OPTIMIZATION OF WATER MANAGEMENT IN POLDER AREAS
SOME EXAMPLES FOR THE TEMPERATE HUMID AND THE HUMID
TROPICAL ZONE

Optimization of Water Management in Polder Areas

Some examples for the temperate humid and the humid tropical zone

DISSERTATION

Submitted in fulfilment of the requirements of
the Academic Board of Wageningen University and
the Academic Board of the UNESCO-IHE Institute for Water Education
for the Degree of DOCTOR
to be defended in public
on Friday, 29 April 2005 at 15:30 h in Delft, The Netherlands

by

PREECHA WANDEE

born in Petchaburi, Thailand

CRC Press
Taylor & Francis Group
Boca Raton London New York

CRC Press is an imprint of the
Taylor & Francis Group, an **informa** business

CRC Press
Taylor & Francis Group
6000 Broken Sound Parkway NW, Suite 300
Boca Raton, FL 33487-2742

First issued in hardback 2017

© 2005 Taylor & Francis Group pic, London, UK
CRC Press is an imprint of Taylor & Francis Group, an informa business

No claim to original U.S. Government works

ISBN-13: 978-0-4153-7555-9 (pbk)
ISBN-13: 978-1-1384-7474-1 (hbk)

Visit the Taylor & Francis Web site at
http://www.taylorandfrancis.com

and the CRC Press Web site
http://www.crcpress.com

Contents

Acknowledgement

This work would never have materialized without the contribution of many people to whom I have the pleasure of expressing my appreciation and gratitude.

First of all, my deepest gratefulness is due to my promoter Prof. E. Schultz, PhD, MSc, who gave me guidance, encouragement and regular discussions, which have been of great value to me. His continued willingness to listen and discuss helped me to produce this dissertation in an appropriate way. I would like to thank my co-promoter Prof. Tawatchai Tingsanchali PhD, for his valuable discussions and guidance to fulfil the part of the work in Thailand for this study.

I extent my deepest gratitude to H. Depeweg, MSc and F.X. Suryadi PhD, MSc for their guidance and suggestions during my study. I appreciate the staff of the Royal Irrigation Department: Mr. Teerapong Pongsawang, Mr. Vooravut Boonthong, Mr. Somboon Munkwamdee, Mr. Kosit Lorsirirat, Mr. Veeraphol Dethai, the Drainage Division of the Department of Public Works and Town & Country Planning, Mr. Chayya Joemjutitam and the Department of Drainage and Sewerage, Bangkok Metropolitan Administration, Mr. Wirat Thamthong, for their contribution to the fieldwork, information and providing data in Thailand.

I extend my gratitude to the team water management of the Principal Water-board of Delfland, Mr. Job van Dansik, Mr. Kick Bouma, and Dr. Viana A. Achthoven, for providing me data and information for the rural polder case studies in the Netherlands. I appreciate Mr. Sjaak Clarisse of Delft municipality for providing me information about the urban polder case study and Mr. Marcel Haaksma of the Civil Engineering Division of Rijkswaterstaat for giving me the information on economic data in the Netherlands.

During my study at UNESCO-IHE, I enjoyed the friendly atmosphere in this institute, especially in the Core Land and Water Development of the Water Engineering Department. I would like to acknowledge the pleasant attitude of all staff members of UNESCO-IHE, especially Student Affairs.

I would like to express my deepest appreciation to my mother, my brothers, my sisters and my Thai friends in the Netherlands for their encouragement. I would like to give special thanks to my wife and my mother in-law for taking care for my family and support during my study in the Netherlands.

Finally, I am grateful to Rijkswaterstaat, for their financial support for the study. Thanks to the Royal Irrigation Department and the government of Thailand for allowing me to study at UNESCO-IHE.

Summary

Weather and climate dictate which crops can be grown in a region and are mainly responsible for the yearly variation in yields. Related to water management the climate in the world can be divided into three main agro climatic zones, which are: temperate humid, humid tropical and arid and semi-arid.

The Netherlands has a temperate humid climate, with a rather even distribution of rainfall over the year. The mean annual rainfall is about 785 mm. The rate of evaporation from open water varies from 0 mm/day in winter to 3 - 4 mm/day in summer. During summer there is a rainfall deficit. On the other hand during winter most of the time there is a rainfall excess that needs to be drained.

Thailand is a humid tropical country. The climate is monsoon, marked by a pronounced rainy season lasting from about May to September and a relatively dry season for the remainder of the year. The annual rainfall in the central area of Thailand is about 1,200 mm. The evaporation from open water varies between 4 and 5 mm/day the whole year round. In general in the rainy season the average rainfall is more than enough for growing rice or other crops. On the other hand, during the dry season, there is a very small amount of rainfall. Hence the primary function of water management is combined irrigation and drainage.

The demand of land for producing food, for urban development such as housing, industry, shopping areas, infrastructure and also for recreation has increased during the history of mankind. This has, among others, resulted in the reclamation of swamps, floodplains, tidal areas and even lakes by impoldering. A polder is a level area, which has originally been subjected, permanently or seasonally, to a high water level (groundwater or open water) and is separated from the surrounding hydrological regime, to be able to control the water levels in the polder (groundwater or open water) independently.

The main components of the water management systems in rural and urban polders are at present as follows:
- for rural areas:
  - distance between the subsurface or open field drains;
  - depth of the subsurface or open field drains;
  - field drain capacity;
  - percentage of open water;
  - water level below the surface;
  - pumping capacity;
- for urban areas:
  - cross section of the sewer pipes;
  - distance between transport canals or transport pipes;
  - percentage of open water;
  - water level below the surface;
  - discharge or pumping capacity.

The main objective of this study was to evaluate present design practices for water management systems in polders in the temperate humid zone - the Netherlands - and the humid tropical zone - Thailand - and to formulate recommendations for a new water management policy for polder areas in present and future situations, such as recommendations for design and operation and maintenance.

To get insight in the system behaviour for different land use and soil composition under temperate humid and humid tropic conditions, model simulations can be used. The existing software package OPOL, based on a non-steady model, was further developed to the version OPOL5 for the simulation of hydrological conditions and optimization of the main components of water management system in polder areas in the temperate humid and humid tropical zone. The model can show system's behaviour and also the effects of variation in the main components of the systems to the overall costs. In this way, the designs of pumped drainage systems in polder areas can be optimized by varying the main components until the annual equivalent costs are minimum. A GIS tool has been used to complement OPOL5 for the simulation of the real situation in an area such as land use, damage, topography, and soil type.

Simulation for the rural area

The aim of the water management system in rural areas is to create good growing conditions for the crops and good drainage conditions for buildings and infrastructure. In order to determine the optimal values for the main components of water management systems for agriculture, the design periods have been used to determine the following from the transformation of precipitation into evapotranspiration and discharge:
-    changes in moisture content in the unsaturated zone;
-    fluctuations in the groundwater table;
-    fluctuations in the open water level;
-    fluctuations in the ponding open water in a rice field;
-    irrigation water supply in humid tropical zone conditions.
   By using the relationship between the conditions above and agricultural criteria, the optimal value of the main components of the water management systems can be determined.

Simulation for the urban area

The urban area has generally one water management system or is part of the polder as sub area. The aim of water management for this area is to provide good drainage and discharge. In order to determine the optimal values for the main components of water management systems for urban areas, the design periods are used to determine the following from the transformation of precipitation to discharge:
-    occurrence of water at the streets;
-    groundwater rise, under influence of rainfall and water level fluctuation in the urban canals;
-    fluctuation of the open water level.
   At first the values for the main components of the water management system in the urban area are determined, assuming free discharge over a weir or through a pumping station. In this way optimal values for the main components of the water management system for the urban areas are found. Dependent on the situation it has to be analysed whether these values are still optimal when the whole system is being considered.

Case studies in the Netherlands and in Thailand

In order to analyse the applicability of the method and the results of it compared to the present conditions the method has been applied to several case studies in the Netherlands and in Thailand. For the Netherlands the case studies involved:

- Schieveen, a clay polder;
- Duifpolder, a peat polder;
- Hoge and Lage Abtswoudse polder, an urban polder;
  For Thailand the case studies involved:
- a rice polder under rainfed conditions;
- a rice polder under irrigated conditions;
- a dry food crops polder;
- a polder between Chao Phraya river mouth and Bangpakong river mouth in the eastern part of Bangkok, a rural polder;
- Sukhumvit polder, an urban polder.

Data were collected in these areas for the calibration of the parameters, which have the most influence to groundwater, open water and discharge.

The polders for the case studies in the Netherlands were located in the area of the Principal Water-board of Delfland. Delfland is located in the western part of the Netherlands. The area is bounded in the West by the North Sea and in the South by the Nieuwe Maas and the Nieuwe Waterweg. Delfland is a typical polder area, which incorporates a large number of areas with different water levels. The total surface is approximately 40,000 ha, which includes 60 polders.

For the case studies in Thailand first of all an experimental plot with rainfed and irrigated rice, as well as dry food crops in the Central Plain has been selected. The experimental area was located at Samchuk district in Suphanburi province at about 208 km to the northwest of Bangkok. The terrain of Suphanburi province consists mostly of low river plains, with small mountain ranges in the North and the west of the province. The southeastern part consists of the very low plain of the Tha Chin river, where is the paddy rice farming area.

The case study for the rural polder region was located in the southeastern part of Bangkok and the central part of Samut Prakan province. It is bordered by the Klongpravetburirom canal to the North, the Chao Phraya river dike to the West, the King Initiated dike to the East, and the Klongchythale canal to the South. This area consists of flat lowland and is situated between the Chao Phraya river mouth and the Bangpakong river mouth. The area is sloping to the Gulf of Thailand.

The case study for the urban polder was located in the eastern part of Bangkok. The eastern part of Bangkok is divided into 10 polders and the polder Sukhumvit, one of the inner urban polders of the Bangkok Metropolitan Administration, has been selected for this study.

Results of the case studies

The clay polder Schieveen in the Netherlands

The polder Schieveen lies within the municipality of Rotterdam. The area of Schieveen polder is 584 ha. Most of the area is pasture (81.3%). The soil in this area is clayey peat for topsoil and heavy clay for subsoil. The present values for the main components of the water management system are as follows: polder water level during winter is 0.71 m-surface; open water area is 5.6%; drain depth is about 0.50 m-surface; distance between the drains is about 20.00 m and pumping capacity is 13.4 mm/day. The optimal values for the main components of the water management system that were found in this study would be: polder water level during winter 1.51 m-surface; open water area 1.9%; drain depth 1.30 m-surface; distance between the drains 23.19 m and pumping capacity 16.2 mm/day. Drain depth has the most influence on the annual equivalent costs in this area. It could be considered to keep the polder water level during the summer higher and the drain depth could be lower than at present.

The peat polder Duifpolder in the Netherlands

The Duifpolder is part of the region Midden-Delfland and lies in the municipalities Schipluiden and Maasland. The area of the Duifpolder is 370 ha. Most of the area is pasture (88.8%). The soil in this area is clayey peat for topsoil and peat with non-decomposed plants for subsoil. The present values for the main components of the water management system are as follows: polder water level during winter is 0.78 m-surface; open water area is 4.4%; drain depth is about 0.35 m-surface; distance between the drains is about 10.00 m and pumping capacity is 14.9 mm/day. The optimal values for the main components of the water management system that were found in this study would be: polder water level during winter 1.35 m-surface; open water area 1.9%; drain depth 1.15 m-surface; distance between the drains 26.50 m and pumping capacity 4.9 mm/day. Drain depth has most influence on the annual equivalent costs in this area. The pumping capacity in the area may be too high compared to the real requirement. It could be considered to keep the polder water level during the summer higher and the drain depth could be lower than at present.

The urban polder Hoge and Lage Abtswoudse polder in the Netherlands

The case study on an urban polder included the total area of the Hoge Abtswoudse polder and the urban part of the Lage Abtswoudse polder. The area that is drained by the pumping station Voorhof has a total area of 713 ha, which includes 216 ha of the Hoge Abtswoudse polder and 497 ha in the Lage Abtswoudse polder. The model application assumes discharge of the Hoge Abtswoudse polder over a weir to the lower area in the Lage Abtswoudse polder. The present values for the main components of the water management system are as follows: sewer diameter is generally 0.30 m; distance between the transport canals is between 175 and 350 m; canal water level is 0.55 m-surface; open water area is 5.0% and pumping capacity is 8.5 mm/day. The optimization was done for the Lage Abtswoudse polder, the results were as follows: sewer diameter 0.30 m; distance between the transport canals 1,870 m; canal water level 0.87 m-surface; open water area 0.2% and pumping capacity 5.0 mm/day. Variations in sewer diameter and pumping capacity have most influence on the annual equivalent costs in this polder area. These parameters have most influence because the rainfall in the area is not so intense and only daily rainfall was used for the optimization. May be this result is not so accurate. Lowering of the water level in winter could be considered. It was found that the open water area and the pumping capacity are more than enough. However, the percentage open water at optimal conditions in this study is very small because only daily data were available for the simulation, therefore the peak discharge due to flash floods is not included in this simulation.

A rice polder under rainfed conditions in Thailand

In practice a rainfed rice polder may not exist in Thailand because farmers always pump water from the canal to irrigate the area. The polder water level at optimal conditions is 0.09 m+surface. At this level the rice in the plot can benefit from water that flows to the plot during a dry period and excess water can be drained by gravity during a wet period. Open water is 1.1%, field drain capacity is 6.7 mm/day and pumping capacity is 7.5

mm/day. The polder water level has most influence on the annual equivalent costs in this area. The deeper the polder water level in the rainfed rice polder the more water flows through the cracks to the canals, therefore the crops will have more damage during dry periods. On the other hand the loss of this water may sometimes benefit the farmer because there is less water in the field in case of heavy rainfall. While a shallow polder water level will increase the probability of inundation of the crops and overtopping of the bund, which may create large damage to the area.

A rice polder under irrigated conditions in Thailand

The general values for the main components of the water management system under conditions of an irrigated rice polder are as follows: polder water level is 0.80 m-surface; open water area is 2.4%; field drain capacity is 46.0 mm/day and pumping capacity is 26.0 mm/day. Based on the model simulation the optimal values for the main components of the water management system would be as follows: polder water level 0.84 m-surface; open water 2.5%; field drain capacity 44.8 mm/day and pumping capacity 32.9 mm/day. At optimal conditions the water has to be pumped for irrigated rice during dry periods because the polder water level is lower than the ground surface. In the irrigated rice polder system most of the water for irrigation was supplied into the polder area through the dike, therefore the model assumes that throughout the year the polder water level was kept at the preferred water level. The pumping capacity would have to be higher than in practice while the other components are more or less the same. Pumping capacity and polder water level have most influence on the annual equivalent costs in this area. Increase of the pumping capacity could be considered.

A dry food crops polder in Thailand

The general values for the main components of the water management system for a typical dry food crops polder are as follows: ditch water level is 0.60 m-surface; distance between the ditches is between 6 and 8 m; polder water level is 0.50 m-surface; open water area is 1.1%; field drain capacity is 46.0 mm/day and pumping capacity is 26.0 mm/day. The simulated optimal values for the main components of the water management system would be as follows: ditch water level 0.74 m-surface; distance between the ditches 8.45 m; polder water level 0.06 m+surface; open water area 2.4 %, field drain capacity 34.8 mm/day and pumping capacity 36.5 mm/day. At optimal conditions the polder water level is above the ground surface, which means that water for irrigation can flow to the plot by gravity. The ditch water level and distance between the ditches have most influence on the annual equivalent costs in this area. It could be considered to lower the ditch water level at first priority for existing areas. For new developments distance between the ditches may be increased from 6 to 8.5 m. An increase in the pumping capacity could also be considered. The polder water level, especially during the dry period, can be kept higher than the ditch bed level to enable the taking of water to the plot by gravity.

The rural polder in Thailand

The total area of the rural polder is 18,760 ha which is composed of 7,390 ha urban area and 11,370 ha agricultural area. The rice area occupied 12.8% and fishponds 87.2%. Water management in the area is called 'water conservation'. Water is being stored in the main and lateral canal before the end of the rainy season for irrigation and other purposes. The farmers normally pump the water from main canals to lateral canals and then pump to irrigate the crops. The values for the main components of the water management system for a rural polder are as follows: polder water level in the rainy season is 0.80 m-surface; open water area is 3.3 %; field drain capacity is 46.0 mm/day and pumping capacity is 5.3 mm/day. The simulated optimal values are as follows: polder water level 0.24 m-surface; open water area 2.0%; field drain capacity 62.9 mm/day and pumping capacity 6.3 mm/day. The polder water level and pumping capacity have most influence on the annual equivalent costs. It may be considered to increase the pumping capacity. The water level during the storing period may be kept higher to get more water stored in the canal systems and reduce the costs for pumping of water into rice fields and fish ponds. Fishponds perform as storage of excess water; the adaptation of agricultural practices such as changing the time schedule of fishponds to store water during the wet period will improve optimal conditions, which need further study on technical and socio-economic aspects.

The urban polder Sukhumvit in Thailand

The study area for the urban polder is located in the urban area in Bangkok Metropolitan at the eastern part of Bangkok. The sub-polder DF of Sukhumvit polder, which has a clear boundary and for which data are available, was selected. This sub polder has a total area of 368 ha which is composed 248 ha paved area and 120 ha unpaved area. The values for the main components of the water management system for the urban sub-polder DF at present are as follows: diameter of sewers is 0.60 m; distance between the transport pipes is 600 m; canal water level during the wet period is 2.08 m-surface; open water area is 0.6% and pumping capacity is 137 mm/day. The simulated optimal values are as follows: sewer diameter 1.00 m; distance between the transport pipes 1,480 m; canal water level 2.56 m-surface; open water area 1.4% and pumping capacity 117 mm/day. Canal water level has most influence on the annual equivalent costs. Lowering of the water level in the wet period and increase of the sewer diameter can reduce damage due to water at the street. It could also be considered to increase the open water area to enable more storage of water and decrease rapid water level rise.

Evaluation of the results

Capillary flow and soil moisture in the unsaturated zone

After a polder has been reclaimed cracks can be formed due to horizontal shrinkage and compaction takes place due to vertical shrinkage. In the top layer the soil is mixed due to agricultural activities and cracks disappear, but cracks under the top layer may stay open. Therefore for the simulation the soil profile was schematised in three layers as follows: layer I ploughed layer or puddled layer in rice, layer II layer with cracks and layer III semi pervious layer.

The capillary flow in the unsaturated zone in or near the root zone is simulated in the model based on the soil moisture tension at the boundary conditions, which depend on the groundwater depth, soil properties, etc. In the temperate humid zone the soil moisture above the field drain level is above field capacity during winter, so there is no capillary rise. In this period the soil moisture content in layer II is higher than in layer I, because in layer I the model computes the soil moisture balance with evapotranspiration, while there is a small capillary rise from layer II. During the summer the soil moisture is below field capacity, therefore the capillary rise plays an important role. In the summer period the soil moisture content in layer I in the clay and the peat polder in the Netherlands decreases faster than the moisture content in layer II. This may be caused by the fact that the net result of the capillary rise from layer II to layer I, infiltration to layer I and the moisture taken by the crops takes more water than the net result of percolation from layer I to layer II and capillary rise from layer III to layer II supplies. Soils in the humid tropical zone can be drier than in the temperate humid zone due to the high evapotranspiration throughout the year. During dry conditions most of the clay in the humid tropical zone will crack and when sudden rainfall occurs water will be lost from the field to the canal system, especially in rainfed rice conditions. The capillary flow to the topsoil is also small under rainfed conditions during dry periods because of the deep groundwater table. Under irrigated rice conditions the soil moisture is almost saturated, therefore there will be no capillary rise. Deep percolation can play a role in the water management of irrigated system. With dry food crops the soil moisture was almost all the time at field capacity due to the irrigation water that was sprayed to the beds. The soil was only dry during a short period after harvesting. The capillary rise in dry food crop conditions can be of benefit to the crops and the farmers, while less pumping will be required for spraying of water during dry periods. During wet periods capillary rise may be small because the soil is almost above field capacity.

Soil storage

Soil storage may play an important role with respect to the water level fluctuations in the polder canal system, because the soil storage may be very high above the groundwater table, especially in peat soil. Soil storage is dependent on the soil moisture conditions and the soil properties. The soil storage in the humid tropical zone, especially with dry food crops plays an important role for the water level in the plot, because the lower the storage the higher the possibility of inundation of crops in the bed system. Soil storage also plays an important role in rainfed rice conditions because soils may be almost dry and it requires quite some rainfall to fill up this storage.

Groundwater fluctuation

Most polder areas are flat and the hydraulic gradient is negligible. Therefore to model the groundwater zone, a conceptualisation as reservoir storage and discharge to subsurface drains as a function of actual storage can be applied. For the temperate humid zone the discharge as non-linear relation with the storage in layer II can be used. The computed groundwater table in the clay and the peat polder is fluctuating above the field drain depth. In reality the groundwater, especially in peat soil, can drop in summer below the field drain depth. Therefore the lack of soil moisture for grassland in summer due to a low capillary rise may be more severe than in the simulation. In the humid tropical zone the water level involves both groundwater and open water. The observed groundwater table or open water level is needed to compare it with the simulated level. In the case studies in Thailand the results fit well with the observed groundwater table or open water level, especially in case of the dry food crops polder and the irrigated rice polder. The simulated groundwater tables in rainfed conditions were quite different from the observed data due to fluctuation of the water level in the river near the plot. In the experimental area in Thailand, the groundwater responds very fast to rainfall because the soil has a high permeability. Moreover, the groundwater in case of irrigated rice and dry food crops was affected by the water level in the lateral canal as well. In irrigated rice not only the fluctuation in the groundwater table but also in the open water level may have a significant influence on the adjacent area because of loss of water through the bunds and deep percolation.

Runoff relation with rainfall in the urban area

The two main functions of the sewer system in the urban area are disposal of storm water and wastewater from households and industries. The rainfall runoff relation is different for the temperate humid and the humid tropical zone, because of differences in temperature and rainfall pattern. In the case studies on the urban polders in the Netherlands as well as in Thailand, if the sewer diameter increases during optimization there will be a larger discharge to the canal system and a higher possibility to get damage due to too high water levels in the urban area. But in the Netherlands rainfall is not severe, so it is difficult to find a good rainfall runoff relation and damage due to water at the streets. Flash floods and water at the street can much more frequently occur in Thailand due to severe rainfall in a short time.

Water level in the main drainage system

The model assumption of a storage reservoir for the main drainage system during the computation of the water level is acceptable for a flat polder area. In the temperate humid zone the water level in the main drainage system can affect the flow from the subsurface field drainage system in case the water is higher than the subsurface drain depth. It may affect the soil moisture flow in the upper layers. However, a water level above the subsurface drain depth has little effect if the duration is short. In the humid tropical zone the water level in the main drainage system is affected by the discharge capacity from the plots for rice and dry food crops. During high water levels in the main drainage system in the wet season pumping will be needed, because excess water in the field cannot flow through the culvert and damage to the crops may occur. Moreover, the possibility to overtop the dike or bund increases if the water level in the main drainage system is high. During the dry season a high water level in the main drainage system is good for irrigation by letting water flow through the culvert by gravity or a smaller head for pumping. The water level in the main drainage system in polders in the humid tropical zone can be kept high at the end of the rainy season to store water for irrigation in the dry season.

Open water area

The open water area is directly concerned with storage of water in the polder. The smaller the open water area the less storage volume will be available. Normally reducing storage area will cause high water levels in the polder under the same drainage conditions. The possibility of damage to urban properties increases where the open water area is reduced. Also more pumping will be required for both rice and dry food crops due to higher water levels at the same drainage conditions. Moreover, the possibility to overtop the dike or bund will increase if the open water area in rice and dry food crop polder is reduced.

Discharge capacity

The discharge of polders is realised through pumping, or gravity flow through a culvert or over a weir. The higher the discharge capacity the less the polder water level rises up. In the Netherlands the discharge is mostly realised by pumping. In the humid tropical zone the discharge capacity can be divided into two categories: discharge from the field and discharge from the canal system. A lower discharge capacity from the field for rice will be more harmful due to more frequent high water levels in the field. A lower discharge capacity from the plots in case of dry food crops will result in a higher possibility of inundation of the beds in the plots and damage may be very high. Moreover, in a dry food crops polder a high water level in the ditches can create high groundwater tables that may damage the crops.

Closing remarks

The OPOL package is a useful tool for determining optimal values for the main components of the water management system in a polder and also for understanding the effects of changes in the values for the main components of the water management system to the costs and damage in a polder. The values for the main components of the water management system as resulting from the simulations in this study are indicative and have to be determined in practice under real conditions such as physical conditions, operation and maintenance practices, land use, agriculture practices, policy, technical aspects, soil conditions, environment and landscape conditions. The particular data have to be based on local conditions.

The annual equivalent costs as found in this study are rather low. Therefore it is recommended to include a safety margin in the optimal values for the main components of the water management system, which are obtained from the OPOL package. The return period, flood risk and risk costs may be included in further studies and analyses on the main components of the water management system.

The annual equivalent costs increase tremendously when the values of the main components of the water management systems are taken smaller than the optimal values. Therefore the risk of under design of the main components has to be seriously considered. The design would also have to take into account future use, lifetime of assets, growth of population, urbanisation, advance of technology, change of environment, economic and social conditions, statistics in hydrology, risk analysis, etc.

The drain depth has most influence on the annual equivalent costs in the rural area in both the clay polder and the peat polder in the Netherlands. In Thailand the polder water level has most influence on the annual equivalent costs in the area of rice for both rainfed and irrigated conditions. In irrigated rice conditions not only the polder water level but also the pumping capacity has most influence on the annual equivalent costs, while the ditch water level has most influence on the annual equivalent costs in the dry food crops polder. The urban area is very sensitive to the canal water level in Thailand, but not so much in the Netherlands.

In this study it is shown that the optimal values for the main components of water management systems in polder areas for the temperate humid and the humid tropical zone can be determined.

1 Introduction

1.1 General

'A polder is a level area, which has originally been subjected, permanently or seasonally, to a high water level (groundwater or open water) and is separated from the surrounding hydrological regime, to be able to control the water levels in the polder (groundwater or open water) independently' (Segeren, 1982).

The water management in polder areas involves water quantity and quality control. This study mainly focuses on water quantity. The aim of the water management system in terms of quantity in polder areas is to create good growing conditions for crops in rural areas and to prevent water nuisance in urban areas. In tropical areas there is often high evapotranspiration and unreliable rainfall whereas in temperate areas there is low evapotranspiration with moderate rains. The water management in tropical areas therefore often involves both irrigation and drainage, whereas in temperate conditions most of the time drainage will be required.

The main components of the water management systems in rural and urban polders at present are as follows:
- for rural areas:
  - distance between the subsurface or open field drains;
  - depth of the subsurface or open field drains;
  - field drain capacity;
  - percentage of open water;
  - water level below the surface;
  - pumping capacity;
- for urban areas:
  - cross section of the sewer pipes;
  - distance between transport canals or transport pipes;
  - percentage of open water;
  - water level below the surface;
  - discharge or pumping capacity.

To get insight in the system behaviour for different land use and soil composition under temperate humid and humid tropic conditions, a model simulation can be used. The model can show system's behaviour and also the effects of variation in the main components of the system to the overall costs. In this way, the designs of drainage systems in polder areas can be optimized by varying the main components until the annual equivalent costs are minimum.

In order to investigate optimal conditions for water management in polder areas hydrological simulations of the non-steady rainfall runoff relations have been made for a series of years. In economical computations, the cost of constructions, operation and maintenance of the drainage and irrigation system, investments for crops, the value of crops, buildings, infrastructure and damage were calculated. Optimizing of such systems means to determine the main components of the system in such a way that the whole system has minimum annual equivalent costs.

The existing software package OPOL, based on a non-steady model, was further developed for the simulation of hydrological conditions and optimization of the main components of water management systems in polder areas (Schultz, 1982 and Wandee,

2001). Complementary to OPOL5 a GIS tool has been applied to simulate the real situation in an area such as land use, damage, topography, and soil type.

Simulation for the rural area

The aim of the water management system in rural areas is to create good growing conditions for the crops and good drainage conditions for buildings and infrastructure. In determination of the optimal values for the main components of water management systems for agriculture, the design periods are used to determine the following from the transformation of precipitation into evapotranspiration and discharge:
-     changes in moisture content in the unsaturated zone;
-     fluctuations in the groundwater table;
-     fluctuations in the open water level;
-     fluctuation of ponding open water in a rice field;
-     irrigation water supply in the humid tropical zone conditions.
Using the relationship between the conditions above and agricultural criteria, the optimal values of the main components of the water management systems can be determined.

Simulation for the urban area

The urban area has generally one water management system, or is part of the water management system in a polder as sub area. The aim of water management for this area is to provide good drainage and discharge. In determining the optimal values for the main components of water management systems for urban areas, the design periods are used to determine the following from the transformation of precipitation into discharge:
-     occurrence of water at the street;
-     groundwater rise, under influence of rainfall and water level fluctuation in the urban canals;
-     fluctuation in the open water level.
At first the values for the main components of the water management system in the urban area are determined assuming of free discharge over a weir or through a pumping station. In this way optimal values for the main components of the water management system for the urban areas are found. Dependent on the situation it has to be analysed whether these values are still optimal when the whole system is being considered.

Case studies in the Netherlands and in Thailand

In order to analyse the applicability of the method and the results of it compared to the present conditions the method has been applied to several case studies in the Netherlands and in Thailand. For the Netherlands the case studies involved:
-     Schieveen, a clay polder;
-     Duifpolder, a peat polder;
-     Hoge and Lage Abtswoudse polder, an urban polder;
      For Thailand the case studies involved:
-     a rice polder under rainfed conditions;
-     a rice polder under irrigated conditions;
-     a dry food crops polder;
-     a polder between Chao Phraya river mouth and Bangpakong river mouth in the eastern part of Bangkok, a rural polder;
-     Sukhumvit polder, an urban polder.

Data were collected in these areas for the calibration of the parameters, which have the most influence to groundwater, open water and discharge.

The polders for the case studies in the Netherlands were located in the area of the Principal Water-board of Delfland. For the case studies in Thailand first of all an experimental plot with rainfed and irrigated rice, as well as dry food crops in the Central Plain were selected. The case study for the rural polder region was located in the southeastern part of Bangkok and the central part of Samut Prakan province. The case study for the urban polder was located in the eastern part of Bangkok.

1.2 Objectives of the study

The main objective of this study was to evaluate present design practices for water management systems in polders in the Netherlands as example for the temperate humid zone and in Thailand as example for the humid tropical zone. The underlying objectives of the study were:
- updating of factors involved in optimization of water management such as land use, damage, construction, operation and maintenance cost;
- evaluation of the physical dimensions of the polders, which affect costs and damage related to performance of the water management systems, such as ground level and landfill, parcelling, discharge structure and location;
- evaluation of the effects of development trends related to optimization in water management, such as urban development and multiple land use;
- comparison of water management issues in polders in the Netherlands and in Thailand;
- to formulate recommendations for optimal design, operation and maintenance of water management systems in such polder areas.

1.3 Outline of the thesis

This thesis starts with the introduction and description of the objectives of study. In Chapter 2 general information will be given on the of worlds' population, agro climatic zones and water management, polders in the world, in the Netherlands and in Thailand will be described. In Chapter 3 water management and land reclamation in the temperate humid zone and the humid tropical zone, including some aspects of the history of water management and land reclamation in the Netherlands and in Thailand will be reviewed. In Chapter 4 and 5, water management in polders in the Netherlands and in Thailand will be described. In Chapter 6 the model description will be given, which consists of a hydrological computation, an economic computation and an optimization technique. In Chapter 7 the field investigations will be described, as well as the application of the model for the case studies, calibration, validation, analysis and discussion of each case study. In Chapter 8 an overall discussion of the case studies, an evaluation of the model, recommendations for optimal conditions for design, operation and maintenance of polders in the temperate humid zone and in the humid tropical zone and in Chapter 9 evaluation and closing remarks will be given.

2 General background

2.1 Population growth and urbanization

By 2002, the world's population reached about 6.2 billion, it has been estimated that the world's population will be about 7.8 billion in 2025 and 9.1 billion in 2050 (Figure 2.1).

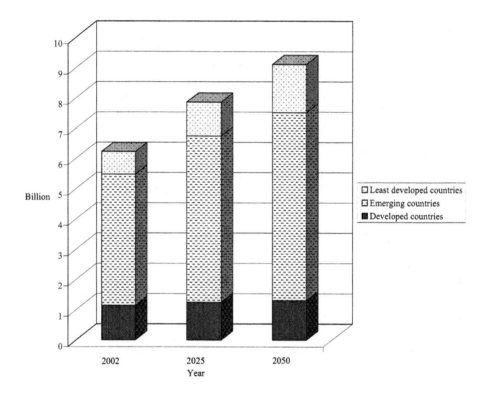

Figure 2.1 World population and growth in least developed countries, emerging countries and developed countries (Schultz, Thatte and Labhsetwar, 2005)

Of special interest in Figure 2.1 is the distinction in least developed countries, emerging countries and developed countries. The majority of the worlds' population lives in the emerging countries. This category comprises most of the Eastern European countries (including Russia), most of the countries in Central and South America, most of the countries in Asia (including China, India and Indonesia), and several countries in Africa. From Figure 2.1 it can be further derived that population growth will take place in the least developed countries and the emerging countries. In the developed countries a slight reduction of the population is expected (Schultz, Thatte and Labhsetwar, 2005).

In recent decades population growth and rapid urbanization, which started after the Second World War, has resulted in a large increase of population in settlements in lowland areas. This process, which was initiated in part by decolonisation and which has accelerated in the past quarter century, due to many factors including industrialization, is mostly due to natural population growth and limitations in rural areas to absorb high rates of population increase. Another interesting aspect related to population growth is the migration from rural to urban areas. The expectation is that due to these developments the population in the rural areas in the least developed countries and the emerging countries will more or less stabilise and that the growth will be concentrated in the urban areas in these regions (Schultz, 2001).

According to the United Nations estimates, the world's urban population will surpass the 60% mark by the year 2050, of which, according to some researchers, approximately one half will be living within a 60 km from seashores (United Nations, 1994). Most of the inhabited areas in this belt will be lowlands. Human population is attracted to coastal zones to a greater extent than to other regions. Urbanisation and the rapid growth of coastal cities have therefore been a dominant population trend over the last decades, leading to the development of numerous mega cities in coastal regions around the world. Of the people living in the coastal zone it is estimated that 600 million live in the area with a chance of flooding of more than once per one thousand years (Nicholls and Mimura, 1998). Collectively, this is both placing growing demands on coastal resources as well as on the increasing people's exposure to coastal hazards.

In historic times, but even more pronounced in recent years, coastal populations around the world have suffered from serious disasters caused by storm floods and related wave and wind attack and precipitation. A dramatic example could be seen in the coastal region of eastern India (State of Orissa), where a tropical storm in October 1999 caused at least ten thousands of deaths and displacement and impoverishment of about 774,000 houses destroyed in a large coastal area (http://news.bbc.co.uk/hi/english /static//2000/ dealing_with_disaster/orissa.stm.).

Recently at Indian Ocean earthquake was occurred under the sea on 26th December 2004. The earthquake generated a tsunami that was among the deadliest disasters in modern history. At a magnitude of 9.0 Richter magnitude scale, and tied for fourth largest since 1900. The earthquake originated in the Indian Ocean just north of Simeulue island, off the western coast of northern Sumatra, Indonesia. The resulting tsunami devastated the shores of Indonesia, Sri Lanka, India, Thailand and other countries with waves of up to 15 m high, even reaching the east coast of Africa, 4500 km west of the epicentre. At about 228,000 to 310,000 people including tens of thousands missing were though to have died and over a million left homeless as a result of the tsunami (http://en.wikipedia.org/wiki/2004_Indian_Ocean_earthquake).

2.2 Agro climatic zones and water management

Weather and climate dictate what crops can be grown in a region and are mainly responsible for the yearly variation of yields. There is a difference between weather and climate: weather means the state of the atmosphere at a given point in time at a given location; climate is a synthesis of weather at a given location over a period of about 30 - 35 years (Ayoade, 1983). In modern dictionaries climate is usually defined as 'average weather' or composite physical state of the atmosphere at a specific locality for a specified interval of time (Sanderson, 1990). The World Meteorological Organization (WMO), which coordinates world climate information, defines this time period as 30 years. Related to water management the climate in the world can be divided into three main agro climatic zones, which are: temperate humid, humid tropic and arid and semi-arid.

About 7% of earth's total land surface is in the temperate humid zone. Some 20% of the earth's landmass is in the humid tropical zone and some estimates suggest that roughly 35% of the earth's total land surface is in the arid and semi arid zone (Andreae, 1981). Although temperate zones account for a small portion of the world's land surface, they are by far the most popular areas to live in and give home to around 40% of the earth's population. This is largely due to the mildness of the climate, which prevents conditions from becoming too harsh, a plentiful supply of rain and generally very fertile soils (http://www.bbc.co.uk/weather/features/weatherbasics/zones_ Temperate.shtml, http://www.unu.edu/unupress/unupbooks/uu24ee/uu24ee0r.htm).

Temperate humid zone

The temperate humid zone generally prevails pole ward of latitudes 30° north or south of the equator. The weather and climate in this zone are dominated by pressure systems, which control daily and seasonal variability of the weather. There are, however, marked spatial variations in the climate of the temperate humid zone arising from the effects of the following factors:
-   location relative to the ocean;
-   latitudinal location, which determines the radiation regime and the length and severity of the cold season;
-   topography, which influences both temperature and precipitation;
-   degree of influence of ocean currents, warm and cold.

The temperate climate varies from the warm type through the cool type to the cold type. According to varying precipitation moist or dry parts can be classified. Areas of the world having a temperate climate are the following (Ayoade, 1983): Europe, most parts of North America, Asia excluding southern and southeastern Asia; New Zealand and eastern parts of Australia; southern part of South America and northern and southern fringes Africa.

Humid tropical zone

The humid tropics occupy a belt roughly 10° north and south of the equator (Figure 2.2). On the whole they are well watered with over 2,000 mm/year rainfall (in some regions 10,000 mm/year), and have year round temperatures typically about 27 °C (except above 900 m altitude). Precipitation is generally at least equal to the potential evapotranspiration, i.e. there is no marked dry period during which vegetative growth is inhibited. There are, however, regions described as humid tropical where precipitation is less than the evapotranspiration for up to five months of the year and most of the humid tropics have at least one such month in a year (Ruthenberg, 1980).

The sub humid tropics are characterized by a wet season of four and a half to seven months, during which precipitation exceeds evapotranspiration, i.e. they are seasonally well watered (Balek, 1977 and Ruthenberg, 1980). In Central and South America the sub-humid tropics lie between roughly 5° South and 25° South, in Asia between 3° North and 10° North. Two generalized precipitation regimes may be recognized: a double rainfall maximum separated by a dry season during the low-sun period, but only a slight dip in rainfall during the high-sun period since the equatorial through would not be far away, and a single rainy season/single dry season regime (Jackson, 1977). Part of the sub-humid tropics, notably India, Pakistan, Bangladesh, Sri Lanka, Southeast Asia, southern China and Japan (Roughly between 7° North and 40° North) receive much of their rainfall as a consequence of seasonal reversals of winds. These 'monsoon lands' are notoriously subject to seasonal and year-to-year fluctuations in time of onset, quantity and distribution of rainfall (Figure 2.3). Consequently there has been much irrigation development in these regions designed to combat rainfall uncertainty, rather than to boost crop yields.

Arid and semi arid zone

The arid and semi arid zone lies mainly between $23°$ North and $35°$ south of the equator. Since solar radiation is higher and longer than in the humid tropics, if moisture is available the tropical dry lands are likely to produce more crops than the area more close to the equator.

2.3 Polders in the world

In this study the optimization of water management systems in polders in the Netherlands as representative for the temperate humid zone and in polders in Thailand as representative for the humid tropical zone concerns the cases for which comparisons and evaluations will be given.

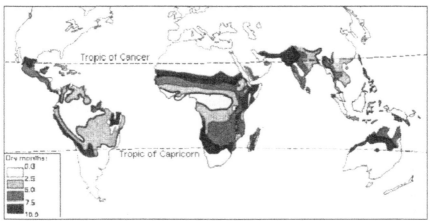

Figure 2.2 The tropics subdivided according to the number of months during which evapotranspiration exceeds precipitation (Harris, 1980)

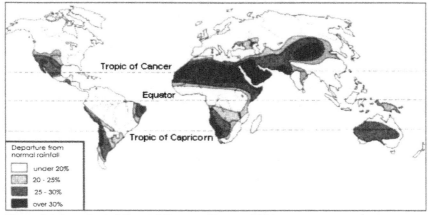

Figure 2.3 Annual rainfall variability map (Barrow, 1986)

The demand of land for producing food, for urban development such as housing, industry, shopping areas, infrastructure and also recreation areas had increased during the history of mankind. This has, among others resulted in reclamation of swamps, floodplains, tidal areas and even lakes by impoldering. With regard to polder development, the polders in the world can be distinguished into three different types (Segeren, 1983):

- polders in already densely populated areas, with mostly small farmers, working under bad conditions of water management, soil and infrastructure. These polder areas are mainly situated in Asia and Europe, for example the polders in Bangladesh (Ali, 2002). The development in these areas concerns not solely improved land and water management but it requires integrated rural development such as education, extension services, credit and marketing facilities, agro processing industries and so on;

- polders in sparsely populated areas in densely populated regions. They are mainly found in South Asia and in some areas in North, Central and South America. A main characteristic of these areas is inaccessibility. The reclamation in these areas is generally undertaken to relieve over population in the surrounding region as well as for the production of food, mainly for the region itself. Large-scale engineering works are often required, as well as on-farm development to prepare the reclaimed land for many small farmers. This will always call for strong government involvement at the initial development stages. Also an integrated approach for the rural area is required;

- polders in sparsely populated areas in sparsely populated surroundings. These are mainly found in Central and South America, parts of Africa and Australia. These areas may produce for the world market. The costs of the reclamation and production have to be recovered by prices of agricultural products on the world market. It is implied that the reclamation and production costs have to be low. The main aim is for food production with minimum costs in contrast to the previous types of polders, where social considerations often play an important role. These low costs are possible because in these areas a high degree of mechanization can be applied for both reclamation activities and agriculture itself.

To gain the full benefits from the reclamation, polders will have to be well planned and executed; therefore detailed data must be available. Also water management systems will have to be well designed and implemented. It is obvious that all aspects of the development in the polder area must be considered in an integrated manner such as socio economic, climate, topography, geohydrology, agricultural practices, irrigation and drainage systems, fresh water reservoirs, farm size, land parcelling, villages, parks, forest areas, and urban areas, planning and implementation. Hence water management in the polder areas will be different from place to place. The same considerations apply improvements to be implemented in reclaimed polder areas.

2.4 Background of the Netherlands

2.4.1 General

The Netherlands has a total area of 41,526 km^2, of which 33,939 km^2 is land surface, some 20,000 km^2 is reclaimed land (Schultz, 1983), or polder. The country borders the North Sea in Western Europe (Figure 2.4). Approximately 25% of the country lies below mean sea level.

The Netherlands has about 16 million inhabitants. The population growth rate is about 0.7%. The population density is 430 per km², making the Netherlands the most densely populated country in Europe. The population is not evenly distributed over the country. There are rural provinces such as Zeeland, with a population density of 198 per km² and there are urban provinces such as Zuid-Holland, with a population density of 1,122 per km².

Figure 2.4 Map of the Netherlands
        (http://www.amsterdam-netherlands.info/img/netherlands_small.gif)

The per capita Gross National Income is € 19,436. Most of the people are working in the service sector and industry, only 5% in agriculture. Although agriculture provides employment for a small portion of the population, it is nevertheless a very important economic sector. Agricultural activities, in a broader sense the so-called agro-business, provide employment to over 11% of the working population (http:// www.nationmaster .com/graph-T/eco_gro_nat_inc_cap and http:// www.library.uu.nl/wesp/populstat/ Europe/netherlg.htm).

The agricultural sector accounts for 17% of imports and for 23% of exports. The export surplus is almost € 7 billion per year. Hence agriculture is one of the most important pillars of the Dutch economy. Therefore, water management in the country to a large extent aims at optimal conditions for agriculture.

2.4.2 Hydrology and land use in the Netherlands

Meteorology

The Netherlands has a temperate maritime climate, with a rather even distribution of rainfall over the year. The mean annual rainfall is about 785 mm. The rate of reference evapotranspiration varies from 0 mm/day in winter to 3 - 4 mm/day in the summer, the mean annual reference evapotranspiration being about 550 mm. The mean monthly variations in rainfall and evapotranspiration are shown in Figure 2.5.

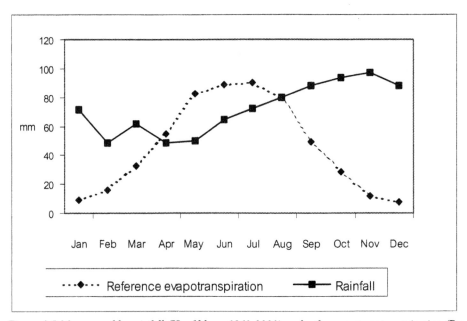

Figure 2.5 Mean monthly rainfall (Hoofddorp, 1960-2001) and reference evapotranspiration (De Bilt, 1960-2001)

Owing to the relative even distribution of rainfall over the year, the increased evapotranspiration during the summer and the low evapotranspiration in the winter, there is a rainfall deficit in the summer months amounting to 82 mm under average conditions and a surplus of rainfall in winter amounting to on average 394 mm.

Average monthly data on temperature, relative humidity, hours of sunshine and wind velocity are given in Table 2.1. In Table 2.1 it can be seen that the temperature and the hours of sunshine remain relatively low throughout the year. The humidity is high during the whole year. It may be concluded that the Netherlands has cool summers and mild winters. Therefore it can be derived that the evapotranspiration in the Netherlands is relatively low.

In Figure 2.6 it can be seen that during the summer there is a certain rainfall deficit based on the average conditions. On the other hand during the winter (Figure 2.7) most of the time there is rainfall excess that needs to be drained.

The depth duration frequency curves for the rainfall at Hoofddorp are shown in Figure 2.8.

The rainfall in the Netherlands has increased during the past decades, which can be seen from the average 20 years annual rainfall at Hoofddorp for the years 1890 to 2001 (Figure 2.9).

Table 2.1 Average temperature, relative humidity, hours of sunshine and wind velocity at De Bilt for the period 1971 – 2000

Month	Temperature in °C	Relative humidity in %	Hours of sunshine per day	Wind velocity in m/s
January	2.8	88	1.7	4.1
February	3.0	85	2.8	3.9
March	5.8	81	3.7	3.9
April	8.3	76	5.3	3.6
May	12.7	74	6.6	3.2
June	15.2	77	6.2	3.1
July	17.4	77	6.3	3.0
August	17.2	78	6.2	2.7
September	14.2	84	4.4	2.8
October	10.3	86	3.4	3.1
November	6.2	88	2.0	3.6
December	4.0	89	1.4	4.0
Annual average	9.8	82	4.2	3.4

Source: http://www.knmi.nl/voorl/kd/lijsten/normalen71_00/c-stationsgegevens/stn260/4-normalen/260_debilt.pdf

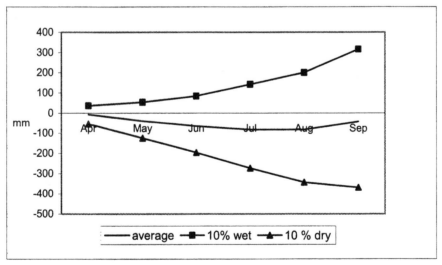

Figure 2.6 Accumulation of the rainfall surplus during the summer period for average, 10% wet and 10% dry conditions in the Netherlands, Hoofddorp (1960 – 2001)

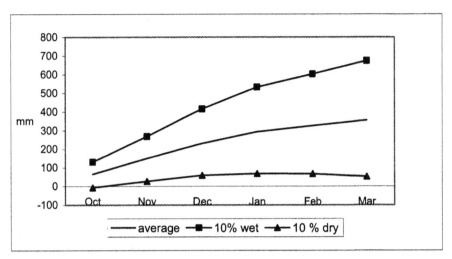

Figure 2.7 Accumulation of the rainfall surplus during the winter period for average, 10% wet and 10% dry conditions in the Netherlands, Hoofddorp (1960 – 2001)

Land use

The Netherlands is a small country lying at the mouth of three rivers: the Rhine, the Meuse and the Scheldt. Large tracts of land have been reclaimed in this land that would have been otherwise periodically or permanently under water. The Dutch have a long history of dealing with floods by constructing earth dikes and channelling surplus water through networks of drains to sluices, which discharge the water when the outside water level is low. In this way the Dutch protect those lands that are lower than the water level outside. With windmills several areas were reclaimed in the North and West by draining the lowlands and even many lakes were drained to gain land (Figure 2.10).

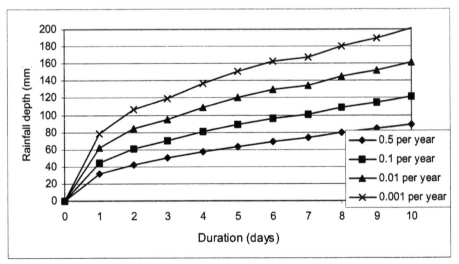

Figure 2.8 Rainfall depth-duration-frequency and chance of occurrence per year according to the Gumbel distribution at Hoofddorp, the Netherlands, period 1867 to 2001

In eight centuries of human activities swampy regions and natural lakes and streams were changed into an irregular pattern of small and large polders, each having its own embankment and a preferred water level at elevations depending on the elevation of the polder surface, soil type and land use, which was mainly agriculture (Figure 2.11).

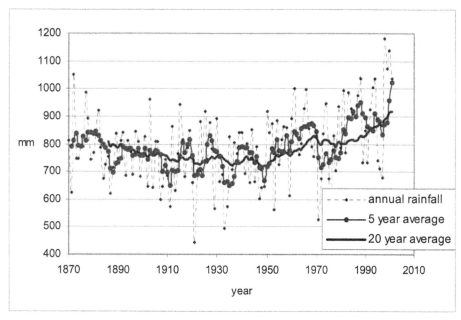

Figure 2.9 Rainfall trend in the Netherlands at Hoofddorp station, period from 1867 to 2001

About 68% of the entire surface is used for agriculture, which area is composed of 32% grassland, 27% arable land and 9% is forest area. Uncultivated land is 4%. The urban and industrial areas take up 16%. Land use in the Netherlands may be classified as follows: agriculture 56.0%, nature 11.7% urban 8.8%, recreation 2.1% inland water and coastal 18.7% and other 2.7% (http:// www.cbs.nl/en/publications/ articles/general/ statistical-yearbook/a-3-2004.pdf).

2.5 Background of Thailand

2.5.1 General

Thailand is situated in the heart of Southeast Asia. It covers an area of 513,115 km². It is bordered by Laos in the Northeast, Myanmar in the North and the West, Cambodia in the East, and Malaysia in the South (Figure 2.12).

Thailand is normally divided into four regions: the North, the Central Plain or the Chao Phraya basin, the Northeast or the Korat Plateau, and the South or the southern Isthmus. The North is a mountainous region characterized by natural forest, ridges and deep narrow, alluvial valleys. The Central Plain, the basin of the Chao Phraya river, is a lush, fertile valley. It is the richest and most extensive rice-producing area of the country and is often called the 'Rice Bowl of Asia'. Bangkok, the capital of Thailand, is located in this region. The Northeast region, or Korat Plateau, is an arid region characterized by rolling surface and undulating hills. The violent climatic conditions often result in floods and droughts. The southern region is hilly to mountainous, with thick virgin forests and rich deposits of minerals and ores. This region is the centre for the production of rubber and cultivation of the other tropical crops.

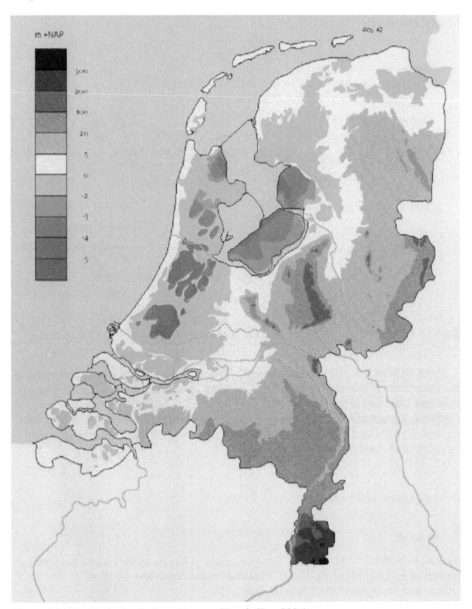

Figure 2.10 The Netherlands elevation map (Van de Ven, 2004)

The population of Thailand is about 62 million. With a growth rate 1.2 to 1.4% per year the population is projected to reach 70 million by 2010. The population density is 120 per km^2. About 38 million or 63% of the population lives in the rural area; about 92% of them are farmers (World Bank, 2000).

The per capita National Income is 1,573 € (National Economic and Social Development Board, 2001). Most of the population is working in agriculture as farmer and herdsman. Agriculture's role in economy has shrunk over the last two decades, in line with the increase of investment in the manufacturing and service sector. Nonetheless, the sector still accounts for about 50% of the total labour force and 25% of the total export value (http://www.boi.go.th/english/Thailand/index).

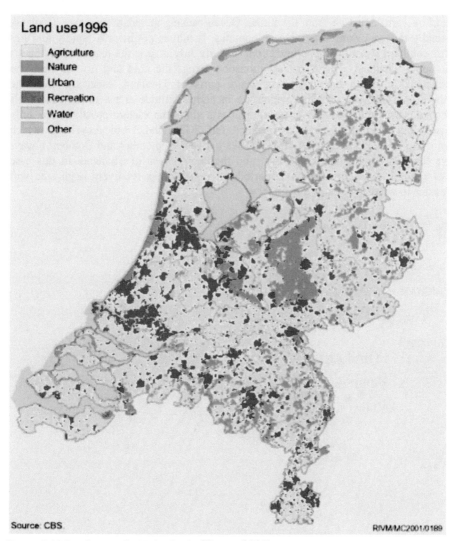

Figure 2.11 Land use in the Netherlands (Visser, 2000)

Thailand is a humid tropical country. The climate is monsoon, marked by a pronounced rainy season lasting from about May to September and a relatively dry season for the remainder of the year. Due to the climate conditions, rice forms a staple part of the Thai diet (35% of GDP), while it is still the basis of the rural economy, followed by rubber (20%), tree crops (15%) and vegetables (10%).

2.5.2 Hydrology and land use in Thailand

The climate of Thailand is tropical and therefore warm throughout the whole year. At the same time it is dominated by the monsoon winds that bring about the seasonal changes from wet to dry, creating three seasons in the North and the central areas and two in the South. The onset of monsoons varies to some extent. Southwest monsoon usually starts in mid-May and ends in mid-October while Northeast monsoon normally starts in mid-October and ends in mid-February. The Southwest monsoon brings a stream of warm moist air from the Indian Ocean causing abundant rain over Thailand, especially the windward side of the mountains. It moves northwards rapidly and lies across southern part of China around June to early July, that is the reason of dry spells over upper Thailand. The Northeast monsoon brings the cold and dry air from the anticyclone in China mainland over major parts of Thailand, especially over the northern and northeastern parts, which are in higher latitude areas. In the South this monsoon causes mild weather and abundant rain along the eastern coastline. Not only monsoon but also tropical cyclones cause rainfall in Thailand. Tropical cyclones, which often occur in the southern part of China Sea between September and October, always move towards the Northeast, the East and the central area of Thailand. In this case heavy rainfall and long continuous periods of rainfall in a relatively large area will occur due to wind and clouds disturbance.

Figure 2.12 Map of Thailand (http://www.nilkhosol.com/thailand/1/map.jpg)

The annual rainfall in the central area of Thailand is between 1,000 and 1,400 mm and slightly more in the Northeast, but in the Southwest and Southeast part it is much higher, about 2,000 mm. As mentioned above the distribution of rainfall is not spread all over the year (Figure 2.13) and the rainfall is also unreliable.

The reference evapotranspiration varies between 3 and 4 mm/day the whole year round. In the dry periods from November to April, the evapotranspiration is exceeding the rainfall (Figure 2.13). In the rainy season there is a rainfall surplus of 745 mm but the rainfall deficit is around 313 mm in the dry period. It can be concluded that Thai farmers will face with too dry and too wet conditions for crops water management.

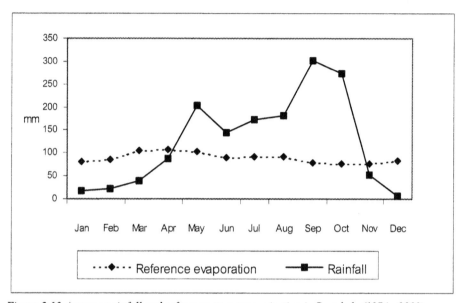

Figure 2.13 Average rainfall and reference evapotranspiration in Bangkok, (1974 - 2000)

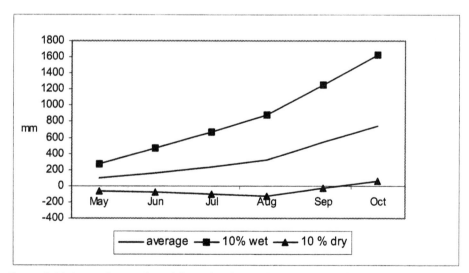

Figure 2.14 Accumulation of rainfall surplus during the rainy season for average, 10% wet and 10% dry conditions in Thailand, Bangkok (1974 - 2000)

From Figure 2.14 it is clear that a rainfall deficit can occur at the probability of 10% in the rainy season, but rainfall excess at the probability of 10% is very high. Therefore irrigation water has to be supplied to meet the crop water requirement when rainfall deficit occurs. On the other hand the drainage system is very important because high amounts of excess rain always occur.

In the dry season usually a rainfall deficit occurs (Figure 2.15). At this time if the farmers grow second crops, such as rice, irrigation is necessary to meet the crop water requirement. On the other hand during the rainy season (Figure 2.13) most of the time there is rainfall excess that needs to be drained.

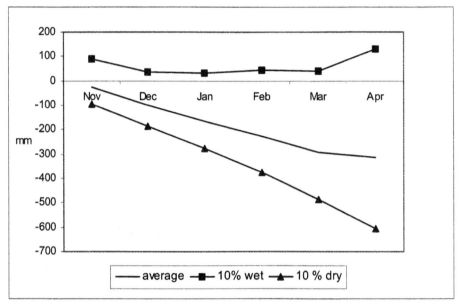

Figure 2.15 Accumulation of the rainfall surplus during the dry season for average, 10% wet and 10% dry conditions in Thailand, Bangkok (1974 - 2000)

Extreme rainfalls are quite high (Figure 2.16) due to tropical storms, which have enormous influence on drainage requirements.

Rain in Thailand has not clearly increased with time as can be seen in Figure 2.17. May be because rainfall in Thailand is influenced by monsoon and tropical storms.

In Table 2.2 it can be seen that the temperature is high almost throughout the year, high hourly sunshine per day, low humidity during the dry period and high humidity during the rainy season. It may be concluded that Thailand has hot and warm weather throughout the year. Therefore it can be derived that the evapotranspiration in Thailand is high throughout the year.

Land use

Overall land use in Thailand may be divided as follows: arable land 34%, permanent crops 6%, permanent pastures 2%, forest and wood land 26% and other 32%. (http://www.photius.com/wfb2000/countries/thailand/thailand_geography.html).

The total area of agriculture land is about 1,900 million ha. Rice is the major crop, it covers about 55.4% of the agricultural land. Others are as follows: field crops 21.3%, permanent crops 8.1%, rubber 8.0%, forest 1.5%, pasture 1.2%, vegetable crops flowers, ornament plants 0.9%, wasteland 2.0% and others 1.6%.

Land use in the lower deltaic area in the Central Plain (1998) is composed of urban area 14.1%, rice 56.5%, fruit and trees 13.1%, vegetables 0.8%, field crops 4.2%, forest 0.1% and others 11.2%.

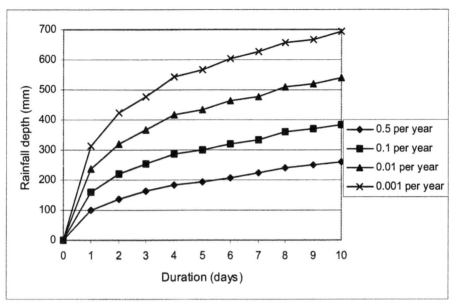

Figure 2.16 Rainfall depth-duration-frequency and chance of occurrence per year according to the Gumbel distribution at Bangkok, Thailand, period 1974 to 2000

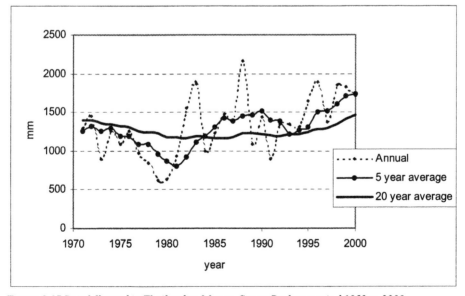

Figure 2.17 Rainfall trend in Thailand at Muang, Samut Prakan, period 1952 to 2000

Table 2.2 Average temperature, relative humidity and wind velocity in Bangkok for the period
*   1981 to 2000*

Month	Temperature in °C	Relative humidity in %	Hours of sunshine per day	Wind velocity in m/s
January	26.6	69	7.4	1.8
February	28.2	72	7.9	2.4
March	29.3	72	7.8	2.7
April	30.3	72	7.4	2.8
May	29.8	75	6.0	2.4
June	29.3	74	4.7	2.4
July	28.9	75	4.5	2.4
August	28.6	76	4.1	2.1
September	28.1	80	4.2	2.0
October	27.8	79	5.1	2.5
November	27.4	71	6.3	1.8
December	24.7	67	7.2	1.7
Annual average	28.2	74	6.1	2.3

Source: Meteorological Department, Thailand

3 Water management and land reclamation

This chapter the focus will mainly be on water management and land reclamation of lowland areas, while these are areas where the polders are located.

3.1 Review of water management and land reclamation in the Temperate Humid Zone

3.1.1 Water management and land reclamation in Europe

It is said that around 500 BC, the Etruscans reclaimed the Pontine marshes in Italy. This can be considered as the first impoldering activities in Europe. The Etruscans were renowned as great engineers and may have acquired the science of hydraulics through trading connections with Mesopotamia. The works fell in disuse after 300 BC and were not restored during the Roman Empire (Wagret, 1959). Between Rome and Pescara is the plain of Avezzano situated at 650 m+MSL (Mean Sea Level) with intensive used agricultural areas. Central-Italy's greatest lake with an extension of 140 km² and a depth of 20 m was located at this place (Its Latin name: Lacus Fucinus, the Italian name: Lago di Fucino). Lake Fucin was notorious for its level risings from 12 to 16 m), leading to a surface extension up to 75 km². These sudden risings had devastating consequences for the deeper based areas surrounding the lake and occurred after long periods of low water. Villages and arable land were given up several times, which caused an emigration of the rural population; but the soils' high quality in the inundated area made them resettle again and again. Very early, people began to search for a possibility to drain the lake or at least to prevent the rising of the water level. The only solution was a 5.7 km long tunnel (emissar), which was completed under the regentship of emperor Claudius after 13 years of construction in the year 54 AD. With this measure the periodical risings of the water level should be prevented and new agriculture areas were obtained. The building rotted in the past-roman time and collapsed in the early Middle Ages. The periodical overflows began again and the water level rose up to its old level. Even though there were several efforts to reconstruct the tunnel in the centuries from 1240 to 1835, but without any success. Only when several extremely high water events occurred - around 1815 and 1860 - the lake was drained completely by a new tunnel and the plain was used agriculturally since 1875 (Döring, 2000).

Long before the start of the Christian era, people settled on the coastal marshlands of the Netherlands. Like in other parts of Europe, the population increased considerably in the period 800 to 1250. In many places this process entailed the reclamation of coastal marshes, inland marshes and river plains. At that time, the first works consisted of a primitive form of land drainage, with hardly any protection from flooding by the sea or the rivers. In the coastal areas, artificial hillocks or dwelling mounds were built, partly through accumulation of trash, partly on purpose, as a defence against the sea floods. No continuous dikes existed in the river areas; only locally some low dikes offered some sort of protection for the lowest patches. Continuous and encircling embankments were not built before 1100. In the period 1150 to 1250, many encircling dikes were built in the southwestern coastal areas as a reaction to the storm surge of 1134, which caused great losses of land. In the river area this did not take place before 1250. Also in the northern coastal area, continuous dikes were only built in the second half of the 13th century. In course of time new techniques (windmills, steam engines) and capital made it possible to realise large reclamation projects in the Netherlands, which lead to its present shape (Van de Ven, 1993). In conclusion it can be stated that the first polders of the Netherlands date from around 1200.

The development of coastal areas in Europe did benefit from the know-how and experience acquired in the Netherlands, due to a certain professional emigration in this particular field. The first mention in historical records of the involvement of the Dutch in reclamation dates from the year 1113 (Van de Ven, 1993). The ruler-bishop of Bremen then concluded a treaty with a few Dutchmen, coming from the newly reclaimed peat area east of the city of Leiden, concerning the reclamation of an area in the valley of the Weser (Figure 3.1). This treaty is virtually identical to the 'cope' agreements the count of Holland had with the reclaimers of his wilderness. The size of the homesteads and the rights and duties of the new village community were laid down. In the treaty only watercourses are mentioned, whereas no mention is made of the construction of dikes. This points to reclamation by drainage. Apparently the venture was a successful one, for soon much new reclamation followed. The emigration towards the East went on, and it extended to low-lying parts of Germany, Poland and Russia.

Figure 3.1 Statue of priest Heinricus, leader of a group Dutchmen, who in 1113 initiated large scale reclamations in Dutch style in the North German plain

In a later stage (after 1550), Dutch experts worked in practically the whole of Europe (Figure 3.2). Three typical representatives of this group are Leeghwater (1590 - 1685), Vermuijden (1575 - 1650) and Bradley (Netherlands Organisation for Applied Scientific Research TNO, 1989).

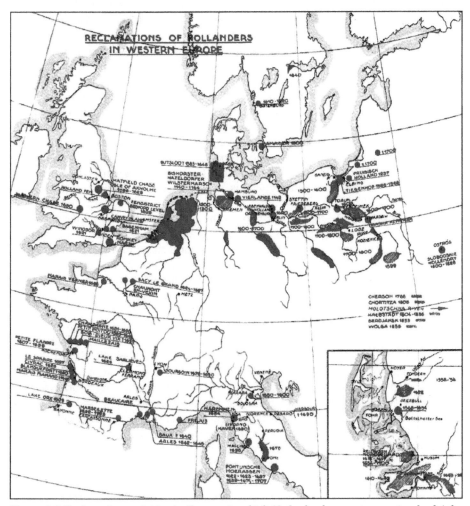

Figure 3.2 Reclamation activities in Europe, in which Netherlands experts were involved (after Van Veen, 1962)

In the Netherlands, Leeghwater, was for long time considered the authority on draining lakes and marshes by pumping. He also did consultancy work in the Baltic region, Denmark, England, France, and Germany. But actually he was in the first place a windmill specialist, and windmills were applied for draining large lakes to the north of Amsterdam (Beemster, 1612; Purmer, 1622; Wormer, 1625; Schermer, 1631) and did not play a leading role in the above mentioned reclamations (Van de Ven, 1993).

The Dutch expert, Vermuijden, did most of his engineering work in England (TNO, 1989). His main achievement was the reclamation of the Yorkshire marshes and the Fens of the Wash.

One of the difficulties of the reclamation of marshes or lakes, which lead to opposition of the inhabitants of the surrounding areas, was the drainage of these areas after reclamation. The solution commonly applied by the Dutch engineers consisted of dredging a catch canal along the ridge of the adjacent areas, to cut off the drainage water from the new lands. This diversion, however, might worsen the drainage conditions because of the decrease of the flow gradient. Vermuijden's idea consisted of increasing the general gradient of the drainage by means of one shortcutting channel running towards the Wash. By 1655 the planned works were finished and soon the benefits of the work became apparent. In a later stage difficulties were experienced with the subsidence of the peat, resulting in a hampered drainage.

The life of Humphrey Bradley is not well known. His original name may have been Braat, but perhaps he Anglicised his name after a short stay in England. Most of his professional career was in France, where he successfully reclaimed many marshes. His general design principle was the same as that of Leeghwater and Vermuijden, and included the peripheral canal to the sea to divert water from the surrounding catchment areas away from the reclaimed low-lying areas ('Ceinture des Hollandais' in the Petit Poitou). This was only completed after his death because of hostility of the local population (TNO, 1989).

Albert Brahms (1692 - 1758) was born in the coastal area in the northern part of Schleswick-Holstein. Through self-study he became familiar with mathematics, geography, history, Latin, French and Dutch. The storm surge of 1717, which caused much damage along the German and Dutch coasts, made that he became involved in coastal and polder engineering, when he was made responsible for the repair works. The quality of his work was at such a level, that he was appointed inspector and judge for the Public Works water affairs.

Through centuries, hydraulic engineering was based on experience from earlier work and empirical facts. Around 1700, a tendency in Europe developed to include the knowledge of applied mathematics and mechanics. In the Netherlands, Simon Stevin, his son Henric, Van Bleiswijk, Van Velzen and others were involved. As Brahms also believed in this new approach, he visited hydraulic engineering works in the Netherlands. He compared, among others, the coastal defences of the Netherlands and Germany. In 1769, his work on the fundamentals of hydraulic engineering was published. Undoubtedly he learned a lot from his Dutch colleagues, but he surpassed them in the field of data collection and measurement techniques. He was the first who investigated the relation between the wave height in front of a dike and the wave run-up.

3.1.2 Water management and land reclamation in the Americas

Argentina

In the delta of the Parana river there are 440,000 ha with good possibilities for economic development. The area is a vast concave flatland with a scant Southeast slope of low height compared to the surrounding rivers. The river levees are rather extensive and occupy some 15% of the total area. Their height is about 1.5 m above the surrounding terrain and their width varies from 10 to 100 m. This configuration hampers drainage of the land, which reduces the agricultural development. Other natural elevations are formed by the sand dunes, rising up to 10 m, which are of great importance for other infrastructural works, like the Urquiza railway line.

About 14,000 inhabitants live in the area. In the western part the main productive activity is cattle breeding and some agriculture. In the eastern part the predominant activity is extensive forestry, with cattle breeding and some supplementary agriculture.

The soils are of the alluvial type, supported by a base of marine banks. The upper 0.7 m layer is composed of organic matter, clay, lime, and fine sand, and has a rather low permeability. The chemical characteristics of the soils make them fit for the development of vegetable species, pasture, and forestry. Below 1 m the soil consists mainly of sand.

The weather is temperate humid, without dry season. Temperatures are moderate, due to the proximity of the ocean and the river environment. Rainfall is regular with an annual average of 1,000 mm.

The zone is subject to relatively frequent and extended flooding, originating mostly from the rise of the Parana River. Local rains may worsen the situation, whereas the southeasterly winds produce floods from the River Plate and delay drainage. The floods cause social damages:

- a forced exodus of the inhabitants out of their houses or isolation;
- suspension of schooling activities;
- interruption of the family income derived from lack of economic activities and the economic losses of young crops, forest plantations, and cattle;
- lack of productivity of fields during and after the floods;
- suspension of industrial and commercial activities;
- isolation of the area resulting from a blocked infrastructure.

Diminishing the negative effects of the floods would enable the maintenance of the area's current function, whereas a greater development could be promoted. Impoldering of the whole delta of the Parana River is not possible at present. There is however, a plan, which consists basically of the construction of a dike, running in a North-Northwest direction on the current sand dunes line where their height decreases, and then Southwest up to the Parana Ibicuy river (Figure 3.3).

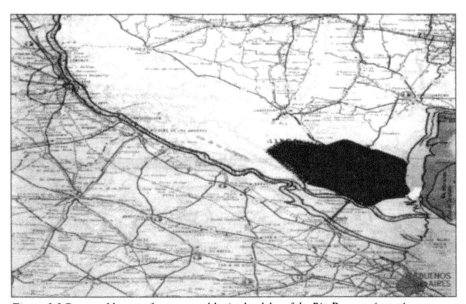

Figure 3.3 Proposed layout of an open polder in the delta of the Rio Parana, Argentina

This dike would have a lateral drainage canal, which could serve for providing filling material. The length of the dike would be about 50 km, with a maximum height of 6 m and 2.50 m width at the top. The dike would prevent the inflow of water from the Parana and Gualequay rivers. To the West it would leave an expansion area of 90,000 ha for storing the floods. During flooding the cattle in the protected area could be moved to the protected side of the dike.

The protected zone can be considered to be an open polder. To promote the outflow of rainwater and floods the project includes the construction of two auxiliary canals with a length of about 20 km, connecting the Paranacito river with the Uruguay and Parana rivers.

United States

Millions of hectares of land in the United States have been embanked and drained for agriculture and settlements, which include the reclamation of bays and riverbeds. Several hundreds of thousands hectares of impoldered farmland below sea level exist, particularly in tidal regions, like in the Sacramento Delta and in the coastal lowlands of North Carolina (Figure 3.4).

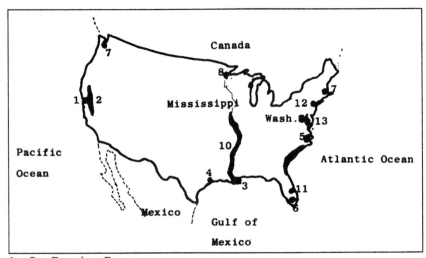

1 = San Francisco Bay
2 = Sacramento – San Joaquin Delta
3 = Louisiana: city of New Orleans
4 = Texas: land surrounding Houston
5 = North – Carolina: peat land around Albemarle and Pamlico
6 = Florida: agricultural land in the southern part
7 = Massachusetts and Washington: bays have been diked and managed for cranberry culture
8 = Minnesota: peat bay has been drained for agriculture
9 = Louisiana and some other states: diked and managed for rice cultivation
10 = Lower Mississippi Valley (Illinois to Louisiana)
11 = Gumbo Island
12 = Hackensack Meadow
13 = Chesapeake Bay

Figure 3.4 Polder development in the United States (Stoutjesdijk, 1982)

The following types of impoldering can be found:

- *impoldering of areas in the sea.* Impoldering activities in these areas started in the 17th century. The reclaimed land is mainly used as grassland for the production of hay. A relatively small area along the coast of the United States can be impoldered, as the major part consists of rocks and sand;
- *reclamation of marshes and swampy areas.* The reclamation of many low-lying marshes is economically feasible. The most important work is the construction of dikes, the digging of main ditches and the construction of outlets;
- *reclamation of lakes by pumping.* An example of this type of polder is the Lake Mattamuskeet in North Carolina. The area of the lake is 20,000 ha and was reclaimed around 1915. Around 1930 the project became unprofitable and pumping was stopped. The area became a lake again.

The most likely impoldering activities in future will concentrate along the seacoast to protect residential, industrial and agricultural areas from flooding. More river bottomland will be drained and a part of this area will be impoldered through embanking and pumping. Considerable acreage of low-lying peat deposits may be impoldered to facilitate the peat mining (for fuel) and farming.

Embanking and filling activities have reduced the San Francisco Bay's area by 30%, i.e. about 50,000 ha. The low lying land at the junction of the Sacramento and San Joaquin rivers was reclaimed by earth levees built by Chinese labourers a century ago. Farm tracts bordered by the levees are increasingly vulnerable to large-scale flooding as the land subsides and the dikes age and crumble.

The total deltaic area of the Sacramento - San Joaquin Delta comprises 300,000 ha. The lands of the delta are highly productive. The principal crops cultivated are asparagus, corn, alfalfa, sugar beets, tomatoes and melon. The polder area, situated between 1.50 m+MSL and 4.50 m-MSL, amounts 170,000 ha. The reclamation of the delta islands was initiated at about 1850. By 1870 some 6,100 ha were reclaimed by constructing low embankments. In 1960 the total agricultural land was 245,000 ha. Because of the industrial and recreational activities the agricultural land is expected to decrease to 239,000 ha.

An estimated amount of 400,000 ha from more than 6 million ha of river bottomland in the Lower Mississippi Valley (Illinois to Louisiana) was embanked and pumped dry.

Chesapeake Bay is the largest estuary on the East coast. It extends about 350 km, with a width varying from 6 to 50 km. The tidal shoreline totals more than 12,000 km. During the last century many areas were reclaimed. However, the rate of erosion is increasing. During the same period 18,000 ha were lost, due to erosion.

Canada

Because of the harsh northern climate, only 12% of the land in southern Ontario is suitable for agriculture. This land has the lowest elevation. It has been estimated that Canada has one-seventh of the world's fresh water. In addition to the Great Lakes, which it shares, with the United States, Canada has many large rivers and lakes (Figure 3.5). (http://www.infocan.gc.ca/facts/geography_e.html)

The region of southern Quebec and Ontario also has prime agricultural land, for example, the Niagara Peninsula. The large expanses of the lakes Erie and Ontario extend the number of frost-free days, permitting the cultivation of grapes, peaches, pears and other fruits. Two-thirds of the Great Lakes coastal wetlands have been drained or reclaimed for land development purposes including the need for prime farmland, new harbour facilities, and urban expansion.

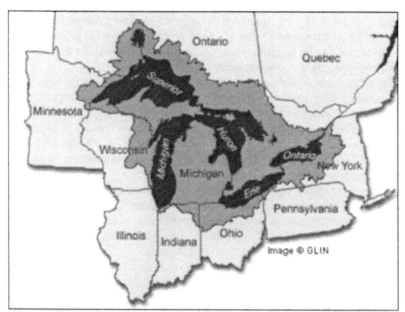

Figure 3.5 The Great Lakes basin in Canada (http://www.great-lakes.net/lakes/)

Water level fluctuations affect most of the 40 million people who live within the catchment boundaries either directly or indirectly. High water levels are of serious concern to those who own property and live on Great Lakes shoreline as serious flood and erosion damages can occur during storm conditions.

There have been extended periods of wet or dry years in the Great Lakes basin, which were responsible for extremely high lake levels in the early 1970s and in 1986, or low water levels of the 1930s, mid-1960s and 1998 - 1999. The full extent of these ranges is not seen during any one year. However in 1998, Lake Ontario saw a drop of 1.2 m from April to December. Spring lake level rises due to heavy precipitation and snowmelt runoff can also be dramatic. (http://www.on.ec.gc.ca/wildlife/factsheets/ fs_sustaining-e.html),(http://www.on.ec.gc.ca/glimr/water-e.html)

Mexico

Around 1200 the Aztecs entered in Central America, where they faced the problem that all the good land was already occupied and that no city allowed them to settle in its surroundings. They were actually shunted from place to place. The Aztecs, however, tried to settle, which resulted in a brief war. They were quickly driven into the marshes of Lake Texcoco, where they found shelter amongst the reeds. Then, according to the chronicles, their God advised them to build their new city Tenochtitlan on a small-uninhabited island in the reed marshes. After some time, the Aztecs realized that their new location had a number of advantages:
-    there were no land claims by third parties;
-    the island was completely surrounded by marshes, which gave the inhabitants an important defensive advantage during wartime;
-    three mainland cities, which offered good marketing possibilities, were located nearby;

- there was a large expansion potential. Floating gardens (chinampas) were constructed from silt mixed with reeds and refuse, whereas the marshes were gradually filled with dirt and rocks (Figure 3.6);
- the surrounding water offered them good transportation facilities (Aztecs did not know wheeled vehicles).

Figure 3.6 Aztec farmer cultivating his land (source: Luijendijk, 1987)

From this location, the Aztecs expanded their empire and reached a high level of civilisation. Some examples of their civil engineering skills were (Figure 3.7):
- the use of canal irrigation;
- the construction of a dike with a length of 16 km, which prevented flooding and allowed for a water management independent of the surrounding saline marshes;
- the use of control gates in the dike;
- the construction of a 3 km long aqueduct for drinking water supply with double pipelines;
- the construction of three causeways;
- dredging activities, from which the spoil was used for further land reclamation.

In 1519, an invasion by the Spaniards put a sudden end to this previously unknown civilisation, whose achievements compelled an enormous admiration from the entire western world of that era. Nowadays, Tenochtitlan is a suburb of Mexico City.

Figure 3.7 Tenochtitlan, the polder city of the Aztecs (Source: Luijendijk, 1987)

3.2 Review of water management and land reclamation in the Humid Tropical Zone

3.2.1 Water management and land reclamation in the Middle East

The oldest reclamation of marshy lowlands was carried out by the Sumerians in Mesopotamia some 6,000 years ago. It was in the area between the rivers Tigris and Euphrates, where according to the Koran and the Bible paradise must have been situated.

The reclamation works of King Menes form another example of ancient impoldering activities. He built his new capital Memphis on the old fertile riverbed, for which according to the historian Herodotus he constructed a dam in the river Nile some 20 km south of Memphis at Kosheish. The course of the river was diverted to a channel, which was excavated between two hills. The gravity dam is supposed to have had a maximum height of 15 m and a crest length of some 450 m. In a later stage king Menes excavated a lake northwest of the new town and dug a canal to connect it with the river Nile. The system of watercourses, viz. the lake, the canal, and the river, served as a moat to protect him from his enemies. The dam had to be guarded and maintained carefully, because in case of a breach the entire city of Memphis would have been flooded. When Herodotus visited Egypt some 2,500 years later, the dam was still guarded with greatest care by the Persians (Biswas, 1972). However, this narration of Herodotus has been contested by Norman Smit (1971), who stated that 'it is incredible that around 3000 BC Egyptian civil engineering had developed to the point where a river of the size of the Nile could be dammed, even temporarily - let alone be permanently - as the story implies'. Whatever is true, such a narration indicates the relative importance of these civil engineering activities on society. Nowadays the 'lost city of Memphis', the remnants of palaces, temples and houses, can be found in the cultivated area east of the Necropolis, all buried under the Nile deposits. The size of Memphis is deducible from a strip of over 30 km on the western Nile bank (Figure 3.8).

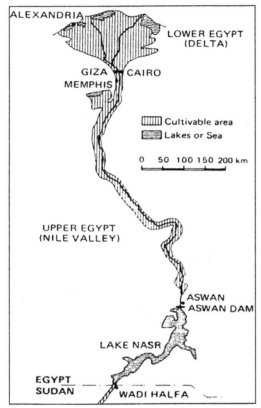

Figure 3.8 Location of the lost city of Memphis, Egypt (Maaskant, et al., 1986)

3.2.2 Water management and land reclamation in Asia

Bangladesh

From 1200 to 1757, the Mughal rulers constructed canals along the rivers and embankments besides the rivers. The low dikes and wooden box sluices were constructed and maintained by them for protection against saline water intrusion and floods during this period.

From 1757 to 1947, under the British regime flood control, drainage and irrigation has been abandoned because the ruler was more concentrated on navigation for trade. The development and maintenance of flood control and road communication networks was suddenly stopped.

Large-scale land and water development schemes began in early 1950s. After the country had suffered from unprecedented floods in two consecutive years 1954 and 1955, embankments and drainage sluices were constructed along the Bay of Bengal to increase agricultural production and reduce flood damages in 1986.

In 1974 the flood control and drainage improvement project (FCD) was reviewed and implemented by the government again. The areas for the FCD Schemes are developed in the lowlands of Bangladesh. Two types of floodplain can be distinguished, namely river floodplains and deltaic floodplains (Ali, 2002).

China

In China, reclamation of lowlands through drainage and irrigation has a long history. According to a Chinese legend, about 2,200 BC, the Yellow River was tamed by Emperor Yu of the Xia Dynasty, henceforth surnamed the Great. For nine years, men worked under his supervision to clean up the silt deposited in the river channels to lead it into the sea. In this way large areas of reclaimed land were made available to farmers (Framji, et al., 1981). About 300 BC, the Wei State diverted water from the Zhang River in the present Hebei province to reclaim saline land and grow paddy rice.

Since the Sui and Tang dynasties, the middle and low reaches of the Yangtze River begun to be developed and controlled. The Taihu lake basin was the most representative example. From 770 to 476 BC, people in the Taihu lake basin began to build dikes to reclaim the lake. During the Tang dynasty (618 - 907), polder construction developed rapidly and problems concerning flood, drought and waterlogging were tackled in a compressive way. By the northern Song dynasty (960 - 1125), basin wide networks of rivers, river outlets, lakes and ponds were basically formed. The protective embankments in lake and riverside areas along the middle and lower reaches of the Changjiang river started in the southern Song Dynasty (1127 - 1279). In areas along the southeastern coast impressive sea walls were constructed (Figure 3.9).

Figure 3.9 Seawalls in Zhejiang, China

Today the flood control systems in China safeguard 32 million ha of farmland (Cai Lingen, 2001). The total waterlogging removal area is about 20 million ha, accounting for 82% of that prone to waterlogging and the total reclaimed area of saline and alkaline lands is about 5.4 million ha, which accounts for 70% of the total cultivated saline lands. In addition in South China, improved area (groundwater table control) of waterlogged lowland is about 3.1 million ha, accounting for 31% of the total.

India

In India there are several reclaimed areas, which will be briefly reviewed.

The Kuttanad polders are situated in the southern part of India. The area, which is 1.5 metre below sea level, is approximately about 10,000 ha. Near by an impoldered area of 50,000 ha can be found in the Lagoon of Koshina (Kerala District).

The total area of the Sunderbans within West Bengal is 800,000 ha. Out of this area about 360,000 ha was reclaimed until 1982 for which about 3,400 km of bunds had been constructed to protect the land against frequent flooding by saline water (Stoutjesdijk, 1982). However, the protection by bunds is very limited. There were some suggestions to construct a closure dam across estuaries and interconnecting this dam by strong dikes encircling the entire western Sunderbans.

The total area of the salt lakes is approximately 12,000 ha. Most of the time these lakes were covered by 0.6 m of water. Some 7,000 ha have been reclaimed for agriculture in 1953. The area has been surrounded by embankments and is separated into two parts: 3,600 ha of northern Salt Lakes and 3,400 ha of southern Salt Lakes.

The Bhal area represents a former tidal flat that was abandoned by the sea. It is an alluvial coastal plain, situated in the Sasuashtra or Kathiawar Peninsula of which the tidal area is 36,000 ha. The reclamation of land in this area was done by constructing an earthen 42 km dike to protect 22,000 ha against the sea. The area is not quite successful because the saline soil problem has not been solved. A pilot polder of 1,500 ha was constructed in 1959/1961 but this project was not successful because of shortage of fresh water, drainage was not sufficient due to silting up of canals and low permeability of the Montmorillonite clay, permitting insufficient salt percolation. By 1982 some 50,000 ha of land were protected and used for cultivation. By the end of 20th century all potential saline area of 80,000 ha were reclaimed.

The Mahanadi catchment extends over an area of 14 million ha. The flood problem in this area is largely confined to the lower reaches. The worst flood condition in the delta occurs when there is simultaneously rainfall in the various sub catchment areas. During such floods, there is a prolonged inundation of vast agricultural fertile land, also resulting in sand deposition damage. There is a plan to construct storage dams in the upstream and strengthen the existing embankments in the delta.

The Godavani river has been embanked in the delta region with a total irrigated area 400,000 ha. Along both bank of the Krishna River, embankments can be found from the Vijaywada barrage to the sea. The embankment and drainage systems in the Cauvery delta are constructed in the area to prevent flood and reclamation land.

Irrigation in India dates back since prehistoric times as the records of old scriptures/literature, centuries old irrigation structures still exist in different parts of the country. However, the modern irrigation with storage dams, river diversion structures and canals started in the 19th century during the British regime in the wake of several famine and drought years. Irrigation induced waterlogging and salinity in some of the command areas there by 10% of the net irrigated area. Waterlogging of irrigated fields was first noticed during 1850 in the State of Punjab (Baig, 1997). In 1885, a rise in the groundwater table and salinity problems developed in Karnal of present Haryana State under the Western Yamuna canal system. Then, in 1892 the lower Chenab canal system followed by the other areas in the State of Punjab, similar problems were noticed.

Maharashtra also had the problem of waterlogging and salinity after introduction of irrigation in the Nira valley.

Drainage problems during 1870s and 1880s could be tackled with moderate canal realignment and improvement of drainage systems. But in the present day drainage has to deal with much more complex problems with vast extension of irrigates areas, intensive irrigation practices with multiple cropping, adoption of high water consuming crops and irrigation management involved with socio-economic conditions. Moreover, various infrastructure developments such as roads, railways, canals, and urban growth have also added to the problems by way of obstruction of natural drain lines.

In 1995 - 19998 the Gulf of Khambhat Project was planned to store fresh water to be used for irrigation, water supply and industrial requirement in the Saurashtra region. Operation levels of the fresh water reservoir range between 5 m-MSL to 5 m+MSL. Between these levels some 12,000 million m^3 of fresh water can be stored. The land reclamation and drainage improvement of salty mud and saline soil amount to 225,000 ha. This area consists of 105,000 ha between 8 m+MSL to 11 m+MSL, which will be used for dry land farming and the remaining area of 120,000 ha between the maximum reservoir level and 8 m+MSL will be mainly wasteland and lowland use of the area.

Indonesia

There is 43 million ha of lowland in Indonesia, mainly in the coastal area, of which 10.5 million ha has potential for agriculture (Stoutjesdijk, 1982). Some 7 million ha are in the tidal zone. Both spontaneous and government organizes tidal land development since more than 100 year ago covering in total an area of several hundreds of thousands of ha, polders have been made among the others, as follows.

Sisir Gunting polder in the northern part of Sumatra has been constructed in 1924. The total area of the polder is 3,000 ha. After 1975 - 1976 the dikes and sluices gradually deteriorated to such extent that more than 1,000 ha became unused.

Polder area near Kupang (Timor Island) lies in the Nusa Tenggara Timur province. The area lies at the Bay of Kupang and covers approximately 3,500 ha, which has been reclaimed from swamps. 10 km dike, a pumping station, a gate and canals have been constructed.

Setjanggang polder is situated on the Northeast coast of Sumatra, near Medan. It is a pilot polder, with an area of 3,600 ha.

The Rawa Sragi Swamp Reclamation project is situated along the downstream reach of the Way Sekampung province. It covers an area of approximately 34,000 ha, some parts have been reclaimed in earlier days. A polder of 7,400 ha has been constructed for 4,000 farmers. Kramat and Pisang swamps of 8,600 ha have been reclaimed from 1982 to 1984.

At Kalimantan, a floodplain area of 25,000 ha along the Negare and Martapura river has been protected by a dike.

The delta of the Kalibrantas, situated in eastern Java, is a small area, which has been impoldered.

The city of Jakarta is situated on low-lying land near the sea. The large and densely populated areas are becoming more and more liable to frequent inundation. The lower part of the city on the West and the East side near the coast require pumped drainage from polder areas, protected against the inflow from higher grounds. This area includes the Sunter West polder, Pademangan polder, Pluit polder, Tumang Barat polder, Grogol polder, Muara Karang polder and Kapuk polder.

Japan

Swampy land in its natural state was used as paddy field and irrigation was not always applied. Consequently large numbers of paddy fields suffered bad harvests as a result of floods and persistent dry spells. In the Middle Ages (800 - 1600) this type of paddy field was improved and new paddy field development was carried out. Flood protection works, like dikes and diversion weirs, were constructed. The shift from centralised to water management at village level guaranteed a continued expansion of the water management system, even during periods of political instability.

With its large population and relative small agricultural area (only 15 to 20%), Japan's land reclamation through impoldering has always played an important role. However, little is known about the origins. Around 1200 (the Minamoto Shogunate) there was not much need for reclaiming tidal foreland and coastal marshes. This changed when Toyotomi Hideyoshi established the Tokugawa Shogunate in 1603. It marked the end of a long period of unrest and local wars between the 'dyamyos' (feudal lords). Hidoshi, who was a daymyo himself, unified and pacified Japan. From then on, the daymyos could only increase their power by peaceful conquest of land on the sea. A famous engineer of that time was Kato Kyomasa, who is well known in Japan's history as a general and an architect of castles. He built the dikes of his polders like castle walls with steep facing. In 1639 the Tokugawa Government closed the country for foreigners and Japan went into a period of nationalisation. Yet the Japanese polder technology developed independently of the West. An example of the reclamation of an inland marsh of that period is the drainage and irrigation of the Minumadai area, which commenced in 1727 (Figure 3.10).

The completion of flood control works in the large rivers resulted in a rapid decline of damage caused by floods, especially in the low-lying lands in the midstream and downstream river reaches. The next important item was the water control of these embanked lowlands. With the establishment of drainage canals, pumped drainage started. The Azumi Irrigation Project, which aimed at the development of approximately 5,000 ha of the Azumi plain, can be considered the first project within this framework.

Before the modern times or before Meiji Restoration (1868), all governors such as feudal Lords tried to construct not only irrigation facilities, but also drainage facilities to protect land against flood disasters and reclaim agricultural land. Drainage improvement of marsh or swamp area resulted not only in improvement of agricultural land conditions but also in expansion of the area of paddy fields by simultaneous reclamation. A matter of their main concern was to reclaim arable land as wide as possible and to reduce flood damage in order to maximize their tax collection from farmers. At this time, not only irrigation but also drainage facilities such as new flood diversion channels, major drains with gates, and on-farm ditches had been developed.

As there was a strong need for more arable land, the Japanese irrigation, drainage and reclamation techniques rapidly developed during the period 1868 - 1945, and included the reclamation of tidal forelands, drainage of lakes and inland marshes, reclamation, flood protection and drainage of floodplains, enclosure of estuaries and lagoons with partial reclamation, etc. It was in this period that the western hydraulic engineering technology reached Japan, a large group of Dutch engineers stayed many years as advisors on various hydraulic projects.

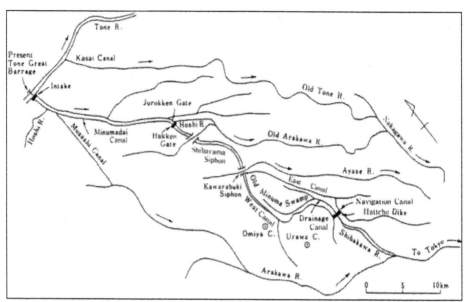

Figure 3.10 Situation sketch of the Minumadai Irrigation Project, Japan

At the end of the 19th century more attention was paid to the drainage and irrigation conditions of existing agricultural lands and to reclamation works in new areas. Especially the areas in the proximity of large rivers were poorly drained. In the period 1923 - 1945 many activities focused on these areas, covering an area of 759,000 ha. Drainage improvements were carried out in some 25% of the existing schemes.

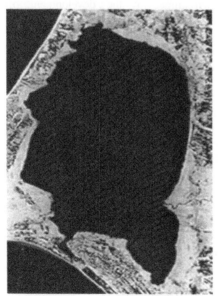

Hachiro Gata Lake, before reclamation   Newly born Land after reclamation

Figure 3.11 Hachiro Gata Reclamation Project, reclaimed in 1957 and finished in 1977
(http://www.ogata.or.jp/english/outline/polder/polder.htm)

From the post-war period, the Japanese Hachiro Gata Reclamation Project, reclaimed in 1957 and finished in 1977, must be mentioned. Hachiro Gata was the second largest lake in Japan, after completion it became new land of 17,203 ha (Figure 3.11).

Malaysia

During the last 70 years many lowlands and swamps located along the coasts have been reclaimed and converted to highly productive paddy polders. Paddy polders generally range from 1,200 ha to 4,800 ha. Individual polders are surrounded by an embankment and provided with drainage systems. Main drains are usually designed to have a total length of less than 5 km to the outfall. From 1932 to 1975, a total of 155 schemes, covering a gross area of 340,000 ha have been completed. Polders with an area larger than 1,000 ha are as follows:
-   Muda Irrigation Scheme covers, 96,000 ha of paddy land;
-   Muda River Scheme, 6,500 ha of paddy land;
-   Krian Irrigation Scheme, 6,500 ha of paddy land;
-   Trans-Perak Scheme, 300,000 ha of agricultural land, which comprises of 9,500 ha of paddy and 215,000 ha of trees crops;
-   Bagan Datoh Reclamation Scheme, 2,500 ha of trees crops;
-   Northwest Selagor Agricultural Development Scheme, 63,000 ha of agricultural land, which comprises of 20,000 ha of paddy and 43,000 ha of tree crops;
-   Klang-Kuala Langat Scheme is reclaimed from swampy coastal of area, 15,000 ha for tree crops;
-   West Johore Agricultural Development scheme, 146,800 ha of tree crops;
-   Tumpat-Pengkalan Kubor Drainage Scheme, 7,000 ha of tree crops;
-   Balai Besar Irrigation Scheme, 1,200 ha of paddy land;
-   Norok Reclamation Scheme, 9,000 ha of paddy and coconut;
-   Kabung Nyabur Drainage Scheme, 4,500 ha of coconut;
-   Dao-Loba Balu Drainage Scheme, 5,000 ha of paddy and coconut;
-   Klias Drainage Scheme, 5,000 ha of oil palm.

The Philippines

The stonewalls, canals, dams and reservoirs of the Igorots can also be considered as a type of architecture, or at least stone engineering. The amount of stones used by the Igorots in their hydraulic engineering works is estimated to far exceed in bulk of those used in building the Pyramids or the Great Wall of China. Many of these walls and canals are thousands of years old and have withstood countless typhoons and the effects of sun, wind and time.

Of the approximately 3.4 million ha of rice lands, and some 2.1 million ha (61%) are irrigated, 1.2 million ha (35%) are rainfed lowland, and 0.07 million ha (2%) are upland. Much of the country's irrigated rice is grown on the Central Plain of Luzon, the country's rice bowl. The rest comes mainly from various coastal lowland areas and gently rolling erosion plains e.g. in Mindanao and Iloilo (http://pne.gsnu.ac.kr/riceipm /philippi.htm).

Republic of Korea

Concentrated efforts have been made to reclaim one million hectares of upland and 600,000 ha of tidal land to expand the limited land resources of the nation and to support country's aim to become self-sufficient in meat and dairy products

(Stoutjesdijk, 1982). Extensive tidal lands on the South and the West coast are very favourable for reclamation, because of the occurring continental shelves with low depth, extremely indented shored line and wide tide. The gross area is approximately 180,000 ha, measured along the low-water line. To develop, it must be protected from the twice-daily inundation. The Mokpo area in the southwestern part of the country has been partly reclaimed and other parts are considered to be most suitable for reclamation because of good soil for agriculture. From 1930 - 1940 some reclamation was done in the Mokpo -Yongsan area. By closing the estuaries, three lakes were formed with a total area of 10,700 ha and tidal area situated above mean sea level of 22,275 ha. Chan Po polder, a pilot polder of 228 ha, has been reclaimed. The two pilot polders Kanghwa and Kwang Yang of each 120 ha, were planned to be reclaimed in that area. Due to lack of funds only 60 ha of the Kanghwa polder has been constructed on the Kwangwa Island, near Onsu-ri village in 1961.

From 1970 - 1977 the Pyongtaek project in the northern part of the country 2,682 ha of the total area 18,419 ha has been reclaimed for rice. From 1974 - 1979 the Gyehado project has been executed. By connecting the island and the shoreline with two sea dikes the area between the two dikes became a polder. From 1978 - 1986 the Yong San Gang area has been reclaimed for rice by closing off estuaries, the total area is 20,700 ha. From 1982 - 1986 the Da Ho development project was executed. A tidal area has been reclaimed for 3,700 ha.

Comprising the two free-flowing estuaries of the Mangyeung and Dongjin rivers, the Saemangeum area is being considered for reclamation. Saemangeum comprises some 30,000 ha of tidal-flats (being up to 25 km wide in some stretches) and 10,000 ha of shallows. An entire area of 40,100 ha is envisage to be reclaimed in the world's largest known ongoing coastal reclamation project. The first phase of the project, the construction of an outer seawall wide enough to take two lanes of traffic and over 5 m high, started in 1991, and is still ongoing with over 90% of the 33 km long wall already completed. As proposed the wall will eventually dam off both estuaries in order to create 28,300 ha of rice-field and industrial land, and 11,800 ha of barrage lake. As proposed, it will likely be completed by 2009, with the gradual conversion of the tidal-flats to rice-fields following a few years later (http://www.wg-sdta.icidonline.org/contents.pdf).

Republic of Singapore

Between 1965 and 1987, a series of large scale land reclamation projects were undertaken, enlarging the coastal zone by 50 km, creating Southeast Asia's largest industrial complex, the Jurong Industrial Estate. More than 300 factories, deepwater wharves and the Changi International Airport were built largely by filling in the marsh and shallow waters of the straits (http://www.firstlegoleague.org/default.aspx?pid=7830).

In 1999, the plan was developed to reclaim land for enlarging Changi Airport, Jurong and Pasir Panjang. Singapore is some 3,300 ha bigger than it used to be - 5% of its total land area - and it aims to get even bigger yet. Predictions are that it will have grown to 82,000 ha by 2010 (http://www.atimes.com/atimes/Southeast_Asia/EG31Ae01.html).

Sri Lanka

Records of a drainage project exist dating back to the 15th century. The marshy land of Muthurajawela, north of Colombo, 2,400 ha in extent, is known to have been cultivated with paddy before 1518. The Muthurajawela for paddy cultivation does not exist anymore due to high salinity problems and flooding. Drainage for agricultural development has been undertaken in 1940. During the Second World War the construction of salt exclusion works with pumps for drainage had been developed and between 1964 - 1965 these areas did extent to 16,600 ha of land in ten sites of low lying lagoons and marshes along South, Southwest and Northwest coastal zones.

By 1967 reclamation of land along the West and Southwest coastal belt was proposed. The project area consisted of 16,400 ha along 240 km length of coastal belt from Chilaw to Matara. The purpose of the project was to develop low lying flat coastal swamps for paddy cultivation by draining out the excess water, partly by pumping and partly by gravity and exclusion of salinity. Siltation and weed growth clogged the internal drains seriously, narrowing the river outfall by sand deposits or frequently complete closure by sand bars due to wave action during low streamflow. The project works composed of clearing existing drainage canals, enlargement of the streams, construction of a new drainage system with pumping facilities and regulators for exclusion of salt and discharge control. Three years after completion it was found that yields and benefits were not equally distributed among the schemes. The upper tracts of land were double cropped and had increased yields, while lower tracts produced low yields and had only a single crop. The upper tracts faced with a lack of adequate and timely water supply and water shortage for homestead and farming of land above 1.3 m+MSL due to improved drainage of land. The lower tracts faced with drainage conditions and salt water intrusion (Weerakorn, 1997)

In 1976 flood protection in the Southwest coast for the river Gin Ganga was completed and in 1983 flood protection for the river Niwala Ganga was completed. These two projects consisted of constructing a dike along the river and drain the area by pumping and gravity. The problems that have occurred after draining the area are land subsidence and acidification due to the drawdown of the groundwater table and exposing the subsurface of pyrite for free oxidation (Weerasinghe, et al., 1996).

Vietnam

The major river in the South is the Mekong River, which flows from China through Laos and Cambodia into Vietnam. It created the Mekong delta and continues to extend it. Many canals have been built for drainage, irrigation and transportation. The main river in the North is the Red River. It flows from southern China into Vietnam. It has created the 1,553,300 ha Tonkin Lowland. For over 2000 years there have been canals, irrigation and dikes built in the lowland. These lowlands belong to the best rice fields in the world.

The water management in the Red River delta is involved with flood control, water supply and drainage. Flooding usually originates in China. Inundation of inland caused by heavy rainfall from storm usually occurs between July and September. In the dry season between November to April rainfall amounts to 15% of the annual rainfall. At this time water levels in all rivers are low and salinity intrudes very far up to 20 - 25 km from the coast (Vinh, 1997). A dike system with a total length 2,100 km and 500 sluices, irrigation and drainage have been constructed over the last centuries, especially since 1960. There are three typical types of drainage in this region:
- by gravity in highly elevated land;
- by using tidal fluctuation, there are 15 irrigation schemes with the total drainage area of about 410,000 ha;
- by pumping, an area of about 400,000 ha is drained by pumps of various types and capacity of each unit ranging from 1,000 m^3/hr to 30,000 m^3/hr (Vinh, 1997).

3.3 History of water management and land reclamation in the Netherlands

During many centuries the inhabitants of the marshy part of the Lowlands lived on natural elevations and later on man-made mounds too. But about 1,000 years ago, the introduction of a sluice or a gate enabled the rather influential local or regional authorities to start embanking of the lowlands that could be drained by gravity. Within a few, but rather turbulent, centuries practically the whole low part of the Netherlands has been endiked. Since then the lowland-Dutch live behind dikes (Figure 3.12) (Schultz, 1982 and Van de Ven, 1993).

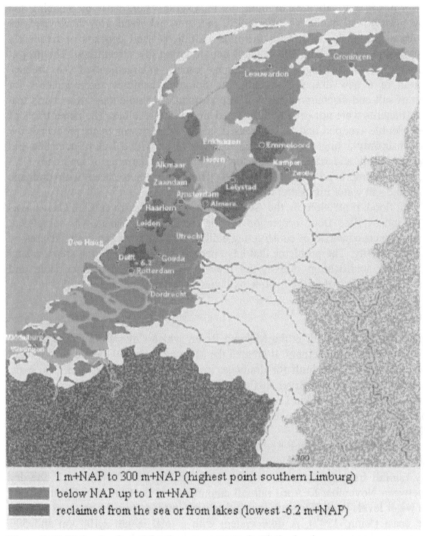

    1 m+NAP to 300 m+NAP (highest point southern Limburg)
    below NAP up to 1 m+NAP
    reclaimed from the sea or from lakes (lowest -6.2 m+NAP)

Figure 3.12 Higher lands and flood prone areas in the Netherlands
(http://www.rabbel.info/images/lowhighNeth.gif)

In the 15th and 16th century existing types of windmills were modified gradually in order to get a usable tool for lifting water. The improved windmills could work in all directions of the wind and arranged in series – made the drainage of rather deep lakes possible (Figure 3.13). Striking in this development is the fact that the first application of a series of windmills was a very big land reclamation project namely the reclamation of the Beemster, being a lake of 7,100 ha, drained between 1608 and 1612 with 15 series of three windmills (Schultz, 1992).

In the first half of the 17th century 28,000 ha of lake bed was brought under cultivation. In the same period around 40,000 ha of coastal accretion were embanked. It was in this time that the Netherlands became a powerful and rich nation. The combination of self-confidence, financial means and technical innovations prompted Leeghwater, to write in 1641 his famous book on draining the Haarlemmermeer by means of 160 windmills, a huge lake near Amsterdam on the bottom of which nowadays Schiphol International Airport is situated. This book became a bestseller and went through 17 editions (Leeghwater, 1641).

Probably even more interesting was that in 1667 Henric Stevin launched the idea of closing of the Zuiderzee. This Zuiderzee was originally a lake, or a set of lakes connected to the sea by means of a natural channel. The channel widened and from the beginning of the 13th century the lake became an estuary of the sea. Henric Stevin pointed out that an enclosure of this estuary would eliminate the storm floods and the venom of the salinization from the heart of the country and would make much land suitable for embanking. In view of technical means of that time the idea was impracticable and it did not draw much attention (Stevin, 1667).

Figure 3.13 Windmills arranged in series of three to pump a polder dry

After the 'Golden Age', Dutch history shows a duller period. It is clear that a small country like the Netherlands could not afford the power needed to keep a leading position during a longer time and it slipped back in the ranks of European nations. Also, the extension of agricultural land slowed down. However, not only a failing spirit, other factors too hampered the rate of reclaiming new land. The growth of the coastal accretions decreased and the draining of most of the remaining, rather deep, lakes had to wait for the new tools. The new tools came, 125 years after the boom-period of the draining of lakes.

In 1776 the first steam engine was imported from England to be installed in a pumping station. Then, in particular after the Napoleonic time the tide turned: industry developed, initially strongly influenced by English products and enterprises, the leading positions in the country shifted to other countries. In this period the old idea of the enclosure and partial reclamation of the Zuiderzee revived too. This revival was stimulated by two events: A disastrous storm flood in 1825 sweeping the eastern shores of the Zuiderzee and the technically successful draining the Haarlemmermeer between 1840 and 1852. The way of draining of the Haarlemmermeer was again a striking example of the push to new developments given by a big land reclamation project. Since 1776 several steam pumping stations had been placed in polders, but always together with windmills. The Haarlemmermeer was the first large lake in the Netherlands drained with only steam power. After the successful draining of the Haarlemmermeer lots of windmills were replaced by steam engines (Schultz, 1992).

The other area was the Zuiderzee and as soon as the reclamation of the Haarlemmermeer had been executed, serious plans started to reclaim land in the Zuiderzee. A disaster was necessary to give the final push to start the works. On 13th January 1916, a gale caused flooding in the areas around the Zuiderzee. During the execution of the works in this area new ideas about the land use in the polders have been developed (Schultz, 1983). So in the recent polders, apart from the use for agricultural purposes, land is also used for town building, recreation areas and nature reserves. In February 1953, in the southwest of the Netherlands 195,000 ha were flooded from the sea and almost 2,000 people drowned. In 1958 the Delta Act was accepted, giving way to the building of large dams and other hydraulic constructions and the raising of existing dikes along the North Sea and main rivers.

The past decade has shown a similar reaction. In 1993 and 1995 the whole country panicked when very high water levels in the rivers occurred. For years plans had been made to reinforce the river dikes, but neither government and nor the public were aware of the fact that large river discharges formed an actual threat to the safety of the country. Therefore, after 1995 the river dikes were reinforced at high speed.

An excess of water due to heavy rainfall in 1998 also surprised many people. It was thought that the capacity of the canals and pumping stations were able to match all weather conditions. After all, much money had been invested to that effect in the second half of the twentieth century.

Both the high water levels in the rivers in 1993 and 1995 and the excess of water in 1998 clearly showed how vulnerable the Netherlands still are where water is concerned, which came as a surprise to the government, too. It was therefore decided to institute a State Committee, 'The committee for water management in the 21st century'. This committee published its proposals in August 2000 to decrease the vulnerability of society to floods and excess of water. The committee did not only take into account infrastructural measures such as the heightening of dikes, the widening of rivers and waterways, and the construction of even bigger pumping stations; research was also done on whether the Netherlands can be made less vulnerable by taking measures in the field of physical planning.

3.4 History of water management and land reclamation in Thailand

In order to understand the history of water management and land reclamation in Thailand it may be divided into 5 topographical regions as shown in Figures 3.14 and 3.15.

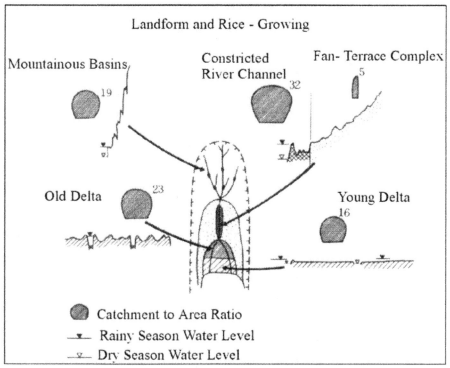

Figure 3.14 The five topographical regions of the Chao Phraya basin (adapted from Yoneo Ishii, 1975)

Three types of plateau lands can be distinguished: constricted river channels, hilly and level. Topographically, the constricted river channel type is identical with the region of the same name in the Chao Phraya basin. The hilly type is intermediate between the fan-terrace complex area and the mountainous basin, and finally the level type is similar to the fan-terrace complex area but smaller catchments to the area ratio, which means that the water conditions are poorer. These types of plateau topography are common in the Indian subcontinent as well as in the northeast of Thailand.

Considering the conditions in Thailand, it reveals that the climate is not the only factor for cultivation but also the topography of the land. Under the humid tropics, weathering and erosion are continually in the flood-free upland that comprises the arable land of the humid tropics. Under flat topographic conditions where erosion is negligible, weathering progresses steadily and eventually produces laterite.

Due to the topographical conditions the soil in Thailand can be divided into three types, namely groundwater soil, intermediate soil and upland soil (Yoneo Ishii, 1975). The groundwater soil is at least part of the year strongly influenced by groundwater. This soil is found in the topographically low place, namely the young delta and constricted river channel area.

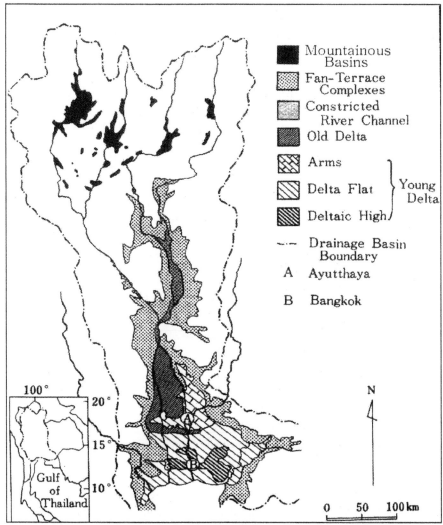

Figure 3.15 Location of regions in the Chao Phraya basin (Yoneo Ishii, 1975)

Water collects here from surrounding regions bringing silt and nutrients and the soil is generally wet and fertile. Upland soil is found on higher grounds, where drainage is good but the water tends to carry soil particles and nutrient elements away to lower grounds. That is why this soil is highly susceptible to weathering and erosion. It is found mainly in the higher parts of the fan-terrace complexes and mountainous basins and tends to be dry and infertile. The intermediate soil lies between the two extreme soils: the land is intermediate in both topography and degree of fertility. Its distribution coincides with the old delta and the lower parts of the fan-terrace complexes. In the higher parts of this region upland soil prevails and in the lower parts semi groundwater and groundwater soils are predominant.

The groundwater soil region, with exception of vegetable and fruit gardens on polder land around Bangkok, is used exclusively for lowland rice cultivation, whereas the upland soil region has been left mostly shrub and has occasional dry fields, mainly maize and cassava. In the intermediate soil region, flood paddy fields occupy the lower parts and dry fields in the higher parts and besides maize, cassava, and sugarcane are sometimes grown.

Development of rice growing in Thailand

Archaeological remains suggest that rice was cultivated on the higher grounds, on the hills and alluvial fans, where it must have relied solely on rainwater (Yoneo Ishii, 1975). Therefore the upland soil region and the higher parts of the intermediate soil region are thought to have been the scene of the earliest upland rice cultivation. The first cultivators, as present day's rice farmers, probably cleared the lands by cutting back the forest and burning dried vegetation and then planted rice seeds in freshly cleared land after the start of the rains. Topsoil unlike today's upland soil, would have been rich in nutrients under the forest cover and yields must have been good. However, the same land could not be cultivated for more than a few years, since upland rice cultivation may affect soil sickness and furthermore the nutrients absorbed from the soil by the crops would not have been replenished from external sources. Thus the farmers had to open new land by clearing more forest, in practice known as shifting cultivation, or slash-and-burn agriculture. Under the climatic conditions of Thailand, land abandoned after erosion does not readily revert to forest, but becomes covered with shrub, and does not again develop the fertile topsoil of the forest.

The destruction of the forest caused by shifting cultivation of upland rice, and the accompanying transformation of the soil into the groundwater soil type, upland soil type and intermediate region, eventually undermined the basis for cultivation of upland rice and brought an end to the upland rice. Thus people started to settle in alluvial fans and valleys where water supply was good, and cultivated lowland rice there. As the population increased lowland rice growing expanded downstream, eventually into large catchments. The young delta was brought under cultivation between the middle of the 19th century and the beginning of the 20th century. Later, particularly after the Second World War, new fields were opened on higher grounds with less favourable conditions. Nowadays rice lands in Thailand can be classified as irrigated, rainfed lowland, deepwater and upland ecosystems (Kupkanchanakul, 1999).

Irrigated ecosystem

Less than 20% of the rice area is irrigated. Irrigated land is the most favourable environment for the rice growth. The purpose of the irrigated ecosystem is to increase rice production in the rainy season and the dry season. In the crop year 1997/1998, the irrigated ecosystem accounted for 32% of the annual rice cultivated area. Average yield in the irrigated area was about 4.31 ton/ha in the dry season and 3.55 ton/ha in the rainy season in the crop year 1997/1998. Major production constraints in this area are water scarcity in the dry season, insect pests and rats.

Rainfed lowland ecosystem

Rainfed lowland is the predominant rice ecology in Thailand. In the crop year 1997/1998, rainfed lowland accounted for about 75% of the rainy season rice area and accounted for about 49% of the annual rice production. Under rainfed conditions rice is usually grown only once a year in the rainy season, where the monsoon rain is the single source of water supply for the rice cultivation. Average yield in the rainfed lowland was extremely low, about 1.87 ton/ha in the crop year 1997/1998, which needs to be improved. Major production constraints in this area are rainfall variability, drought, submergence and inherent low soil fertility, especially in the northeastern part of Thailand. Infrastructure at farm level in most rainfed lowland rice is very poor and cannot support a high level of rice production (Kupkanchanakul, 1999).

Deepwater ecosystem

Vast rice growing areas in Thailand are subject to long periods of deep flooding annually. The deepwater ecology in the Central Plain has changed due to land modification and infrastructure made to control floods which resulted in reduction of the deep water area (Molle and Keawkulaya, 1998). The harvested area under the deepwater ecosystem was estimated at about 0.16 million ha or about 1.9% of the rainy season rice area in crop year 1997/1998. The average yield in the deepwater ecosystem is low, about 1.95 t/ha. Drought at the early vegetative phase, long term deep flooding at the late vegetative phase to early ripening and weed competition are the most important rice production constraints in this area.

Upland ecosystem

Upland rice is the smallest rice ecosystem in Thailand. In the crop year 1997/1998, upland rice accounted for less than 1% of the area and production. Drought, poor soil fertility and weed competitions are the most important constraints in this area.

Land reclamation in the Chao Phraya basin

To clear the wasted land and cultivate rice in the Chao Phraya basin in the former times, the two basic conditions were access to the land and water for cultivating the adjoining land. Therefore many canals were constructed in this area between the 18th and 19th century. The rice cultivation in this area depends on the water supplied by gentle natural flooding of the Chao Phraya river.

However in the lower delta area, downstream from Ayuttaya, where very flat lowlands are located, there is inundation, which is usually caused by overflow from upstream and overtopping and breaches along the Chao Phraya river and the Tha Chin river. The inundation water spreads over a wide area and most of the water is detained in this area for a period of 2 or 3 months and then naturally or artificially drained into the river or directly to the sea. In this area, major flood damage occurs to both agricultural and urban areas. Providing dikes along the river, including drainage systems and pumping facilities created polders. The mode of irrigation in the Chao Phraya basin is shown in Figure 3.16.

Nowadays most of the canals still exist, but in some areas the canals were filled and the width of some of the canals was reduced because of changes in the canal boundary, as a result of there is a reduction of drainage capacity. Also the mode of transportation was changed from boat to car or bus. So many roads have been constructed. Whereas the urban area has increased and the ground surface has become lower due to settlement which was caused by groundwater extraction. Therefore pumping stations and regulating gates have been built along the edge of the Chao Phraya river to evacuate and manipulate the excess rainwater in the polder along the riverside. However, the increased capacity of pumping may not be sufficient to solve the drainage problems because the limitation in water carrying capacity of the canals to the pumping station, if the width and depth of the canals will not be increased. Moreover, according to traffic congestion, the boat transportation has been selected for mass transportation, which is one of the factors constraining drainage management.

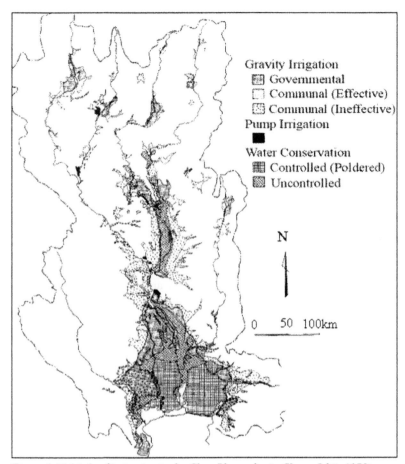

Figure 3.16 Mode of irrigation in the Chao Phraya basin (Yoneo Ishii, 1975)

4 Water management in polders in the Netherlands

4.1 General water management objectives

In the polders in the Netherlands the primary function of the water management system is drainage, but more precisely to maintain the desired groundwater table in different parts of the area in such a way that the area can function in the desired manner. Initially the polders were intended for agriculture. The design of the main drainage system was mainly based on agricultural requirements. Later land was used for other purposes such as urban areas, recreation and nature.

With the above objective the systems were designed to collect the excess rainfall, temporary store it in the open canal system and eventually discharge it out of the polder by discharge sluices or pumping (Figure 4.1). Due to the subsidence that has occurred in the polders in the Netherlands, almost all the polders are nowadays drained with pumping stations. With respect to the objective of drainage, off-season drainage is required in the Netherlands, while rainfall excess generally occurs in the non-cropped season. During the cropping season the groundwater is sufficiently deep, but high groundwater tables in the off-season may adversely affect the soil structure, soil workability, soil temperature and accessibility of the land and thus indirectly the development of the following crop.

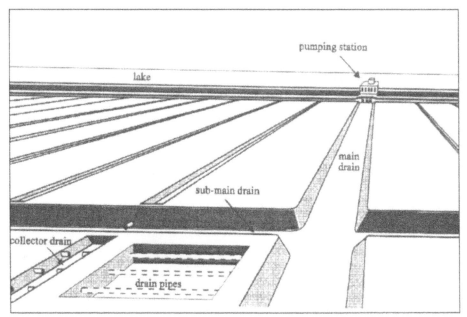

Figure 4.1 Scheme of a typical clay polder in the Netherlands (after Luijendijk and Schultz, 1982)

Water management in a polder is in fact groundwater management and urban drainage. Open water can influence the groundwater table. The water level in the open watercourses is kept more or less constant throughout the year to prevent subsidence of the soil, ensure a sufficient water supply for the crops and sufficiently dry conditions during the wet periods.

The main drainage system serves as hydraulic conveyance to the structures and pumps and at the same time as storage reservoir apart from other purposes. The alignment of the system of drains is influenced by the surface level and by the location of the pumps. During extreme rainfall situations there is a need to keep the rise of the water level within tolerated limits. In order to do this an amount of around 1 to 10% of the whole area is provided with open water storage in terms of open drains in the rural areas. Moreover, an open water area of around 3 to 8% exists in the urban areas.

The main drainage system of the polder consists of collector drains, main drains and structures. The agricultural fields of the polder area are normally drained by gravity within the systems. The automatic pumping stations with electrical pumps are installed and operated at the main drains. In pumped drainage the system is generally composed of three major components, which are field drains, open main drainage network and pumping station. Characteristic values for the water management systems in different types of polders in the Netherlands are shown in Table 4.1.

Table 4.1 Characteristic values for the water management systems in different types of polders in the Netherlands (Luijendijk and Schultz, 1982)

Type of polder	Open water in %	Polder water level in m-surface	Pumping capacity in mm/day
Peat polder	5 - 10	0.20 - 0.50	8 - 12
Old clay polder:			
- grassland	3 - 10	0.40 - 0.70	8 - 12
- arable land	5 - 10	0.80 - 1.00	8 - 12
IJsselmeerpolders	1 - 2	1.40 - 1.50	11 - 14
Urban polder	3 - 8	1.50 - 1.80	15 - 30
Greenhouse polder	3 - 10	0.80 - 1.00	20 - 30

The standard drain depth depends on the soil type and on the land use. The standards that are generally used are shown in Table 4.2. The variation depends on the soil type, the drainage system (surface or subsurface), the grown plants and the cultivation method.

Table 4.2 Standard drain depths in m-surface for different types of soil and land use in the Netherlands (Cultuurtechnische Vereniging, 1988)

Land use	Clay	Peat
Urban	0.80 - 1.00	0.60 - 0.80
Pasture	0.60 - 0.80	0.40 - 0.60
Greenhouse	0.60 - 1.00	0.60 - 0.80
Cultivation	0.80 - 1.00	0.60 - 0.80
Nature	0 - 0.30	0 - 0.30

In a flat area like a polder the hydraulic gradient, which is required for flow in the open drains is generated artificially by the drawdown produced by the pumping stations.

Dimensions of drains and pumping capacity

The required pumping capacity to maintain the water level in the main drains depends on the following conditions:

- climate;
- soil;
- land use;
- drainage;
- seepage;
- storage capacity of the open drains.

One of the methods to determine the pumping capacity for a polder is based on the time series of daily pumping amounts in an existing polder. Daily change in the open water storage is added to the pump outflow to obtain the inflow in the main drainage system due to excess water. The inflow is further corrected for seepage and then these data can be used for frequency analysis with Gumbel's extreme value method. The following water balance equation may be used to find the pumping capacity:

$$I + S = Q + B$$

where:

I   = net inflow of excess water into open drains (mm/day)
S   = seepage inflow (mm/day)
Q   = discharge of the pumping station (mm/day)
B   = storage in the open drains (mm/day)

In general depth and distance between field drains determine the capacity of the system. The optimal design may be formulated using certain economic criteria where the net benefits as a result of drainage are maximal. When the economic criteria are translated into hydrologic criteria the design discharge can be determined. The system should be able to discharge this amount with an acceptable rise of the water level during the concerned period. However, the economic optimal drainage may vary for different hydrologic situations and also vary based on design formulation i.e. steady or unsteady state.

The ultimate objective of the design of a pumped drainage system is to optimize the system in such a way that the sum of discounted benefits and of discounted costs is minimal. Such an economic assessment can be coupled with the design of the three major components of the system. The process to design the water management system of a polder in the Netherlands may then be composed of four main parts as follows (Figure 4.2):

- data collection and data analysis;
- hydrological computation;
- hydraulic simulations;
- costs analysis.

Data collection and data analysis

Data on daily rainfall for rural areas and preferably on hourly rainfall for urban areas, evaporation, crop water requirement, data on soil properties, topography, land use and agricultural practices were collected. The data have been analysed for consistency. Data on rainfall have been analysed for probability of rainfall for a certain period, which can be used for the hydrological computation.

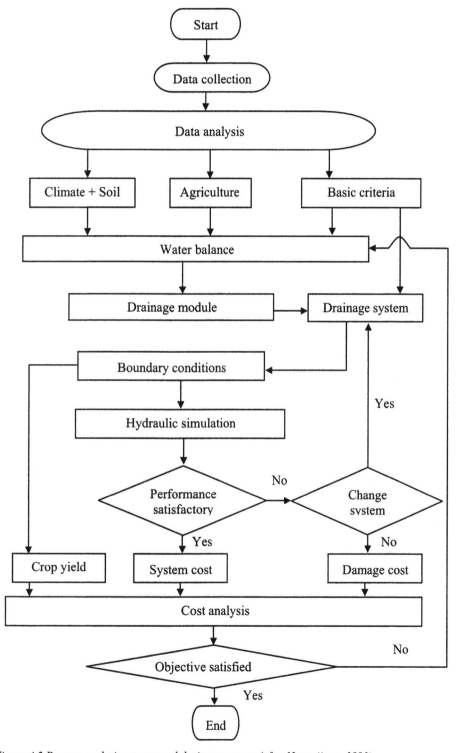

Figure 4.2 Process to design a pumped drainage system (after Notowijoyo, 1991)

Hydrological computation

In the hydrological computation normally the water balance and peak discharges for consecutive days are being determined. The peak drainage discharges have been analysed to determine the return periods.

Hydraulics simulations

In the hydraulics simulations, the hydrodynamic modelling can be used to simulate the water levels in the canal systems at peak drainage discharge, which data input is topography and geometry of the canal systems, layout of the canal systems and control structure. Water levels and discharges at different locations can be used for calibration and verification of the model. Then with the different pumping capacities, canal dimensions and layout, the water level rises can be calculated.

Costs analysis

Based on the results of the hydraulic simulations, the cost of each drainage system for different ground elevations, for the following item have been determined.
- bill of quantities and cost estimates;
- comparison of construction, operation maintenance and damage costs;
- selection of the dimensions that best satisfy the objective.

4.2 Water management in a clay polder

4.2.1 Groundwater table

In the agriculture field a certain groundwater table has to be maintained for good soil moisture conditions in the unsaturated zone, especially in the top layer. This is important as far as the yield of crops is concerned. A very deep groundwater table during the growing season and a very shallow groundwater table during the winter season would adversely affect the soil structure, soil workability, soil temperature and accessibility of the land. It indirectly affects the crop yield reduction. Maintaining an optimum temperature for the crops (15 - 25 °C) is an important reason of draining the soil in spring. By removing water from the soil, the soil is heated faster, cultivation of crops may start earlier and therefore a better yield may be expected. Most plants get oxygen through their roots and aeration of the root zone is improved by draining water from the soil during the growth period. Moreover, draining of water from the soil is necessary for proper development of the roots and the soil structure.

4.2.2 Subsurface drainage

A subsurface drainage system in a clay polder nowadays generally consists of corrugated PVC pipe drains, which lie above the water level in the open drains. In the IJsselmeerpolders the bottom of the open drains lies above the polder water level and there is normally no water retained in the drain. In this area for agriculture the minimum required depth of the subsurface pipe drains at the outlet is 1.10 m-surface. For fruit orchards this depth is 1.30 m. To avoid frequent submergence of the subsurface drain outlets the polder water level should be at least 1.40 m-surface. A freeboard of 0.30 m is sufficient to account for flow resistance and temporarily higher water levels due to wind effects and rainy periods. In old clay polders, the bottom of the open drains lies generally below the polder water level (Table 4.1).

The distance between the subsurface pipe drains is generally designed such that a fixed discharge q is released at a fixed groundwater midway between the drains. Flow to the drains is mainly through the soil above drain level as permeability below the drains is normally very low. The subsurface drains in the rural area are designed for a discharge of 5 - 10 mm/day or slightly more to allow for upward seepage, where this occurs. Depending on the land use, the design discharge is linked to a technically preferential depth of the drainage system and standard convexity of the groundwater between two drains. The major components of a drainage system for a clay polder are:
-   subsurface field drainage;
-   open main drainage network;
-   pumping station.

Ideally the design of these three components should be carried out together. In practice the design of the subsurface field drainage is carried out independent of open main drains and pumping stations. Design of the open main drainage network is based on the assumption of steady flow through the network with attention of temporal changes in runoff and flow through the pumping station. The assumption of steady flow is done while this results in acceptable dimensions in an easy way.

However, to improve the design of a network, it is important to simulate how it performs together with the temporal variation of rainfall, runoff and pump discharge. With this analysis, it is important to know the relation of the duration of the water level exceedence in the canal with decreasing frequency. For this purpose it is important to examine the system behaviour in response to the reduction of pumping capacity for a given condition of extreme rainfall.

In the design of subsurface field drainage for a clay polder there are two distinct criteria:
-   determination of drain depth and distance between the drains;
-   required transport capacity of the drains.

The first criterion is a combination of the required depth of the groundwater table and corresponding discharge rate and is further related to the type of the drainage flow equation, which will be used i.e.:
-   steady state criteria, which assume a constant groundwater table depth and discharge corresponding to an average depth and average discharge under design conditions;
-   non-steady criteria where a certain fall of the groundwater table has to be reached in a certain number of days after rain or irrigation.

4.3 Water management in a peat polder

During centuries the low peat areas in the western part of the Netherlands have mainly been utilized for grassland. Many ditches characterize the old polders in this area with a water level between 0.20 - 0.40 m-surface. Drainage is only possible by extraction of water by pumping stations. A more or less constant open water level is maintained by discharge in the winter and water supply from the surrounding water bodies in the summer.

At high groundwater tables the bearing capacity of the soils may be not sufficient for the mechanization in dairy farming during winter and spring as well as in wet periods in the summer. This frequently impedes the required activities in farming. In the wet periods it is therefore difficult to inject manure slurry into the land. In spring fertilizing and other activities have to wait until the dry periods. Thus, the growing season starts at a later date and this also affects the grazing period. In the wet periods of the grazing season part of the grass production may be lost by trampling and the sod may be damaged, which may result in lower yields in the following periods. In autumn the grazing period may shorter and this results in a reduction of the grazing period as a whole.

4.3.1 Capillary rise above the groundwater zone

The great mass of the roots of grass is in the shallow soil layers. For a good production these roots should be able to absorb sufficient moisture. The groundwater table is able to assist in this as a result of the upward capillary flow of water.

4.3.2 Groundwater table and open water level

Groundwater tables usually vary dependent on rainfall in the area. During the dry period the groundwater table will drop below the water level in the drains. At this time the open water level can be kept high to let water flow to the soil, then crops will benefit. While during the winter period the water level is above the drains therefore at this period there is flow from the soil to the open field drain.

In the experimental field Zegvelderbroek, at an open water level of 0.30 m-surface in winter, the groundwater table could rise to the ground surface under surplus precipitation conditions (De Bakker and Van den Berg, 1982). In summer the groundwater could drop down to about 0.60 m-surface, due to the evapotranspiration surplus. The infiltration from the open drain was very small (about 1 mm/day) because of a low permeability in a narrow strip of about 5 m from the open drain due to the lack of cracks and a compressed side slope by drinking cattle. Beyond this strip the permeability was very high (Table 4.3). Lowering of the drain water level resulted in a lower groundwater table. By lowering the drain water level to 0.50 m-surface, the groundwater table could rise to about 0.40 m-surface in winter and drop to 0.90 m-surface in summer.

Table 4.3 Permeability of peat soil at the experimental field Zegvelderbroek (De Bakker and Van den Berg, 1982)

Depth in m-surface	Permeability in m/day
0.30 - 0.50	1.95
0.50 - 0.70	0.90
0.70 - 1.00	0.18

In case of a continuous dry period all suspended water had been used. The groundwater table was kept not lower than 0.50 m-surface to provide the rising capillary moisture for the grass root zone. In the wet period the groundwater table was kept at 0.40 m-surface to prevent shortage of air with harm to the roots and consequently to the grass. Moreover, it was found that to keep the bearing capacity sufficient the average groundwater depth must not exceed 0.30 m-surface with a range of 0.20 m to 0.50 m-surface, depending on the duration of the high groundwater table. For example in autumn at a groundwater table depth of 0.20 m the bearing capacity can be sufficient, while insufficient at a groundwater depth of 0.40 m in spring because of reswelling of peat in this period.

4.3.3 Open field drain water level

During a dry period the rising of the capillary moisture provides the only water available to the grass. The open field drain water level should be maintained higher than the groundwater table to make the water flow to the soil. In the wet period the normal groundwater table is high, while the rainfall excess infiltrates to the soil. The open field drain water level in this period should be kept below the groundwater table to drain the ground and surface runoff. The open field drain water level in both cases is dependent on the peat permeability and distance between the open field drains. To avoid frequent submergence in the area the polder water level should be at least 0.50 m-surface during wet periods. The major components of a drainage system for a peat polder are:
- open field drains;
- open main drainage network;
- pumping station.

Design of the open field drains and open main drainage network is based on the assumption of steady flow through the network with attention of temporal changes in runoff and flow through the pumping station. In the design of open field drains for a peat polder the only criterion is the required transport capacity of the drains. This criterion is a combination of the required depth of the groundwater table and corresponding discharge rate and is further related to the water level in the main drainage network. The methodology may be summarized as follows:
- select a return period (design period) of a storm event for the analysis;
- from the hydrologic rainfall-runoff model deduce the flows through the network;
- determine operation and maintenance cost of the pumps and drains;
- determine cost of crop yield reduction, damage to buildings, roads and other infrastructure by using suitable damage functions of each alternative;
- optimize the system for minimum discounted cost and maximize benefits;
- repeat the procedure for another alternative;
- select the most economic one.

4.4 Urban drainage

The urban area is generally a part of the water management system in the polder as sub area. The aim of water management of this area is to provide good drainage and discharge. In urban areas a more or less fixed water level is required to avoid possible damage of buildings and infrastructure.

Their paved and unpaved areas characterize urban areas. The paved area generally consists of houses, buildings, streets and squares. The unpaved areas consist of the green areas, parks and gardens. Drainage from the paved area is generally released by means of sewer systems while drainage from the unpaved area is generally realised by subsurface drains (Figure 4.3).

Basically there are three systems of landfill for urban areas in polders, which are as follows:
- no landfill;
- excavation of clay or peat at the roads and refill with sand;
- complete landfill from 1 m to several metres.

Surplus precipitation is generally drained from the covered surface via the storm water sewers into urban canals. Domestic and industrial wastewater is discharged by way of the wastewater sewers. In the urban areas in the IJsselmeerpolders the surface was raised by 1 m of sand and subsurface drainage systems were installed in the covered area. Drains are laid at least 1.30 m-surface. This depth is chosen so as to limit disturbance to a minimum.

In the urban environment excess rainfall is carried to the main drainage system through two paths, the sewer system and the subsurface drains. Both of them discharge through the urban canals and weir/pump to the main drainage system of the polder.

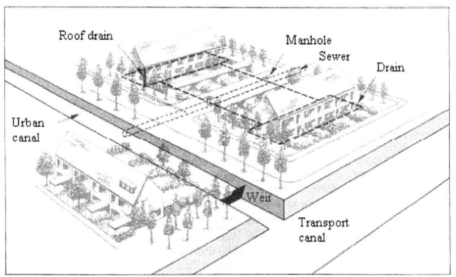

Figure 4.3 Scheme of an urban polder in the Netherlands (Schultz, 1992)

The major components of the drainage system in urban areas in the Netherlands are as follows:
- cross-section of the sewer systems;
- distance between the canals;
- percentage of open water;
- water level below the surface;
- pumping capacity or discharge capacity.

5 Water management in polders in Thailand

5.1 General water management objectives

Thailand has a humid tropical climate. The seasons in Thailand can be divided into rainy season and dry season. In general in the rainy season the average rainfall is more than enough for the growing of rice or other crops in polders. In fact farmers often experience flooding, resulting in the ruin of crops. On the other hand, during the dry season, there is a very small amount of rainfall. Thus the farmers are confronted with two extreme conditions. Hence the primary function of water management is to create prevention of waterlogging and flooding in the rainy season and to create prevention of drought and prevention of salinity by basin irrigation and surface drainage in the dry season. In order to fulfil these functions generally irrigation and drainage are required. The irrigation water supply has to meet the crop water requirements, therefore irrigation scheduling is necessary. Due to the unreliable rainfall during the rainy season and salt accumulation in the dry season caused by high evapotranspiration and saline seepage (in the area near the sea), good quality irrigation water is needed. The drainage has to evacuate excess rainfall in both cropping seasons and where applicable to evacuate polluted water due to salinization, in the dry season.

Therefore in the polders in Thailand the primary function of the water management system is to control the application of irrigation water and drainage. Initially the polders were mainly intended for rice cultivation. The design of the main irrigation and drainage system was mainly based on agricultural requirements. Later land was used for other purposes such as urbanization, shrimp farming, recreation, etc.

During the dry season in most of the polder areas there will be high evapotranspiration from the soil surface and a high upward movement of salt due to saline groundwater (especially in the northeastern part of Thailand). Normally a second rice crop during the dry season can help to minimize salinity problems. The supply of high quality irrigation water over the ground surface is effective in dissolving and carrying away the soluble salts, which finally drain to the surface drainage system.

5.2 Water management in a rice polder

Normally the irrigation water for the polders comes from a diversion or storage dam, barrage or weir, which divert water from the river and feed it to the trunk canals, lateral canals and ditches. Dikes have been constructed along the rivers to prevent inundation of the cultivated land where the major crop is rice (Figure 5.1). Lowland rice is usually grown in a layer of ponded water on the surface of the field. This water is supplied by rainfall and supplemented by irrigation. The drainage system normally is surface drainage, which is composed of field drains and main drains. The drains receive water from the irrigation surplus, excess rainfall in the rainy season and leaching water (if any) in the dry season. Therefore the main objectives of paddy field drainage are the removal of excess surface and subsurface water for leaching conditions in the dry season.

In Figure 5.1 the water management scheme for a rice polder is shown. This is a combined drainage and irrigation system. Farmers can control the water level in the field by pumping water or by gravity flow if the water level in the field is lower than the preferred water level and drain water out if the water level in the field is higher than the preferred water level by pumping or by gravity dependent on the water level in the lateral canal.

The major components of a system in a rice polder are:
-   bund, inlet and outlet for irrigation and drainage;
-   water level in the field and in the canals;
-   open canals network;
-   pumping station.

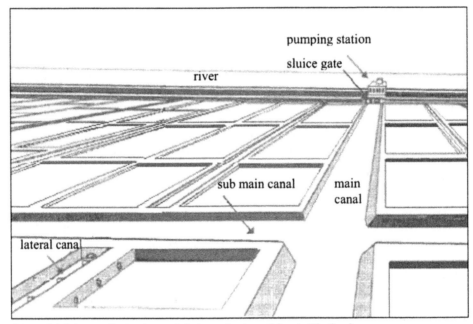

Figure 5.1 Scheme of irrigation and drainage in rice polders in Thailand

Continuous submergence of water throughout the cropping season is the general practice in Thailand. For the irrigation practice water will be drained out from the field within one week after maximum tillering stage (Figure 5.3) to let the soil oxidize in order to reduce toxicity in the root zone. High yields may be expected, if the water depth in the field is suitable for rice at different growth stages (Figures 5.2, 5.3 and Table 5.1). Moreover, water depth control in the field can be an advantage for weed control and optimal time for fertilization.

Due to unreliable rainfall and agricultural practices in Thailand, farmers always grow two or three crops if irrigation water is available. The water for irrigation will be stored in the main canal before the dry season. On the other hand in the wet or rainy season the water level in the main drain cannot be kept too high to prevent damage due to unexpected storms, which may cause heavy rainfall in the area. In the rainy season in case of water needed by the crops farmers usually let the water flow to the lateral canals through the culvert and pump or supply through a culvert to the field dependent on the water level in the lateral canal and the field. But in the dry season the farmers usually pump the water from the main canal to the lateral canal and finally pump it to the field. In case of heavy rainfall farmers let water flow to the lateral canal by a pump or through a culvert dependent on the water level in the lateral canal and the field. In case of harvesting this is always at the same time as the water level in the main canal is high owing to storing water for the second crop in the dry season, farmers always drain water by pumping.

Because lowland rice is mostly grown under conditions of near soil saturation and submergence, loss through percolation is minimal. Continuous submergence with intermittent drainage is the most effective method (Figure 5.3). Due to lack of labour for transplanting and high labour cost, nowadays most farmers have changed the method from transplanting rice to spreading seed. During and immediately after spreading the water is kept at 10 mm for about 1 week. In the following period, submergence is deeper to maximum 80 mm (Table 5.2). Drainage and drying the topsoil about 3 days is practiced in the heading period since rice can tolerate a water shortage and root development is enhanced (Charoenying, 1989) (Figure 5.3). Drainage in this period will drain the toxic substances from the field. Moreover, the oxygen in the soil surface, which can be utilized by the rice roots, increases. The rice shoots will have a good resistance to diseases and strength of the rice shoots increases so it will not fall easily. After spreading fertilizer, adequate water supply from head development through flowering is essential. Because root activity at this time is at maximum, a high water temperature is preferred but below 35 °C (Doorenbos and Kassem, 1986). Continuous flow irrigation or drainage and renewal of water once or twice during this period is sometimes practiced. During the ripening period fields should gradually be drained to facilitate harvest operations.

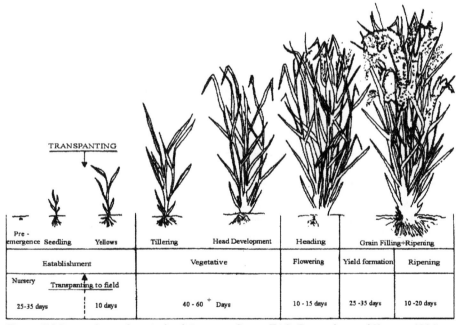

Figure 5.2 Stage of growth periods of rice according to FAO (Doorenbos and Kassem, 1986)

Table 5.1 Stages of growth periods and agriculture practices of rice (Royal Irrigation Department, 1994)

Stages of growth and agriculture practice	Periods in days (approximate)
1 Land preparation	30
2 Establishment	20
3 Vegetative	40
4 Heading + flowering	30
5 Yield formation	20
6 Ripening	10

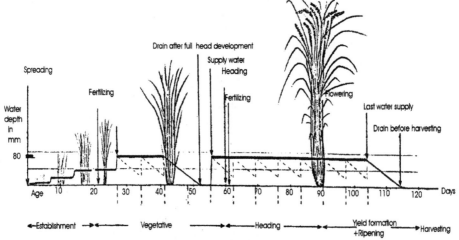

Figure 5.3 Preferred water layer for rice growth stages (after Royal Irrigation Department, 1975)

Table 5.2 Preferred water layer depth for rice (Royal Irrigation Department, 1994)

Day	Stage of growth and agriculture practice	Preferred water layer depth in mm
1 - 7	Soak the soil	30
8 - 13	Soil scaring	20
14 - 30	Let weeds grow and get rid of weeds	10
31 - 37	Establishment	10
38 - 44	Establishment	25
45 - 49	Establishment	50
50 - 56	Vegetative	50
57 - 70	Vegetative	80
71 - 80	Vegetative	80 - 0
81 - 83	Vegetative	0
84 - 109	Vegetative	80
110 - 119	Flowering	80
120 - 134	Yield formation	80
135 - 144	Yield formation	80 - 0
145 - 150	Ripening	0

Land preparation for the rice fields, is composed of soil soaking, soil scaring and let weeds grow and getting rid of weeds. Soaking soil requires 2 - 3 days or more dependent on the type of soil (Charoenying, 1989). Charoenying suggested that in soil ploughing, it requires some water depth to dissolve soil mud at the ploughing disc. After ploughing the field will be left about 2 weeks to let the weed grow during this stage the water depth should be kept at about 1 cm (Charoenying, 1989). After that the water layer on the surface is usually kept at various depths dependent on the crop stage and the agricultural practice (Table 5.2). In practice, when there is a dry spell the farmers will pump water from the drainage canal by trying to keep the water level depth in the field at the preferred water level as shown in Table 5.2. In the wet periods the surface runoff is calculated based on the assumption that the farmers will drain the excess water into the lateral canals if the water level in the rice field is higher than the maximum water level that rice can tolerate as shown in Table 5.3.

Table 5.3 Water layer depth for rice that can be tolerated in the wet periods

Day	Stage of growth and agriculture practice	Maximum water layer depth in mm
1 - 7	Soak the soil	600
8 - 13	Soil scaring	200
14 - 30	Let weeds grow and get rid of weeds	50
31 - 37	Establishment	12
38 - 44	Establishment	40
45 - 49	Establishment	67
50 - 56	Vegetative	102
57 - 70	Vegetative	127
71 - 80	Vegetative	172
81 - 83	Vegetative	271
84 - 109	Vegetative	380
110 - 119	Flowering	482
120 - 134	Yield formation	524
135 - 144	Yield formation	300
145 - 150	Ripening	300 - 0

5.3 Water management in a dry food crops polder

In flat lowland areas in the Lower Central Plain of Thailand especially the area between four main rivers; the Chao Phraya river, the Thachin river, the Meklong river and the Bangpakong river, in the former time the farmers mostly grew rice. The increasing population in Bangkok and vicinity, has increased the need for vegetables and fruits. Moreover, vegetables and fruits can give more benefit compared to rice, therefore farmers have changed and improved their land in order to grow dry food crops by constructing dikes around the area, excavating small ditches and to make use of the soil excavated from the ditches to fill the row in the plot (Figure 5.4) (Charoensiri, 1991). This technique is the so called 'raised-bed system'. It gives a better protection of the land against flooding, salt intrusion and it reduces salinity and acidity in the soil. Therefore many kinds of dry food crops can be grown in this land (Sutti, 1998). The raised-bed system can be found in Bangkok and vicinity. Raised-bed systems are composed as follows (Charoensiri, 1991):
- surrounding dike with a height of around 1 - 2 m or more than the flood level in the past, around the land to prevent flowing of water from outside into the area and also for transporting the product from the area. This dike normally is constructed by earth from the excavation in the area (Figure 5.6);

-   canal or river branch outside the area, which is used for water storage;
-   beds, which are constructed for growing crops. Normally they have 6 - 8 m width dependent on the type of crops. The length is dependent on the geometry of the area. Somewhere the surface of the bed is constructed as convex curve usually for vegetables. Generally there is a small path along the side of the bed, which is used for walking (Figure 5.5);
-   ditches are excavated around the bed in order to collect water and transport water for crops. On the other hand they will be used for draining excess water in the rainy season. The dimensions of the ditches are 1.0 - 1.5 m wide, 0.8 - 1.0 m deep and the bed width is 0.7 - 1.0 m;

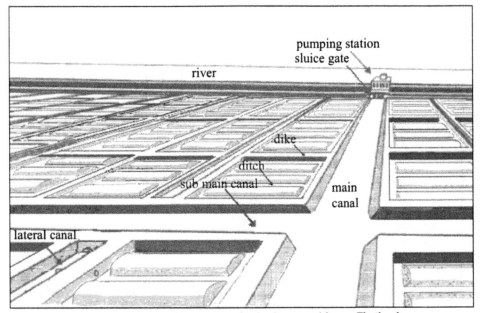

Figure 5.4 Scheme of irrigation and drainage in dry food crops polders in Thailand

-   irrigation, normally in a big area small boats with a pump are used for spreading water to the crops by moving it along the ditch (Figure 5.5);
-   water control normally established by a pump, water wheel, regulating gate, and culverts (Figure 5.6 and 5.7).
    The major components of a system in a dry food crops polder are:
-   bed, ditch and surrounding dike;
-   inlet and outlet for irrigation and drainage;
-   water level in the ditches and in the canals;
-   open canals network;
-   pumping station.

Water management at field level for the raised-bed systems

Water management in the raised-bed systems can be done by two methods, which are gravity and pumping (Sutti, 1998). Depending on the water level in the ditch inside the dike and the water level in the canal outside the dike. Water management can be distinguished into four cases as follows:

- *Case 1. Outflow by gravity.* In case of excess of water in the plot and the water level in the plot is higher than in the canal. The farmers will open the gate of the culvert, and water will flow to the sump. From the sump water will flow out to the canal;
- *Case 2. Outflow by pumping.* In case of excess of water in the plot and the water level in the canal is higher than in the plot, the farmers will open the gate of the culvert, and water will flow to the sump. From the sump water will be pumped out to the canal;
- *Case 3. Inflow by gravity.* When there is need for irrigation water for the crops in the plot and the water level in the canal is higher than in the plot, the farmers will open the gate or culvert in the canal, and water will flow to the sump. Then the gate of the plot is opened where irrigation water is needed. The water will flow from the sump to the plot;

Figure 5.5 Irrigation supply in a raised-bed system for dry food crops in polder areas in the Central Plain of Thailand

- *Case 4. Inflow by pumping.* When there is need for irrigation water for the crops in the plot and the water level in the canal is lower than in the plot, the farmers will open the gate of the culvert, when the level of the culvert is lower or equal to the bed of canal, and water will flow to the sump. Then water will be pumped from the sump to the plot by closing the higher gate, which prevents water to flow back to the canal and open the gate and let water flow to the plot where irrigation is needed.

In some areas in the polders especially for vegetables and fruit trees, combined irrigation and drainage systems can be found. Due to the flat area the drainage network is used for irrigation as well. While the systems of drains have a low level, it is necessary to pump the irrigation water from the main drains to the field or to apply sub-irrigation. The irrigation water can be supplied to the plot from the river or main canal by gravity, or pumped, which depends on the water level in the canal and the river.

Inside the polders, a raised-bed system is applied for fruit and vegetables, in this system the small canals and surrounding dikes were constructed as a network (Figures 5.6 and 5.7). The irrigation water can be supplied by using small portable pumps.

Water level in the ditches and moisture in the soils

The farmers will keep the water level in the ditch all year round more or less constant in the system if water in the canal is available, the water level in the ditch is kept at 0.50 - 0.60 m-surface for vegetables and 0.70 - 0.80 m-surface for fruit trees (Charoensiri, 1991). Most of the farmers control the water level by pumping for vegetables every 2 - 3 days and for every 7 - 10 days for fruit trees.

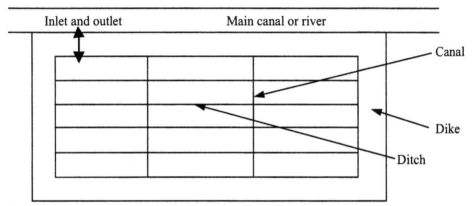

Figure 5.6 A raised-bed system in polder areas in Thailand

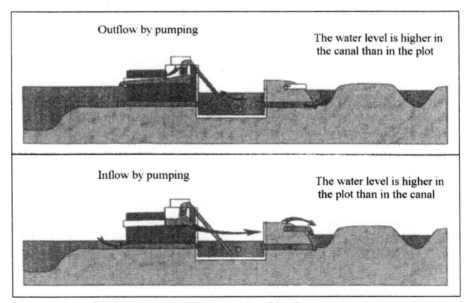

Figure 5.7 Raised-bed system water management in polders by pumping

Most of the time the farmers will keep the moisture at field capacity (-0.033 MPa or -0.33 bar), but sometimes it can be a little lower than field capacity but not lower than a moisture tension of -0.1 MPa or -1 bar for vegetables because the irrigation water supply is enough and there is no limit on water use for the farmers (Sutti, 1998). In case of Chinese kale (Brassica oleracea var. alboglabra) the minimum moisture preferred in the soil is not lower than -0.25 bar (Sanders, 1997).

The top of the saturated zone in soil is about 0.10 - 0.15 m above the water level in the ditches. The water level in the ditches should not be above 0.35 m-surface while otherwise there will be not enough air for the crops (Sutti, 1998) due to high groundwater. However, the moisture in the topsoil will decrease as the distance from the ditch increases, in dry conditions the evapotranspiration is higher than the movement of water to the topsoil, therefore spraying of water is needed to keep the moisture at field capacity.

5.4 Water management in an urban area

In the polders in Thailand, especially in the Chao Phraya river basin, dikes were constructed in the downstream zone to prevent inundation of the cultivated areas. Residential areas were also protected by dikes and polder systems at a higher safety level, together with a number of pumping stations, especially in the most downstream part of the river, Bangkok and its vicinities (Hungsapreg, et al., 1998). Due to urban development, most of the surface areas are covered with houses, roads, or other paved surfaces. Due to little water absorption, rainfall tends to convert almost immediately into runoff flowing to the drainage system.

Components of the drainage systems for urban areas are as follows (Drainage Office, Bangkok Metropolitan Administration, 1998):

- *collector drain pipes.* The collector drainpipes are the pipes, which collect water from wastewater or rainwater from a residential area. The general diameter is between 0.3 - 1.5 m. These pipes usually lay in the foot part or shoulder of the road at 1.5 - 3.0 m depth and the distance between the service manholes is about 12 m. In some narrow roads (width less than 4 m) this pipe is laid at the centreline of the road and there will be a connecting pipe of 0.30 m diameter at every 12 m to the manhole to collect water from houses;
- *transport pipes.* The transport pipes collect rainwater or wastewater from the collector drain pipes and transport the water to a main drain or an urban drainage canal by gravity. The diameter is generally between 1.5 - 2.25 m. They are laid at a depth of more than 3.0 m to avoid crossing with the collector drainpipes. The service manholes are positioned as shown in Table 5.4. The manhole is the point where the collector drain pipes join with the transport pipe. These transport pipes are generally laid at the centre of road;

Table 5.4 Transport pipe diameter and distance between manholes

Transport pipe diameter in mm	Distance between manholes in m
less than 800	80
from 1,000 to 1,500	120
more than 1,500	200

Source: Drainage Office, Bangkok Metropolitan Administration, 1998

- *force mains*. The function of the force main is to transport water from the booster pump to the main drain canal under pressure of the pump, where there will be no service manhole but only a discharge structure at the end of the pipe. These pipes have 1.5 - 2.0 m diameter and are placed at a depth of about 3 - 4 m;
- *boosters pumping stations*. Booster pumping stations may be components of the urban drainage system. They are designed to collect water from the transport pipe and pump the water to the drain. The general dimensions of the pumping station are 3.5 m width, 20.0 m length and 5.0 - 8.0 m depth. The location of the booster pumping stations is at the junction of the main road;
- *pumping station at the end of a road.* At the end of a road and join with a drain, there will be a sump with a depth of 3.0 - 4.0 m to collect water and pump it to the drain;
- *pumping station or sluice at the end of drain.* This pumping station is used for pumping water out of the polder area when the water level in the river is high or drains by gravity dependent on the water level in the drain and the river.

The excess rainfall is carried to the main drainage system through the sewer system and discharges through the urban drains and weir or pump to the main drainage system or the river. In flat areas, like polders the capacity of the drainage system depends highly on the sewer and drain storage. The sewer and drain networks function is not only water conveyance, but also storage.

Due to heavy rainfall characteristics in Thailand, a heavy rainfall can occur in a short time. So the sewer pipe diameters have to be large enough to carry rainwater in a short period, while in the dry period there is only domestic wastewater and the discharge is small. Therefore sewer collection of rainwater and wastewater is normally separated in Thailand. Otherwise there will be some sediment sewer system when there is low flow where there is a very flat slope in the area.

The major components of the drainage system in urban areas in Thailand are as follows:

- cross-section of the sewer systems;
- distance between the transport pipes;
- percentage of open water;
- water level below the surface;
- pumping capacity or discharge capacity.

6 Model description

6.1 General

In order to achieve the objectives of the study the existing software package OPOL was further developed for hydrological conditions and practices of water management for the temperate humid and humid tropical conditions (Schultz, 1992 and Wandee, 2001). The model objective is to find the relationship between rainfall and discharge in polder areas. The first modelling step concerns the determination of the model parameters by calibration. When the parameters of the model have been determined, the model will be validated for a year with available hydrological data. Thereafter the optimal values for the main components of the water management system in polder can be determined by the computation of annual investment, operation and maintenance costs and damage in such a way that the annual damage and investment costs over the period under consideration are minimum. The model involves the following activities:
- hydrological analysis and calibration in order to use it for the case studies and to improve the understanding of the hydrological processes;
- determination of extreme hydrological situations;
- optimization of the main components of the water management system for a given set of extreme hydrological conditions and economic criteria.

6.2 Optimization model for water management in a polder

The OPOL package consists of several programs in FORTRAN language, which are:
For the Netherlands:
- RAINRUR (rainfall runoff relation in a rural area);
- RAINURB (rainfall runoff relation in an urban area);
- OWAP (optimization of water management in a polder).
For Thailand:
- RICERURF (rainfall runoff relation in a rainfed rice rural area);
- RICERURI (rainfall runoff relation in an irrigated rice rural area);
- DRYRUR (rainfall runoff relation in a dry food crops rural area);
- RAINURB (rainfall runoff relation in an urban area);
- OWARF (optimization of the main components of water management system in a rainfed rice polder).
- OWARI (optimization of the main components of water management system in an irrigated rice polder).
- OWAD (optimization of the main components of water management system in a dry food crops polder).

RAINRUR, RICERURF, RICERURI, DRYRUR and RAINURB are hydrological models. OWAP, OWARF, OWARI and OWAD consist of the subroutines for the optimization computations for rural and urban areas.

6.2.1 Hydrological computation for the Netherlands

Rural and urban areas respond differently to rainfall. This response has an impact on the design of the main drainage system in a polder.

In rural clay polders the subsurface drainage systems result in an effective drainage of the excess rainfall and groundwater control.

In rural peat polders the open field drainage systems result in an effective drainage of the excess rainfall and groundwater control.

In urban areas the paved area is nowadays generally provided with a separate sewer system. The unpaved part is provided with a subsurface drainage system. Because of the greater depth of the drains and the sandy cover, the urban subsurface drainage reacts more slowly and smoothly on the rainfall than the rural field drainage system. On the other hand the urban storm water sewer system discharges quicker than the rural discharge and the peak is smoothed out by the open water storage in the urban areas. Therefore the hydrological computations in the two areas are treated by separate sub-modules:

- RAINRUR   for the rural area;
- RAINURB   for the urban area.

It is difficult to model strictly the physical process of the rainfall-runoff in soils where deep cracks occur (Figure 6.1), while the usual groundwater flow equations are not applicable here. Therefore a reservoir model is conceptualised in the above two computations. These are one-dimension lumped models. In the following the process components and steps involved in the computation are described in brief.

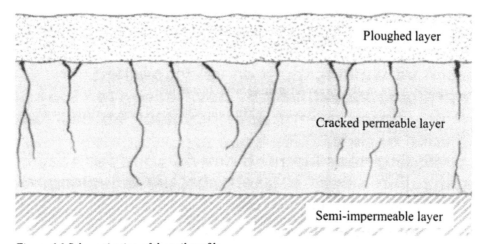

Figure 6.1 Schematisation of the soil profile

Hydrological model for the rural area in a clay polder

In the model for the rural area in a clay polder the hydrological processes involved in the surface and subsurface environment are schematised into four layers where the water balance is determined assuming each as a reservoir (Figure 6.2).

The model components are as follows:

Above the surface:
- precipitation (mm/day)   (P)
- interception (mm)   (E_i)
- maximum infiltration in layer I (mm/m)   (S_{i1})
- open water evaporation (mm/day)   (E_0)

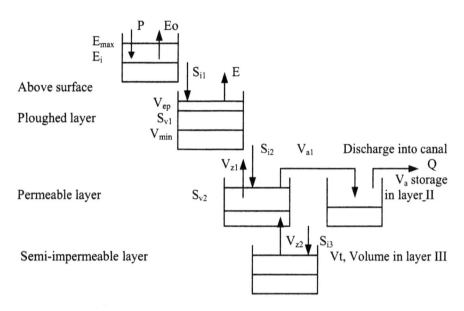

Figure 6.2 Scheme of the hydrological model for the rural area

Layer I (ploughed layer):
- evapotranspiration   (E)
- maximum infiltration in layer I (mm/m) (S_{i1})
- capillary rise from layer II (mm/day) (V_{z1})
- maximum percolation in layer II (mm/m) (S_{i2})
- storage above field capacity in layer I (mm/m) (S_{v1})
- soil moisture at which evapotranspiration is at potential rate (mm/m) (V_{ep})
- soil moisture at which evapotranspiration is zero (mm/m) (V_{min})

Layer II (permeable layer):
- percolation from layer I (mm/m) (S_{i2})
- capillary rise to layer I (mm/day) (V_{z1})
- percolation of surplus water to layer III (S_{i3})
- capillary rise from layer III (mm/day) (V_{z2})
- storage above field capacity in layer II (mm/m) (S_{v2})
- discharge of surplus water to collector drain (V_{a1})

Layer III (semi-impermeable layer):
- maximum percolation from layer II into layer III (mm/m) (S_{i3})
- moisture at saturation in layer III (mm) (V_t)

Open water storage:
- inflow from layer II through the subsurface drain (mm) (V_{a1})
- change in storage (mm) (V_a)
- discharge through pumping (mm) (Q)

The above components will be described briefly.

Interception

Interception depends on the type of vegetation and its coverage, in the model interception at any instant is the precipitation at that instant and the residual interception from the previous time step, which is limited by the maximum interception E_{max}. This upper limit of interception is a calibrated model parameter and usually ranges between 2 - 3 mm/day (Biswas, 1976).

In the polder surface runoff is hardly evident therefore this component in the open water balance has been neglected. However, when the rainfall exceeds the sum of the maximum interception and the water holding capacity in the underlying two layers the surplus is added to the discharge from layer II to the subsurface drain.

$$E_i(t + \Delta t) = E_i(t) + P(\Delta t) \qquad (6.1)$$

where:
$E_i(t)$ = interception (mm)
$P(\Delta t)$ = precipitation (mm/time step)
Δt = time step (hours)

Evapotranspiration

If the interception is less than the maximum (E_{max}) but still greater than zero, evapotranspiration will take place at open water evaporation rate (E_0), leaving the remaining interception to be added to the precipitation in the next time step. Otherwise evapotranspiration is a function of the potential rate (E_p) and the moisture status in the underneath layers. Open water evaporation is expressed as follows:

$$E_o(\Delta t) = \frac{E_o}{N} \qquad (6.2)$$

where:
E_o = open water evaporation (mm/day)
$E_o(\Delta t)$ = open water evaporation (mm/time step)
N = number of time steps per day

As long as the moisture related to the field capacity in layer I is above the upper limit of field capacity in layer I evapotranspiration takes place at potential rate and if it is less than V_{min} it reduces to zero (Figure 6.3). If the moisture status is in between these limits evapotranspiration is linearly interpolated between the potential value and zero.

$$E_p(\Delta t) = f(t) * E_0(\Delta t) \qquad (6.3)$$

where:
$E_p(\Delta t)$ = potential evapotranspiration (mm/time step)
$f(t)$ = crop factor (-)

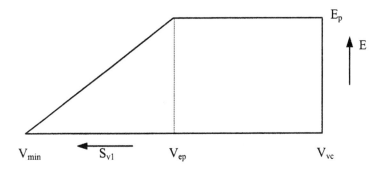

Figure 6.3 Evapotranspiration and soil moisture

$$E_a(\Delta t) = \frac{(V_{min} - S_{vl}(t))}{(V_{min} - V_{ep})} * E_p(\Delta t), \text{ if } V_{min} < V < V_{ep} \qquad (6.4)$$

$$E_a = 0.0, \text{ if } V_{min} \geq V \qquad (6.5)$$

$$E_a = E_p, \text{ if } V \geq V_{ep} \qquad (6.6)$$

where:
$E_a(\Delta t)$ = actual evapotranspiration (mm/time step)
$E_p(\Delta t)$ = potential evapotranspiration (mm/time step)
$S_{vl}(t)$ = soil moisture deficit in layer I related to field capacity (mm/m)
V_{ep} = soil moisture at which evapotranspiration is at potential rate (mm/m)
V_{min} = soil moisture at which evapotranspiration is zero (mm/m)
V_{vc} = soil moisture at field capacity (mm/m)

Consequently the moisture content in layer I is reduced for the next time step as follows:

$$S_{vl}(t + \Delta t) = S_{vl}(t) - \frac{E(\Delta t)}{D_1} \qquad (6.7)$$

where:
$S_{vl}(t)$ = moisture content deficit related to field capacity in layer I (mm/m)
D_1 = thickness of layer I (m)

If the actual evapotranspiration from layer I is more than the potential (E_p) the residual is taken as a loss of moisture from layer II in the same manner as above, resulting in a decrease in the moisture content in layer II. The boundary values for V_{min} and V_{ep} are updated for the layer based on the difference between the wilting point and field capacity in layer II. Hence the total evapotranspiration is the sum of losses from the two layers.

$$E'(\Delta t) = E_p(\Delta t) - E(\Delta t) \tag{6.8}$$

where:
$E'(\Delta t)$ = evapotranspiration from layer II (mm/time step)

Moistening of the layers

The amount of moisture that will penetrate in layer I and layer II depends on the interception at the surface, moisture content of the layer in the previous time step, maximum rate of moistening of the layer and the thickness of the layer. Therefore the maximum amount of moisture that can percolate in the underneath layers is assumed to be increasing from V_{min} with the increase of moisture content up to field capacity.

$$S_{i1}(\Delta t) = \frac{\theta_{v1} + 0.1 * S_{v1}(t)}{\theta_{v1}} * V_{max1} * D_1 * \Delta t \tag{6.9}$$

$$S_{i2}(\Delta t) = \frac{\theta_{v2} + 0.1 * S_{v2}(t)}{\theta_{v2}} * V_{max2} * D_2 * \Delta t \tag{6.10}$$

where:
D_1 = thickness of layer I (m)
D_2 = thickness of layer II (m)
$S_{i1}(\Delta t)$ = maximum amount of moisture that can be stored at any instant in layer I (mm)
$S_{i2}(\Delta t)$ = maximum amount of moisture that can be stored at any instant in layer II (mm)
$S_{v1}(t)$ = increase in storage at instant t in layer I up to field capacity (mm/m)
$S_{v2}(t)$ = increase in storage at instant t in layer II up to field capacity (mm/m)
V_{max1} = maximum rate of moistening in layer I (mm/m/hour)
V_{max2} = maximum rate of moistening in layer II (mm/m/hour)
θ_{v1} = moisture content at field capacity in layer I (%)
θ_{v2} = moisture content at field capacity in layer II (%)

Therefore the maximum volume of water that can be absorbed at any instant is as follows:

$$V_{max}(\Delta t) = S_{i1}(\Delta t) + S_{i2}(\Delta t) + E_{i\,max} \tag{6.11}$$

where:
E_{imax} = maximum interception (mm)
$V_{max}(\Delta t)$ = maximum volume of water that can be absorbed at any instant (mm)

When the interception at any instant (E_i) is greater or equal to V_{max}, there is maximum interception as well as maximum accommodation of water in layers I and II while if there is any surplus it goes to the subsurface drain and main drains.

$$V_{al}(\Delta t) = E_i(t + \Delta t) - V_{max}(\Delta t) \tag{6.12}$$

where:

$V_{al}(\Delta t)$ = surplus to be discharged to the drain (mm)

If, however, interception at any instant (E_i) is less than V_{max}, after meeting the maximum interception on the surface i.e. E_{imax}, the remaining goes to fill the room in the successive layer I and layer II in the following manner:

$$V(\Delta t) = V_{max}(\Delta t) - E_{i\,max} \tag{6.13}$$

$$S_{v1}(t + \Delta t) = S_{v1}(t) + \frac{V(\Delta t)}{D_1} \qquad \text{if } V < S_{i1} \tag{6.14}$$

also:

$$S_{v1}(t + \Delta t) = S_{v1}(t) + \frac{V(\Delta t)}{D_1} \qquad \text{if } V > S_{i1} \tag{6.15}$$

Therefore:

$$S_{v2}(t + \Delta t) = S_{v2}(t) + \frac{V(\Delta t) - S_{i1}(\Delta t)}{D_2} \tag{6.16}$$

Soil moisture tension and capillary rise

The soil moisture tension in layer I is assumed uniform throughout its depth, while in layer II it varies linearly and becomes zero or even positive at its bottom during winter and below the bottom during summer when the water table goes down below the cracks (Figure 6.4).

Soil moisture in the first layer is defined as follows:

$$\theta_1(t) = \theta_{v1} + 0.1 * S_{v1}(t) \tag{6.17}$$

where:

θ_{v1} = moisture content at field capacity in layer I (%)
$\theta_1(t)$ = soil moisture in layer I (%)
$S_{v1}(t)$ = moisture added in layer I (mm/m)

Moisture tension in layer I and layer II is based on an empirical formula as follows:

$$\log p_1(t) = a_1 + b_1 * \ln\left[\left(\frac{\theta_1(t)}{n_1}\right)^{c_1} - 1\right] \qquad (6.18)$$

$$\theta_2(t) = n_2 * \left[e\left(\frac{\log(p_1(t) - a_2}{b_2}\right) + 1\right]^{\frac{1}{c_2}} \qquad (6.19)$$

where:
a_1, b_1, c_1 and a_2, b_2, c_2 are constant values for the soil types in layers I and II as will be described and estimated in Chapter 7.

n_1, n_2 = porosity of soil in layer I and layer II (%)
p_1 = moisture tension in layer I (cm)

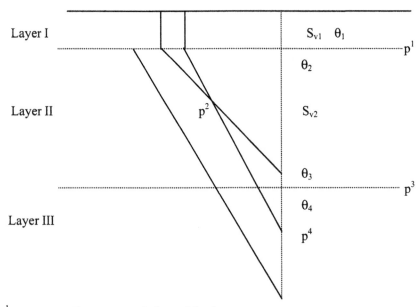

p^1 = suction pressure in layer I (cm)
p^2 = suction pressure in the middle of layer II (cm)
p^3 = suction pressure at the top of layer III (cm)
p^4 = average suction pressure in layer III (cm)
θ_1 = soil moisture in layer I (%)
θ_2 = soil moisture at the top of layer II (%)
θ_3 = soil moisture at bottom of layer II (%)
θ_4 = soil moisture in layer III (%)

Figure 6.4 Schematisation of pressure profiles in the layers

θ_1 and θ_2 are soil moisture content in layer I and layer II at their interface. The soil moisture tension in layer II is based on the moisture content at its boundary, and the moisture content at its bottom, which is computed from the simple geometrical relationship (Figure 6.5) as follows:

$$\frac{\left(\theta_{v2} - \theta_2(t)\right) + \left(\theta_{ve2} - \theta_3(t)\right)}{2} = \frac{S_{v2}(t)}{10} \tag{6.20}$$

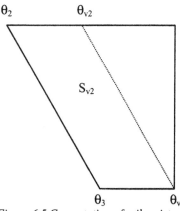

θ_2 θ_{v2}

S_{v2}

θ_3 θ_{ve2}

D_2 = thickness of layer II (m)
S_{v2} = soil moisture deficit in layer II related to field capacity (mm/m)
θ_{v2} = soil moisture at field capacity in layer II (%)
θ_{ve2} = soil moisture at saturation in layer II (%)
θ_2 = soil moisture at the top of layer II (%)
θ_3 = soil moisture at bottom of layer II (%)

Figure 6.5 Computation of soil moisture at the bottom of layer II

$$\theta_3(t) = 0.2 * S_{v2}(t) - \theta_2(t) + \theta_{v2} + \theta_{ve2} \tag{6.21}$$

$$\theta_{2g} = 0.5 * \left(\theta_2(t) + \theta_1(t)\right) \tag{6.22}$$

$$\log p_2(t) = a_2 + b_2 * \ln\left[\left(\frac{\theta_{2g}(t)}{n_2}\right)^{c2} - 1\right] \tag{6.23}$$

$$p(t) = 0.5 * \left(p_1(t) + p_2(t)\right) \tag{6.24}$$

The unsaturated hydraulic conductivity is based on the above averaged soil moisture tension and clay contents in layers I and II which is computed by the following empirical formula (Rijniersce, 1983):

$$k_{ot}(t) = A * p(t)^{-B} \tag{6.25}$$

$$B = 2.07 - \frac{(L_1 + L_2)}{69.6} \tag{6.26}$$

$$A = 10^{(3.21*B-4.78)} \tag{6.27}$$

where:
L_1 = clay content in layer I (%)
L_2 = clay content in layer II (%)
p = soil moisture tension (cm)
$k_{ot}(t)$ = unsaturated hydraulic conductivity (m/day)

Therefore the capillary flow from layer II to layer I is computed as follows:

$$V_{z1}(t + \Delta t) = \frac{1000 * k_{o1}(t)}{O_1} * \left(\frac{p_1(t) - p_2(t)}{50 * (D_1 + D_2)} - 1 \right)$$

(6.28)

where:

O_1 = number of steps in a day
$p_1(t)$ = pressure in layer I (cm)
$p_2(t)$ = pressure in layer II (cm)
D_1, D_2 = thickness of the layers I and II respectively (m)
$V_{z1}(t+\Delta t)$ = capillary rise from layer II to layer I (mm/time step)

The above shows that the capillary rise from the saturated part to the unsaturated part and through the unsaturated zone depends on the metric pressure difference between the boundaries, type of soil and the soil moisture condition in the unsaturated zone. Capillary flow to layer I increases the soil moisture content in layer I as follows:

$$S_{v1}(t + \Delta t)' = S_{v1}(t + \Delta t) + \frac{V_{z1}(\Delta t)}{D_1}$$

(6.29)

However, if the moisture content in layer I is greater than the field capacity (i.e. $S_{v1} \geq 0$), there is percolation from layer I to layer II and the soil moisture content in layer II and layer I in the next time step are calculated as follows:

$$S_{v2}(t + \Delta t) = S_{v2}(t) + S_{v1}(t + \Delta t) * \frac{D_1}{D_2}$$

(6.30)

$$S_{v1}(t + \Delta t) = 0$$

(6.31)

If θ_3 is greater than the moisture content at saturation in layer II, the soil moisture tension at the interface of the layers i.e. P_3 and P_4 is equal to zero and θ_4 is at saturation capacity in layer III. If θ_3 is less than the saturation moisture capacity flow takes place from layer II to layer III that follows the same as the flow computation in layer I.

$$\log p_3(t) = a_2 + b_2 * \ln \left[\left(\frac{\theta_3(t)}{n_2} \right)^{c_2} - 1 \right]$$

(6.32)

$$\theta_4(t) = n_3 * \left[e^{\left(\frac{\log p_3(t) - a_3}{b_3} \right)} + 1 \right]^{\frac{1}{c_3}}$$

(6.33)

$$\theta_4 = MIN(\theta_4, \theta_{ve3})$$

(6.34)

$$\log p_4(t) = a_3 + b_3 * \ln\left[\left(\frac{0.5(\theta_4(t) + \theta_{ve3})}{n_3}\right)^{c_3} - 1\right]$$ (6.35)

$$P_o(t) = 0.5 * (p_2(t) + p_4(t))$$ (6.36)

where:

$P_o(t)$	= pressure halfway between layer II and layer III (cm)
a_3, b_3 and c_3	= constant for soil in layer III (-)
n_3	= porosity of soil in layer III (%)
θ_{ve3}	= soil moisture at saturation in layer III (%)
A and B	= constants based on clay content in layer II and layer III as described earlier (-)

$$k_{o2}(t) = A * p_o(t)^{-B}$$ (6.37)

The effective thickness of layer III is calculated as described in the schematisation in Figure 6.6 in the following way:

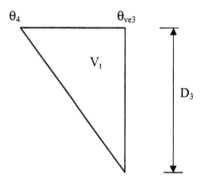

Figure 6.6 Effective thickness and volume to be replenished in layer III

$$D_3(t) = \frac{V_t(t)}{(\theta_{ve3} - \theta_4(t)) * 0.5}$$ (6.38)

Where

$V_t(t)$	= volume to be replenished in layer III (mm)
$D_3(t)$	= thickness of layer III (m)
θ_{ve3}	= moisture at saturation in layer III (%)
$\theta_4(t)$	= moisture in layer III (%)

Therefore the capillary rise from layer III to layer II and the soil status are updated as follows:

$$V_{z2}(\Delta t) = \frac{1000 * k_{o2}(t)}{O_1} * \left(\frac{p_2(t) - p_4(t)}{50 * (D_2 + D_3)} - 1 \right)$$

(6.39)

$$S_{v2}(t + \Delta t)' = S_{v2}(t + \Delta t) + \frac{V_{z2}(\Delta t) - V_{z1}(\Delta t)}{D_2}$$

(6.40)

However, if the soil water content in layer II is greater than the field capacity i.e. S_{v2} > 0, there is percolation to layer III and the soil water (S_{v2}) in layer II is set to:

$$V_{a2}(\Delta t) = S_{v2}(t + \Delta t)' * D_2$$

(6.41)

$$S_{v2}(t + \Delta t) = 0$$

(6.42)

where:

$V_{a2}(\Delta t)$ = volume available to layer III or to be drained (mm/time step)

The net volume of moisture available for the drains and to meet the deficit in layer III is:

$$V_a(t + \Delta t) = V_a(t) + V_{a1}(\Delta t) + V_{a2}(\Delta t) - V_{z2}(\Delta t)$$

(6.43)

If the amount is less than 0 ($V_a < 0$) the net volume in layer III is updated as follows:

$$V_t(t + \Delta t) = V_t(t) - V_a(t + \Delta t)$$

(6.44)

$$V_a(t + \Delta t) = 0$$

(6.45)

If the net volume available is less than the maximum moistening rate into layer III (S_{i3}) then:

$$V_t(t + \Delta t)' = V_t(t + \Delta t) - V_a(t + \Delta t)$$

(6.46)

$$V_a(t + \Delta t) = 0 \text{, if } V_t(t + \Delta t) > 0$$

(6.47)

else:

$$V_a(t + \Delta t) = -V_t(t + \Delta t)'$$

(6.48)

$$V_t(t + \Delta t) = 0$$

(6.49)

If the net volume available is greater than or equal to the maximum moistening capacity of layer III (S_{i3}) then:

$$V_t(t + \Delta t)' = V_t(t + \Delta t) - S_{i3}(\Delta t) \tag{6.50}$$

where:
$S_{i3}(\Delta t)$ = maximum moistening capacity of layer III (mm/time step)

$$V_a(t + \Delta t)' = V_a(t + \Delta t) - S_{i3}(\Delta t), \text{ if } V_t(t + \Delta t)' > 0 \tag{6.51}$$

else:

$$V_a(t + \Delta t)' = V_a(t + \Delta t) - S_{i3}(\Delta t) - V_t(t + \Delta t)' \tag{6.52}$$

$$V_t(t + \Delta t) = 0 \tag{6.53}$$

If there is no room in layer III to accommodate surplus water, then the net surplus goes as discharge to the subsurface drains.

Discharge to the subsurface drains and storage in the groundwater zone

The storage in the groundwater zone is conceptualised as reservoir storage and discharge to subsurface drains as a function of the actual storage in the reservoir. In reality flow to the drains occurs mainly through the cracks (see Appendix I) or pores above the drain level, while the permeability below the drains is generally very low. Therefore the discharge is modelled as a non-linear relation with the storage in layer II (V_a) and introduced with a lagtime of Δt. In the model the actual shape of the phreatic groundwater table is not taken into consideration.

The amount of water that will be discharged to the subsurface drains is the average of the computation during the previous time step and the actual time step at any instant. Thus the discharge is inversely proportional to the distance between the drains and directly proportional to the saturated hydraulic conductivity (k) and net available soil water in layer II to the power n. The parameters k and n in this formula are calibrated parameters to be determined through optimization.

$$Q(t + \Delta t) = k * \frac{V_a(t + \Delta t)^n}{L^2} * \Delta t \tag{6.54}$$

The above non-linear model in estimating discharge has shown a good performance in comparison to the linear model (Ven, 1980).

$$Q_g(\Delta t) = 0.5 * (Q(t) + Q(t + \Delta t)) \tag{6.55}$$

$$V_a(t + \Delta t)' = V_a(t + \Delta t) - Q_g(\Delta t) \tag{6.56}$$

where:
$Q_g(\Delta t)$ = discharge to take place during each time step if the remaining is > 0 (mm)

$$Q_t = Q_t + Q_g(t + \Delta t), \text{ if } V_a(t + \Delta t)' > 0 \tag{6.57}$$

else:

$$Q_t = Q_t + V_a(t + \Delta t) \tag{6.58}$$

$$V_a(t + \Delta t) = 0 \tag{6.59}$$

The above computation is done for each time step (hours) and cumulated for presentation and averaged for each day. The computation of the groundwater table due to a surplus of water in layer II is based on the following schematisation of layer II and the rise in the water level in the cracks (Figure 6.7).

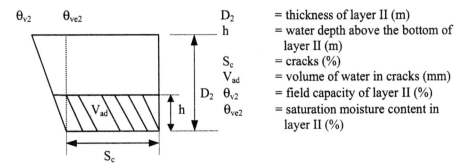

D_2	= thickness of layer II (m)
h	= water depth above the bottom of layer II (m)
S_c	= cracks (%)
V_{ad}	= volume of water in cracks (mm)
θ_{v2}	= field capacity of layer II (%)
θ_{ve2}	= saturation moisture content in layer II (%)

Figure 6.7 Surplus of groundwater in layer II

$$V_{ad} = \sum_{i=1} \frac{V_a}{O_1} \tag{6.60}$$

$$A = \frac{(100 - S_c) * (\theta_{ve2} - \theta_{v2})}{200 * D_2} \tag{6.61}$$

$$A * h(t)^2 + S_c * h(t) - \frac{V_{ad}}{10} = 0 \tag{6.62}$$

where:
O_1 = number of steps in a day
V_a = surplus of water in layer II cumulated (mm/day)

Solving the quadratic equation for h(t) gives the change in the groundwater depth from the bottom of layer II.

Determination of the polder water level in the rural area

Depth of dewatering from the surface is taken as the depth of the drains. However, if the ripening (see Appendix I) is less than the drain depth the dewatering depth is taken as the ripening depth (Figure 6.8).

$D_o = D_d$

$D_o = D_r$ when $D_r < D_d$

For optimization the polder water level is always kept at 0.20 m below the subsurface drain depth. If it is less than the $D_d + 0.2$, then the polder water level is adjusted to $D_d + 0.2$.

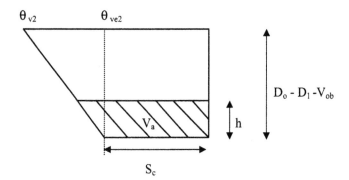

where:
D_o = dewatering depth (m-surface)
D_1 = depth of layer I (m)
h = groundwater table in layer II (m)
S_c = cracks (%)
V_a = saturated volume of groundwater in layer II (mm)
V_{ob} = depth above the dewatering level (m)
θ_{ve2} = moisture content at field capacity (%)

Figure 6.8 Scheme for the computation of the groundwater table

The difference of the water depth between the polder water level and the dewatering depth is as below:

$$V_{op} = PP - D_o \qquad\qquad (6.63)$$

where:
D_o = dewatering depth (m-surface)
D_d = subsurface drain depth (m-surface)
D_r = ripening depth (m-surface)
PP = polder water level (m-surface)

Open water in m^3 from the polder water level up to the level when the pump is stopped is as follows:

$$V_{ow} = P_{ow} * h_p * O_l * 100 \tag{6.64}$$

where:
h_p = lowering of the water level from the polder water level (m)
P_{ow} = open water at polder water level (%)
O_l = total land area (ha)

Groundwater table in layer II

In order to find the relation with the groundwater table in layer II and the volume of water in the layer, the following is assumed:
-   cracks are distributed uniformly throughout the depth;
-   when the drainage level is above the level of the ripening depth, the groundwater table is taken at the same level as the open water level.
From simple geometric similarity the following can be stated:

$$\frac{V_a(t)}{1000} = \frac{S_c * h(t)}{100} + \frac{h(t)}{D_o - D_r - V_{ob}(t)} * \frac{100 - S_c}{100} * \frac{\theta_{ve2} - \theta_{v2}}{100} * h(t) * 0.5 \tag{6.65}$$

$$a = \frac{(100 - S_c) * (\theta_{ve2} - \theta_{v2}) * 0.5}{(D_o - D_l - V_{ob}(t)) * 100} \tag{6.66}$$

$$h(t) = -S_c + \frac{\sqrt{S_c^2 + 0.4 * a * V_a(t)}}{2 * a} \tag{6.67}$$

Therefore the depth of the groundwater table below the surface is as follows:

$$D_2(t) = D_o - D_l - V_{ob}(t) - h(t) \tag{6.68}$$

In the present computation several periods are used to compute the discharges and water levels for given rainfall and evapotranspiration in a similar way as computed in the module RAINRUR5.
The rainfall surplus on the open water is calculated as follows:

$$P_n(t) = \frac{P(\Delta t) - E_o(\Delta t) * O_{wm}(t)}{1000} \tag{6.69}$$

where:
$P_n(t)$ = rainfall surplus over the open water area (m^3/time step)
$O_{wm}(t)$ = area of open water at instant t (m^2)

Cumulative discharge from the five different land use areas into the open drainage system is therefore computed as follows:

$$Q_s(\Delta t) = \sum_{j=1,5} (Q(\Delta t) + K(\Delta t)) * 10 * Q_l(j) \qquad (6.70)$$

where:
Q(j)   = different type of land use in the rural area (ha)
Q(Δt) = discharge into a drain from the land (mm/time step)
K(Δt) = seepage from the bottom (mm/time step)
Q$_s$(Δt) = cumulative discharge (m³/time step)

Discharge into the open water due to seepage is given by:

$$V_0(t) = K(\Delta t) * 10 * O_{wm}(t) \qquad (6.71)$$

where:
V$_0$(t)   = open water volume above polder water level (m³)

Therefore the total open water volume above the polder water level is given by the following formula:

$$V_o(t + \Delta t) = V_o(t) + P_n(t) + Q_s(t) \qquad (6.72)$$

For the computation of the discharge in the main system in the rural area, the urban discharge is added to the total discharge.

Operation of pumps to pump out the excess water

The pump operation to pump out the excess water is handled in the program in the following manner:

When $Y_s(t) < h_a$: $I_w = 0$ and $G_c(t) = 0$, no pumping
When $Y_s(t) > h_a$: $I_w = 1$, start pumping
When $Y_s(t) > h_p$: $I_w = 1$, Gc(t) = capacity pumping (m³/time step)

where:
I$_w$ = indicator of start pumping (-)
G$_c$(t)   = pumping capacity at any instant (m³ /time step)
Y$_s$(t)   = water level above polder water level at any instant (m)
h$_a$ = start level (m+polder water level)
h$_p$ = stop level (m-polder water level)

$$V_o(t + \Delta t)' = V_o(t + \Delta t) - G_c(\Delta t) \qquad (6.73)$$

if $V_o(t + \Delta t) < V_{ow}$

$$V_o(t + \Delta t) = V_{ow} \qquad (6.74)$$

$$G_c(t+\Delta t) = G_c(t) - \left(V_{ow} - V_0(t+\Delta t)\right)$$
(6.75)

where:
V_{ow} = minimum volume in the open water that can be bailed out (m³)

Fluctuation of the water level

If V_o is greater than 0 the rise of the water level in the open drains above the polder water level (Y_s) is computed in the following way:

$$a = \left(\frac{P_{ow}}{100} + \frac{0.5}{L_s}\right) * O_l$$
(6.76)

$$b = \left(2 * \frac{T_s}{L_s} + \frac{T_t}{P_l}\right) * O_l + 0.20 * T_v * (c + d * O_d)$$
(6.77)

$$Y_s(t+\Delta t) = \frac{-a + \sqrt{(a^2 + 0.0004 * b * V_{ow}(t+\Delta t))}}{2 * b}$$
(6.78)

The new open water area in m² due to the rise of the water level is calculated in the following way:

$$O_{wm}(t+\Delta t) = \left(a + b * Y_s(t+\Delta t)\right) * 10,000$$
(6.79)

If V_o is equal to zero then the rise of the water level and open water volume is obtained as follows:

$$Y_s(t+\Delta t) = \frac{V_o(t+\Delta t)}{P_{ow} * O_l * 100}$$
(6.80)

$$O_{wm}(t+\Delta t) = P_{ow} * O_l * 100$$
(6.81)

When the difference depth between the water level and the polder water level is greater than V_{op}.

$$V_{ob}(t+\Delta t) + Y_s(t+\Delta t) - V_{op} > 0$$
(6.82)

The volume to be drained into the main drainage system from each type of land use is calculated as follows:

$$V_a(t+\Delta t) = V_a(t) - 10 * S_c\left(V_{ob}(t+\Delta t) - V_{ob}(t)\right) - 0.5 * A$$
(6.83)

$$A = \frac{(\theta_{ve2} - \theta_{v2})(100 - S_c)(V_{ob}(t + \Delta t)^2 - V_{ob}(t)^2)}{10 * (D_o - D_l)} \tag{6.84}$$

Model efficiency

The optimization model of Rosenbrock has been used to determine the model efficiency of the following parameters (Schultz, 1992):

$$F = \sum_{i=1}^{365} (Q_{obsi} - Q_{comi})^2 \tag{6.85}$$

$$F_0 = \sum_{i=1}^{365} (Q_{ave} - Q_{obsi})^2 \tag{6.86}$$

$$R^2 = 1 - \frac{F}{F_0} \tag{6.87}$$

where:
R   = model efficiency (-)
Q_{obsi} = observed discharge (mm/day)
Q_{comi} = computed discharge (mm/day)
Q_{ave} = average discharge (mm/day)

Hydrological model for the rural area in a peat polder

In peat soil there are no cracks in this type of soil but the basic principle of the model for the clay soil can be applied to peat soil. The only element, which different is calculation of the groundwater table, which is only dependent on pore spaces in the peat. This is simulated by taking cracks to be zero. Therefore in the model for the rural area in a peat polder the hydrological processes involved in the surface and subsurface environment are also schematised into four layers where the water balance is computed assuming each layer as a reservoir in the same way as for a clay polder (Figure 6.2). The hydrological model works in the same way as for the clay polder see Equation 6.1 to 6.53.

6.2.2 Economic computation for the Netherlands

The economic computation is done with the program OWAP. The scheme consists of the subroutines shown in the schematic diagram (Figure 6.9). The economic computation is done separately for the rural and urban area by the subroutines OWARUR and OWAURB. They are coupled with other modules for optimizing the components in the system.

Economic computation for the rural area

The economic computation for the rural area in both a clay and a peat polder use mostly the same formulas. The only differences in the formulas are the constants, which are depending on the different water management practices. The Economic computations for the rural area are described as follows.

The open water area in the rural area is estimated with the following equation:

$$O_w = \left[\frac{P_{ow}}{100} + \frac{0.5}{L_s} + \left(2 * \frac{T_s}{L_s} + \frac{T_t}{P_1} \right) * PP \right] * O_1 + 0.2 * T_v * PP * (a + b * O_d)$$

(6.88)

where:

a, b	= constant (-)
L_s	= distance between the drains (m)
O_d	= total land area (ha)
O_1	= total land for agriculture (ha)
O_w	= open water area (ha)
P_1	= parcel length (m)
P_{ow}	= open water up to water level (%)
PP	= polder water depth from the surface (m)
T_s	= side slope of the drains (-)
T_t	= side slope of the collector drains (-)
T_v	= side slope of the canals (-)

Program   Subroutines   Data files

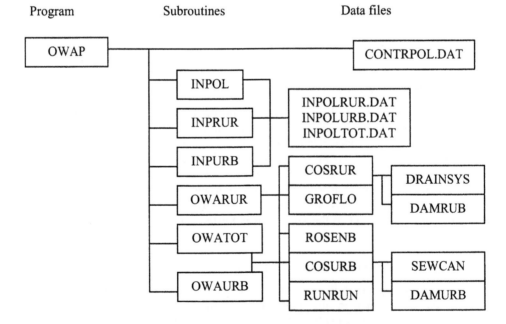

Figure 6.9 Scheme of the model OWAP for the optimization of the main components of water management systems in rural and urban areas in a polder

The area of the infrastructure is taken as a percentage of the net area by the following formula:

$$O_i = \frac{O_l - O_w}{10,000} * G_{oi}$$ (6.89)

where:
G_{oi} = percentage of infrastructure in the net area (%)

The number of farms in the agricultural area is:

$$N_b = \frac{O_l - O_w - O_i}{B_g}$$ (6.90)

where:
B_g = size of the farm (ha)
N_b = number of farms (-)

The area of the building facilities is calculated by using the following formula:

$$O_b = \frac{N_b * (1 + 0.01 * O_v)}{10,000} * O_{bb}$$ (6.91)

where:
O_v = proportion of facilities in the farm area (%)
O_{bb} = area of building and farm (ha)

Therefore the remaining area for agriculture is:

$$O_g = O_l - O_b - O_i - O_w$$ (6.92)

The areas covered by each crop are 'P_{og}' percent of the area 'O_g' ha.

$$O_{gi} = \frac{P_{ogi}}{100} * O_g$$ (6.93)

Costs for the canal system

The costs for the field drains per metre are obtained from the following empirical formula:

$$A_{dr} = (a + b * L) * (1 + c * (D_d - d))$$ (6.94)

where:
a, b, c, d = constant (-)
L   = distance between the drains (m)
D_d = drain depth of subsurface drain or open field drain (m-surface)
A_{dr} = cost of subsurface drain or open field drain (€/m)

The length of the main drains is obtained from the layout of canals in the case study area for both the clay and the peat polder and from literature (Saiful Alam, 1994). The length of the main drain in m can be described as shown in the following empirical formula:

$$V_1 = (a + b * O_d) * 1,000$$ (6.95)

where:
a   = 0.84 for a clay polder and 1.11 for a peat polder
b   = 0.00124 for both a clay polder and a peat polder

The total costs of the subsurface drain or open field drain system at the above rate is:

$$\cos td = \frac{A_{dr} * O_g * 10,000}{L}$$ (6.96)

where:
costd   = total costs of subsurface drains or open field drains (€)

Maintenance costs for the subsurface drains are estimated at 0.07 €/m/year for a clay polder (Rijkswaterstaat, 2004) and maintenance costs for the open field drains are estimated at 0.17 €/m/year for a peat polder (estimated from Van den Bosch, 2003).

$$\cos tdm = \frac{0.10 * O_g * 10,000}{L} \quad \text{for a clay polder}$$ (6.97)

$$\cos tdm = \frac{0.17 * O_g * 10,000}{L} \quad \text{for a peat polder}$$ (6.98)

where:
costdm   = total maintenance costs for subsurface drains or open field drains (€/year)

Earthwork is the sum of the earth movement in the collector drains, sub-main drains and main drains as follows.
Earth movement in the collector drains with a bottom width of 0.5 m (Figure 6.10):

$$E_{xc1} = PP * (0.5 + T_s * PP) * \frac{O_l * 10,000}{L_s}$$
(6.99)

Excavation of the side slopes in the sub-main drains:

$$E_{xc2} = 0.5 * T_t * PP^2 * \frac{O_l * 10,000}{P_l}$$
(6.100)

Excavation of the side slopes in the main drains:

$$E_{xc3} = T_v * PP^2 * O_l * 100$$
(6.101)

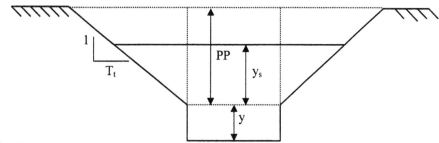

where:
PP   = polder water level (m-surface)
y   = depth of water in the drain (m)
y_s = depth of water from the polder water level (m)

Figure 6.10 Section of a sub-main drain, or main drain

Excavation in between the above side slopes of the sub-main drains and the main drains:

$$E_{xc4} = (y + PP) * P_{ow} * O_l * 100$$
(6.102)

$$G_{ow} = E_{xc1} + E_{xc2} + E_{xc3} + E_{xc4}$$
(6.103)

where:
y   = water depth in the drain (m)
P_{ow} = open water at polder water level (%)
V_1 = length of main drains (m)
P_1 = parcel length (m)
L_s = distance between the drains (m)
PP   = polder water depth from the surface (m)
T_s, T_t, T_v= side slopes of collector drains, sub-main drains and main drains (-)

Construction costs involved in timbering of the sub-main drains and main drains are calculated as follows:

$$K_{xx} = \frac{O_l * 10,000 * K_{obt}}{P_l} + V_l * 2 * K_{obv} \qquad (6.104)$$

where:
K_{obt} = cost of sub-main drains (€/m)
K_{obv} = cost of main drains (€/m)
P_l = parcel length (m)
O_l = total land area (ha)

Therefore the total costs for excavation and timbering of the collector drains, sub-main drains and main drains are:

$$K_{ot} = G_{ow} * K_{og} + K_{xx} \qquad (6.105)$$

where:
G_{ow} = excavation (m³)
K_{og} = cost of excavation (€/m³)

Annual maintenance costs for the collector drains, sub-main drains and main drains are:

$$K_{at} = G_{ow} * K_{gw} + \frac{O_l * 10,000 * K_o}{P_l} + V_l * 2 * K_{ob} \qquad (6.106)$$

where:
K_{gw} = cost of maintenance in earth work (€/m³)
K_o = cost of maintenance in the sub-main drains (€/m)
K_{ob} = cost of maintenance in the main drains (€/m)

Cost of pumping

The required capacity of the pumping station in a polder will be calculated as follows (Schultz, 1992):

$$G_c = \frac{a_{ke} * O_g}{8,640} \qquad (6.107)$$

where:
G_c = pumping capacity (m³/s)
O_g = gross area (ha)
a_{ke} = drainage modulus (mm/day)

The total investment costs for the installation of pumping station is derived from data from Rijkswaterstaat, 2004 and Delfland, 2004 which is expressed with the following formula:

$$A_g = a + b * G_c * h_p \tag{6.108}$$

where:
A_g = total investment costs for pumping station (10^6 €)
a   = constant taken as -0.0495
b   = constant taken as 0.111
G_c = pumping capacity (m^3/s)
h_p = total operating head (m) where $h_p = h_s + h_l$
hs   = pressure head, which is taken as 2.55 m for the clay polder, 2.7 m for the peat polder and for the urban polder 2.3 m
h_l = headloss, taken as 1.5 m

The annual maintenance costs for the pumping station are derived from data from Rijkswaterstaat, 2004 and Delfland, 2004 which is expressed with the following formula:

$$K_{beg} = \frac{A_g}{B} \tag{6.109}$$

where:
K_{beg} = annual maintenance costs for the pumping station (10^6 €)
B   = constant taken as 67

Power required for the station at least favourable circumstances is given by the following formula (Schultz, 1992):

$$G_{cap} = 16.4 * (h_p + 0.5) * G_c \tag{6.110}$$

where:
G_{cap} = power required for the pumping station (KW)

Operation power is based on average annual excess precipitation and seepage (Schultz, 1992):

$$E_{ng} = 0.0454 * (h_p + 0.5) * (P_o + K) * O_d \tag{6.111}$$

where:
E_{ng} = operation power for the pumping station (kWh)
K   = average seepage (mm/year)
P_o = average of excess precipitation (mm/year)
O_d = total area in the polder (ha)

Cost of energy for the pumping (data from Delfland, 2004):

$$K_e = a * G_{cap} + b * E_{ng}$$
(6.112)

where:

a, b   = constants taken as 240 and 0.0738 respectively

K_e = energy costs for the pumping (€/year)

Annual equivalent cost

The total costs for construction may be summarized as follows:

Patt   = field drainage + main drainage + pumping station
(6.113)

Annual maintenance costs are the sum of the following:

Pott   = maintenance of the field and main drainage system + operation and
maintenance cost of pumping station
(6.114)

The annual recovery factor is expressed as follows:

$$V_t = \frac{r * (1+r)^L}{(1+r)^L - 1}$$
(6.115)

where:

r   = annual interest rate, usually 5%

L   = lifetime of the project components (years)

V_t = annual recovery factor (-)

Therefore the annual equivalent costs for drainage are construction and maintenance costs:

$$cdt = V_t * costd + costdm$$
(6.116)

The annual equivalent costs for the main drainage system include excavation, timbering and annual maintenance:

$$cot = V_{t1} * G_{ow} * K_{og} + V_{t2} * K_{xx} + K_{at}$$
(6.117)

where:

V_{t1}, V_{t2} = annual distribution factor for excavation and timbering respectively (-)

Annual equivalent costs for the pump are the costs for installation and operation and maintenance:

$$cgt = V_{t4} * K_{beg} + K_e$$
(6.118)

where:

V_{t4} = annual distribution factor of pump cost of life span 't$_4$' years (-)

The total annual equivalent costs of the water management is therefore:

$$ctt = cdt + \text{cot} + cgt \qquad (6.119)$$

Investment costs for crops, buildings and infrastructure

Maximum and minimum investment for each crop on a hectare of land that possibly can occur during each decade is searched out from a year list of investments for the crops. The value of buildings and other facilities is calculated as follows:

$$wl1 = (1 + 0.01 * O_v) * V1 * \left[\frac{10,000}{O_{bb}} \right] \qquad (6.120)$$

where:
O_{bb} = area of building/farm (ha)
$V1$ = value of building/farm (€)
O_v = % of facilities within the farm (20%)
$wl1$ = value of building and facility in the farm (€/m^2)

Therefore the total costs involved in the farm is the sum of the following:

$$SOMH_{max} = a_1 * wl1 + a_2 * wl2 + \sum_{i=3,5} a_i * wl_{max} \qquad (6.121)$$

$$SOML_{min} = a_1 * wl1 + a_2 * wl2 + \sum_{i=3,5} * a_i wl_{min} \qquad (6.122)$$

where:
wl_{max} = maximum investment costs for crop i (€/ha)
wl_{min} = minimum investment costs for crop i (€/ha)
$wl2$ = value of infrastructure in the farm (€/m^2)

Crop damage computation

Crop damage may be occurred during planting, harvesting and air deficit.

Planting

The reduction of crop yield is computed in the model for certain boundary conditions based on the following parameters:
- soil moisture tension during the execution of tillage;
- date of sowing;
- late date of sowing when crop damage may occur;
- number of days that sowing takes place.

The reduction of crop yield related to planting is based on an empirical relation, based on the available field data, which is given by the following formula (derived from Deinum, 1966):

$$Q_{d1} = a * d^b \qquad\qquad\qquad\qquad\qquad\qquad\qquad\qquad (6.123)$$

where:

a, b   = constants taken as 0.809 and 0.811 respectively (-)
d   = number of days of delay from the last date of sowing (-)
Q_{d1} = crop yield reduction due to late date of sowing (%)

Dates of early planting and late planting as taken in the model are as follows: earliest date for sowing is 15th March and late date of sowing is 15th April (adapted from Deinum, 1966). The period of sowing is taken as 2 days.

Other boundary conditions is the soil moisture tension in the upper peat soil for grass of at least pF 2.0 for a drain water level of 0.20 m to 0.30 m-surface due to the bearing capacity of the peat soil penetration resistance 0.6 MPa (De Bakker and Van den Berg, 1982).

The critical periods in a series of hydrological years have been checked with the above criteria and yield reduction resulting from untimely sowing and unfavourable situations has been computed for each time step and for grass.

Harvesting

The reduction of crop yield is based on an empirical relation, based on the available field data, which is given by the following formula (derived from Deinum, 1966):

$$Q_{d2} = a * d^b \qquad\qquad\qquad\qquad\qquad\qquad\qquad\qquad (6.124)$$

where:

a, b   = constants taken as 0.152 and 1.000 respectively (-)
d   = days from the last limiting date (-)
Q_{d2} = damage due to late date of harvesting (%)

Dates of early harvesting and late harvesting as taken in the model are as follows: earliest date for harvesting is 1st November and late date of harvesting is 1st December (adapted from Deinum, 1966). The period of harvesting is taken as 2 days.

Other boundary conditions is the soil moisture tension in the upper peat soil for grass pF is at least 2.0 in peat soil to be the lowest value at which harvesting can take place for the same reason as mentioned above.

The critical periods in a series of hydrological years have been checked with the above criteria and yield reduction resulting from untimely harvesting and unfavourable situations has been computed for each time step and for grass.

Yield reduction due to air deficit

There may be reduction of yield due to insufficient air entry as a result of a bad soil structure in the upper layer. The relation between the percentage of air in the soil layer and the average groundwater table during the winter is obtained from field data, which uses of the following formula:

$$L_p = a + b * h_g \tag{6.125}$$

where:
L_p = air content in the soil (%)
a, b   = constants equal to 10.7 and 16.3 respectively for clay soil (Boekel, 1974)
           = 1.3 and 32.3 respectively for peat soil (De Bakker and Van den Berg, 1982),
h_g = average groundwater table (m-surface)

The relation between the percentage of air in the soil layer and the average groundwater table during summer in peat land is obtained from Zegvelderbroek 1973 to 1974 field data (De Bakker and Van den Berg, 1982), which assumed to be the same as in the above formulas, but in this case the constants a and b are equal to 1.3 and 32.3 respectively.

A similar relation for the relative yield with the percentage of air is obtained from field data and may be used to calculate the relative yield for a given air percentage (derived from Deinum, 1966):

$$R_o = C + A * B^{(1/L_p)} \tag{6.126}$$

where:
R_o = relative yield (%)
A, B, C = constants equal to 196, 0.001 and -56.2 respectively (-)

6.2.3 Hydrological computation for Thailand

Hydrological model for rice fields in Thailand

In the model for the rural area the hydrological processes in a rice field as applicable to the polders in Thailand involves the surface, subsurface environment, water management at field level and overall water management in the polder areas. These are surface and subsurface drainage systems in rice fields to remove excess open water, to control salinity and to remove toxic by-products such as hydrogen sulphide, soluble manganese, iron or aluminium. Hence in the model for the rural area the hydrological processes in a rice field are schematised into four layers for which the water balances are determined assuming each as a reservoir (Figure 6.14).

Puddling for land preparation in a rice field

Puddling influences pore size distribution and thus water retention and transmission. Puddled soil retains more water at a given soil water potential and may have higher unsaturated hydraulic conductivity, but evaporation losses from puddled soils are low. Rice grown on a puddled soil consequently may suffer less in mild drought than rice on non-puddled soil (De Datta and Kerim, 1974).

Most farmers till a wet field rather than a dry one, because it facilitates transplanting of rice plants, helps level the land, under ploughing weeds and stubble and to improve the soil conditions for plant growth. They first soak the land until the topsoil is saturated, after that they plough the soil until about 0.15 m-surface and harrow or puddle. Repeated ploughing and harrowing or puddling partially or completely destroys soil aggregates, dependent on their stability (Yun-sheng 1983, Sharma and De Datta 1986 and Pagliai and Painuli, 1991) (Figure 6.11).

Cracks in a rice field

Rice is usually grown in clay soil, and alternate soaking and drying produces deep and wide cracks in it. Due to the practice of the rice farmers in both rainfed and irrigated conditions, the rice field will dry due to drought and in dry periods in rainfed conditions, while in irrigated conditions farmers will let the soil dry before harvesting. In both cases cracks can occur.

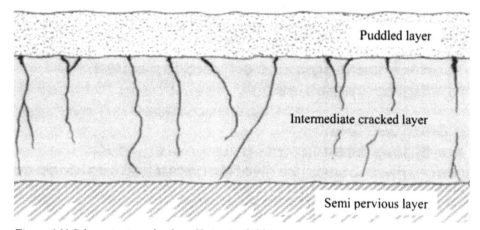

Figure 6.11 Schematisation of soil profile in rice fields

During drying out of clay soil, shrinkage and cracks begin to form at the ground surface, which extend deeper into the soil as the soil becomes drier. When clay first begins to dry out, water is lost between the clay particles causing the soil to shrink by an amount, which is more or less equal to the volume loss of water. The drying process imposes large physical stresses on the soil particles, which forces them together and orientates them parallel to each other (De Crecy, 1982). The shrinkage results in the formation of deep cracks and in the settling of the surface but the main mass of soil remains saturated. This phase is termed normal shrinkage and it may continue up to high suctions e.g. 15 bars or wilting point (Rycroft and Amer, 1995). Beyond this point, shrinkage becomes less than the volume loss of water as air begins to enter the matrix of the clay. Three zones can be recognized: a 'structural' zone at the wet end, a 'normal' and a 'residual' at the dry end where the volumetric change is smaller than the volume change (or nearly so) to the volume of water lost or gained (Figure 6.12).

It can be assumed that almost all cracks will be closed if the soil moisture in the soil fraction elements has reached field capacity (Rycroft and Amer, 1995). In the model for the rural area rice soil is schematised as shown in Figure 6.13.

Modeling the water balance and water transport in swelling and shrinking clay soil

The model FLOCR for the simulation of the soil water balance in swelling and shrinking clay soil was developed by Bronswijk (1988). The soil matrix and cracks are schematised as a layer as shown in Figure 6.13. This model describes moisture transport in a soil matrix and cracks as follows.

When rainfall reaches the surface of a cracked clay soil, part of the water infiltrates into the soil matrix and part of it flows into the cracks. When rainfall exceeds the maximum infiltration rate of the soil matrix at the top layer, water flows into the cracks. In addition a certain part of rainfall falls into the cracks.

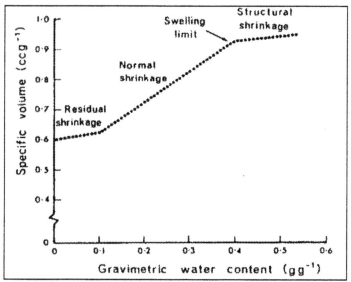

Figure 6.12 Example of a shrinkage and swelling curve (Probert, et al., 1987)

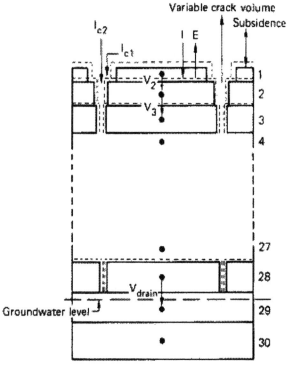

Figure 6.13 Schematisation of soil profile in swelling and shrinking clay soil (Bronswijk, 1988)

In the model, horizontal infiltration into crack walls of water running rapidly downwards along cracks is neglected. However, Hoogmoed and Bouma (1980) showed that this infiltration is small. Moreover, calculations of the water balance of the soil as a whole are not influenced by horizontal infiltration. Bronswijk (1988) has shown that the model in this way was in good agreement with the observed data. For this reason all water infiltrating into cracks is assumed to accumulate at the bottom of the cracks and is added to the moisture content of the corresponding soil layers.

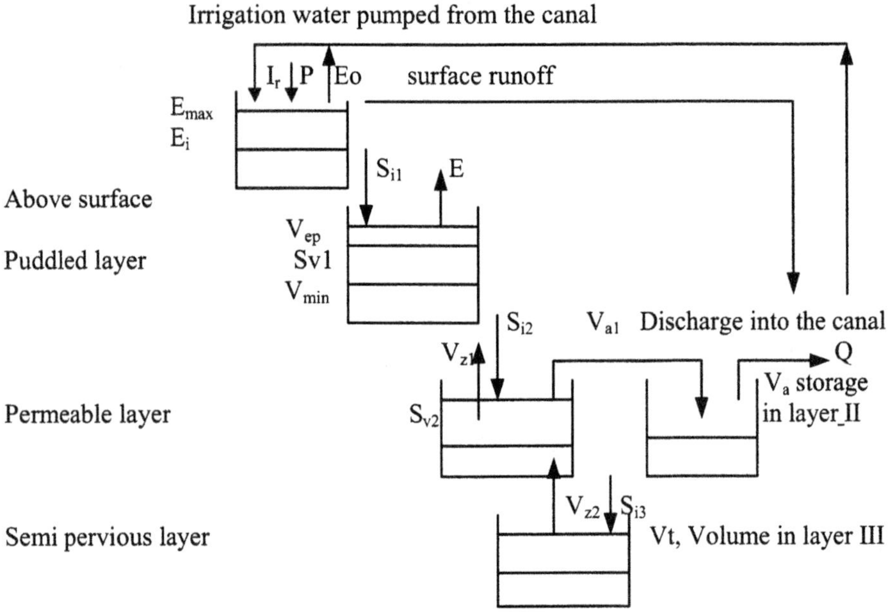

Figure 6.14 Scheme of the hydrological model for a rice polder

The model components are as follows:

Above the surface:
- precipitation (mm/day)   (P)
- open water evaporation (mm/day)   (E_0)
- maximum infiltration in layer I (mm/day)   (S_{i1})
- interception (mm)   (E_i)
- irrigation water which is pumped from canal (mm/day)   (I_r)
- surface drainage discharge (mm/day)   (Q_1)
- water layer depth (mm)   (W_L)

Layer I (puddled layer):
- evapotranspiration (mm/day)   (E)
- soil moisture at which evapotranspiration is at potential rate (mm/m)   (V_{ep})
- soil moisture at which evapotranspiration is zero (mm/m)   (V_{min})
- maximum infiltration in layer I (mm/m)   (S_{i1})
- storage above field capacity in layer I (mm/m)   (S_{v1})
- maximum percolation to layer II (mm/m)   (S_{i2})
- capillary rise from layer II (mm/day)   (V_{z1})

Layer II (intermediate cracked layer):
- percolation from layer I (mm/m) (S_{i2})
- capillary rise to layer I (mm/day) (V_{z1})
- percolation of surplus water to layer III (S_{i3})
- capillary rises from layer III (mm/day) (V_{z2})
- discharge of surplus water to the lateral canal (V_{a1})
- storage above field capacity in layer II (mm/m) (S_{v2})

Layer III (semi pervious layer):
- maximum percolation from layer II into layer III (mm/m) (S_{i3})
- moisture at saturation in layer III (mm) (V_t)

Open water storage:
- discharge from layer II through the lateral canal (mm) (V_{a1})
- change in storage (mm) (V_a)
- discharge through the culvert (mm) (Q_2)
- discharge through pumping (mm) (Q_3)
- irrigation water, which is taken from the canal (mm) (I_r)

The above process components will be described in brief.

Interception

Interception depends on the type of vegetation and its coverage, in the model interception at any instant is the precipitation at that instant and the residual interception from the previous time step (Equation 6.1), which is limited, by the maximum interception E_{max}. The upper limit of interception is a calibrated model parameter. Maximum values may be expected to be the same as described for the Dutch conditions.

Evapotranspiration

If the interception is less than the maximum interception (E_{max}) but still greater than zero, open water evaporation rate (E_0) will take place, leaving the remaining interception to be added to the precipitation in the next time step. Otherwise evaporation from soil is assumed.

When the precipitation and irrigation rate are greater than the infiltration rate of the soil in layer I the evapotranspiration is a function of the potential rate (E_p) and the moisture status in the underneath layers. Research data show that more than one-third of the evapotranspiration water is from the surface of standing water in a rice field (IRRI, 1971). Open water evaporation is expressed as given by Equation 6.2.

As long as the moisture in layer I is above the upper limit of saturation in layer I evapotranspiration takes place at potential rate and if it is less than V_{min} it reduces to zero. If the moisture status is in between these limits evapotranspiration is linearly interpolated between the potential value and zero (Figure 6.15).

In case there is a water layer on the surface of the soil, one-third of the evapotranspiration water is from the surface of standing water in a rice field. Then two-third of the water evapotranspiration is from the root zone. If there is no water layer on the field evapotranspiration will take place from the soil in the root zone.

If WL > 0

$$E_{ps}(\Delta t) = \frac{2}{3} * f(t) * E_0(\Delta t) \qquad\qquad (6.127)$$

If WL < 0

$$E_{ps}(\Delta t) = f(t) * E_0(\Delta t) \qquad\qquad (6.128)$$

where:
$E_{ps}(\Delta t)$ = potential evapotranspiration from the root zone (mm/time step)
$f(t)$ = crop factor (-)

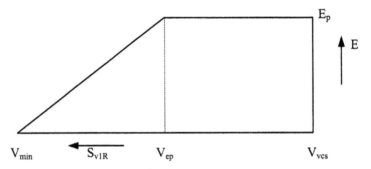

Figure 6.15 Evapotranspiration and soil moisture

$$E_{as}(\Delta t) = \frac{(V_{min} - S_{v1R}(t))}{(V_{min} - V_{ep})} * E_{ps}(\Delta t), \text{ if } V_{min} < V < V_{ep} \qquad (6.129)$$

$$E_{as} = 0.0, \text{ if } V_{min} \geq V \qquad\qquad (6.130)$$

$$E_{as} = E_{ps}, \text{ if } V \geq V_{ep} \qquad\qquad (6.131)$$

where:
$E_{as}(\Delta t)$ = actual evapotranspiration from the root zone (mm/time step)
$S_{v1R}(\Delta t)$ = soil moisture deficit in layer I related to 0.8 saturation (mm/m)
V_{ep} = soil moisture (mm/m) at which evapotranspiration is at a potential rate (0.8 V_{vcs}) (http://www.fao.org/docrep/X0490E/x0490e0e.htm)
V_{min} = soil moisture at which evapotranspiration is zero (0.20 V_{vcs}) (mm/m)
V_{vcs} = soil moisture at saturation (mm/m)

If the actual evapotranspiration from the first layer is more than the potential (E_{ps}) the residual is taken as a loss of moisture from layer II in the same manner as above resulting in a decrease in the moisture content in layer II. The boundary values for V_{min} and V_{ep} are updated for the layer based on the difference between the 0.20 and 0.80 of the saturated moisture content in layer II:

$$E'(\Delta t) = E_{ps}(\Delta t) - E_{as}(\Delta t)$$
(6.132)

where:

$E'(\Delta t)$ = evapotranspiration from layer II (mm/time step)

Consequently the moisture content in layer I is reduced for the next time step as follows:

$$S_{v1}(t + \Delta t) = S_{v1}(t) - \frac{E_{as}(\Delta t)}{D_1}$$
(6.133)

where:

$S_{v1}(t)$ = moisture deficit related to field capacity in layer I (mm/m)
D_1 = thickness of layer I (m)

If the moisture content in layer II is less than 20% of the moisture at saturation rice will die and evaporation will be taken as evaporation from bare soil. The evaporation from bare soil is assumed to decrease from 20% of the moisture content at saturation to zero at wilting point. Hence the total evapotranspiration is the sum of losses from the two layers.

Moistening of the layers

The amount of moisture that will penetrate in layer I and layer II depends on the interception at the surface, moisture content of the layer in the previous time step, maximum rate of moistening of the layer and the thickness of the layer. Therefore the maximum amount of moisture that can percolate in the underneath layers is assumed to be increasing from V_{min} with the increase of the moisture content up to field capacity (Equations 6.9 and 6.10). Therefore the maximum volume of water that can be absorbed at any instant is as follows.

In case there are no cracks occur at the soil surface ($\theta_1 > \theta$ at pF = 2.53)

Equation 6.11 gives the maximum volume of water that can be absorbed. When the interception at any instant (E_i) is greater or equal to V_{max}, there is maximum interception as well as maximum accommodation of water in layers I and II. While if there is any surplus it will be stored at the surface of the rice field and it will be added to the water depth in the field. If the water depth is higher than the maximum allowable depth it will discharge to the main drain.

If $\theta_1(t) > \theta_1$ at pF =2.53

$$V_{af}(\Delta t) = E_i(t + \Delta t) - V_{max}(\Delta t)$$
(6.134)

where:

V_{af} = surplus to be added to the water depth at the field surface (mm)

If however, the interception at any instant (E_i) is less than V_{max}, after meeting the maximum interception on the surface i.e. E_{imax}, the remaining will fill successively the space in layer I and layer II in the following manner.

In case cracks occur at the soil surface ($\theta_1 < \theta$ at pF = 2.53)

When the precipitation is greater than the maximum interception, this part of the water reaches the surface of the cracked clay soil, part of the water infiltrates into the soil matrix and part of it falls into the cracks. When rainfall exceeds the maximum infiltration rate of the soil matrix at the top layer (layer I), water flows into the cracks. In addition a certain part of rainfall falls into cracks.

If $\theta_1(t) < \theta_1$ at pF = 2.53

$$V(\Delta t) = E_i(t + \Delta t) - E_{i\,max} \tag{6.135}$$

$$S_{v1}(t + \Delta t) = S_{v1}(t) + \left(\frac{V(\Delta t)}{D_1}\right) * A_{m1}(t), \text{ if } V * A_{m1} < S_{i1} \tag{6.136}$$

also:

$$S_{v1}(t + \Delta t) = S_{v1}(t) + \left(\frac{V(\Delta t)}{D_1}\right) * A_{m1}(t), \text{ if } V * A_{m1} > S_{i1} \tag{6.137}$$

Therefore, the amount of rainfall into the cracks at layer I is less than the maximum of the moisture that can be stored at any instant in layer I, the increase in storage at any instant in layer II is calculated as follows:

If $V(\Delta t)*A_{m1} < S_{i1}(\Delta t)$

$$S_{v2}(t + \Delta t) = S_{v2}(t) + \left(\frac{V(\Delta t)}{D_2}\right) * A_{c1}(t) \tag{6.138}$$

If $V(\Delta t)*A_{m1} > S_{i1}(\Delta t)$

$$S_{v2}(t + \Delta t) = S_{v2}(t) + \left(\frac{V(\Delta t)}{D_2}\right) * A_{c1}(t) + \frac{A_{m1}(t) * V(\Delta t) - S_{i1}(\Delta t)}{D_2} \tag{6.139}$$

where:
A_{m1} = relative area of the soil matrix in layer I at any instant (-)
A_{c1} = relative area of the cracks in layer I at any instant (-)

In case there is a water layer on the ground surface, the increase in storage at any instant in layers I and II is calculated as follows:

If WL(t) > 0

$$S_{v1}(t + \Delta t) = S_{v1}(t) + \frac{S_{i1}(\Delta t)}{D_2} \tag{6.140}$$

$$S_{v2}(t + \Delta t) = S_{v2}(t) + \frac{S_{i2}(\Delta t)}{D_2} \tag{6.141}$$

For a small depth of the swelling and shrinking clay soil layer, it is assumed that the crack walls are parallel to each other and the soil is isotropic. The relative area of the soil matrix and cracks can then be defined by:

$$A_{m1}(t) = \frac{(D_1 - \Delta D_1(t))^2}{D_1^{\,2}} \tag{6.142}$$

$$A_{c1}(t) = \frac{\Delta D_1(t)^2}{D_1^{\,2}} \tag{6.143}$$

where:
D_1 = depth of the soil in layer I where there is no shrinkage (m)
ΔD_1 = change in depth of the soil in layer I due to swelling and shrinking (m)

Change in depth of layers I and II due to change in the soil moisture can be defined as follows. If the water content of the soil is higher than the water content at a potential of -33 kPa there is no shrinkage. While the water content at a potential of the soil between -33 kPa (pF = 2.53) and -1,500 kPa (pF = 4.18) shrinkage will occur. By assuming linear shrinkage, the following relationship has been proposed (Medina, et al., 1998).

$$L_s = \frac{C_i}{(\theta_{15} - \theta_{1/3})}(\theta_i - \theta_{1/3}) \tag{6.144}$$

where:
L_s = linear shrinkage (-)
θ_{15} = water content in the soil at a potential of -1,500 kPa (pF = 4.18) (%)
$\theta_{1/3}$ = water content in the soil at a potential of -33 kPa (pF = 2.53) (%)
C_i = coefficient of linear extension or linear extensibility (-)

The coefficient of linear extension (C_i) describes the linear change in the volume of the soil at moisture potential from -33 kPa to -1,500 kPa (Grossman, et al., 1968). Values of the coefficient of linear extension are available for many soils from USDA-NRCS (http://soils.usda.gov/).

$$C_i = \left(\frac{\rho_d}{\rho_m}\right)^{\frac{1}{3}} - 1 \qquad\qquad (6.145)$$

where:

ρ_d = bulk density of a dry clod at -1,500 kPa (pF = 4.18) moisture potential (kg/m^3)

ρ_m = bulk density of a moist clod at -33 kPa (pF = 2.53) moisture potential (kg/m^3)

In well-developed alluvial soils, volume changes with changing water contents are isotropic (Brownswijk, 1990). Assuming equidimensional linear shrinkage in the three dimensions of a cube (Medina, et al., 1998), the percentage of volume change is given by:

$$V_s = \left|1 - (1 - L_s)^3\right| * 100 \qquad\qquad (6.146)$$

where:
V_s = volume change in the soil (%)

Change in depth of the soil in layer I due to swelling and shrinking is calculated by assuming that the soil is isotropic as follows (Brownswijk, 1990 and Medina, et al., 1998):

$$\Delta D_1(t) = \left[1 - \left(1 - \frac{V_{s1}(t)}{100}\right)^{\frac{1}{3}}\right] * D_1 \qquad\qquad (6.147)$$

where:
$V_{s1}(t)$ = volume change of soil layer I (%)
D_1 = depth of layer I at no shrinkage (m)
ΔD_1 = change in depth of layer I (m)

Hence the depth of layer I at the next time step is calculated as follows:

$$D_1(t + \Delta t) = D_1 - \Delta D_1(\Delta t) \qquad\qquad (6.148)$$

The percentage of volume change is related to the moisture in soil layer I, which is adapted from Equation 6.146 is as follows:

$$V_{s1}(t) = \left|1 - (1 - L_{s1}(t))^3\right| * 100 \qquad\qquad (6.149)$$

$$L_{s1}(t) = \frac{C_{i1}}{(\theta_{115} - \theta_{11/3})}(\theta_1(t) - \theta_{11/3}) \qquad (6.150)$$

where:
$V_{s1}(t)$ = volume change of soil layer I at any instant (%)
$L_{s1}(t)$ = linear shrinkage of soil layer I at any instant (-)
θ_{115} = water content in soil layer I at a potential of -1,500 kPa (pF = 4.18) (%)
$\theta_{11/3}$ = water content in soil layer I at a potential of -33 kPa (pF = 2.53) (%)
$\theta_1(t)$ = water content of soil layer I at any instant (%)
C_{i1} = coefficient of linear extension of soil layer I (-)

The change in depth in layer II is calculated in the same way as follows:

$$\Delta D_2(t) = \left[1 - \left(1 - \frac{V_{s2}(t)}{100} \right)^{\frac{1}{3}} \right] * D_2(t) \qquad (6.151)$$

where:
$V_{s2}(t)$ = volume change of soil layer II at any instant (%)
D_2 = depth of layer II at no shrinkage (m)
ΔD_2 = change in depth of layer II (m)

Hence the depth of layer II at the next time step is calculated as follows:

$$D_2(t + \Delta t) = D_2 - \Delta D_2(\Delta t) \qquad (6.152)$$

The percentage of volume change is related to the moisture in soil layer II, which is adapted from Equation 6.146 is as follows:

$$V_{s2}(t) = \left| 1 - (1 - L_{s2}(t))^3 \right| * 100 \qquad (6.153)$$

$$L_{s2}(t) = \frac{C_{i2}}{(\theta_{215} - \theta_{21/3})}(0.5 * (\theta_2(t) + \theta(t)_3) - \theta_{21/3}) \qquad (6.154)$$

where:
$V_{s2}(t)$ = volume change of soil layer II at any instant (%)
$L_{s2}(t)$ = linear shrinkage of soil layer II at any instant (-)
θ_{215} = water content in soil layer II at a potential of -1,500 kPa (pF = 4.18) (%)
$\theta_{21/3}$ = water content in soil layer II at a potential of -33 kPa (pF = 2.53) (%)
$\theta_2(t)$ = water content of soil layer II at any instant (%)
C_{i2} = coefficient of linear extension of soil layer II (-)

Soil moisture tension and capillary rise

The soil moisture tension in layer I is assumed uniform throughout its depth, while in layer II it varies linearly and becomes zero or even positive at its bottom during the growing season and below the bottom during the harvesting season when the water table goes down below the cracks (Figure 6.4).

Soil moisture in the first layer is defined by Equation 6.17 and moisture tension in layer I and layer II is based on an empirical formulas by Equations 6.18 and 6.19.

θ_1 and θ_2 are soil moisture content in layer I and layer II at their interface. The soil moisture tension in layer II is based on the moisture content at its boundary, and the moisture content at its bottom, which is computed from the simple geometry relationship as shown in Figure 6.5 and Equations 6.20 to 6.24.

The unsaturated hydraulic conductivity is based on the above averaged soil moisture tension and clay contents in layers I and II which is computed by the empirical formula as in Equations 6.25 to 6.27. Therefore the capillary flow from layer II to layer I is computed as in Equation 6.28.

The above shows that the capillary rise from the saturated part to the unsaturated part and through the unsaturated zone depends on the metric pressure difference between the boundaries, type of soil and the soil moisture condition in the unsaturated zone. Capillary flow to layer I increases the soil water content in layer I as in Equation 6.29.

However, if the moisture content in layer I is greater than field capacity (i.e. $S_{v1} \geq 0$), there is percolation from layer I to layer II and the soil moisture content in layer II and layer I in the next time step are calculated as in Equations 6.30 and 6.31.

If θ_3 is greater than the moisture content at saturation in layer II, the soil moisture tension at the interface of the layers i.e. P_3 and P_4 is equal to zero and θ_4 is at saturation moisture content in layer III. If θ_3 is less than the saturation moisture content flow takes place from layer II to layer III that follows the same as the flow computation in layer I as in Equations 6.32 to 6.37.

The effective thickness of layer III is calculated as described in the schematisation in Figure 6.6 and Equation 6.38. Therefore the capillary rise from layer III to layer II and soil status is updated as in Equations 6.39 to 6.40. However, if the soil water content in layer II is greater than the field capacity i.e. $S_{v2} > 0$, there is percolation to layer III and the soil water (S_{v2}) in layer II is set as in Equations 6.41 and 6.42. The net volume of moisture available for drains and to meet the deficit in layer III is:

$$V_a(t + \Delta t) = V_a(t) + V_{a2}(\Delta t) - V_{z2}(\Delta t) \tag{6.155}$$

If the amount is less than 0 ($V_a < 0$) the net volume in layer III is updated as in Equations 6.44 and 6.45. If the net volume available is less than the maximum moistening rate into layer III (S_{i3}) then Equations 6.46 to 6.49 are applied.

If the net volume available is greater than or equal to the maximum moistening capacity of layer III (S_{i3}) then Equations 6.50 to 6.53 are applicable. If there is no room in layer III to accommodate surplus water, then the net surplus goes as discharge to the surface drains.

In case the groundwater is above the surface only interception, evapotranspiration, deep percolation and discharge to the surface canal are concerned. The deep percolation is reported to be 1 mm/day for heavy clay soil and 3 mm/day for light clay soil (Charnvej, 1999).

Discharge to open field drains and storage in the groundwater zone

Discharge to the lateral canal in case of a water table below the surface

The storage in the groundwater zone is conceptualised as reservoir storage and discharge to lateral canals (Figure 6.16) as a function of the actual storage in the reservoir. In reality flow to the drains occurs mainly through the cracks or pores above the drain level, while the permeability below the drains is generally very low.

The amount of water that will be discharged to the subsurface drains is the average of the computation during the previous time step and the actual time step at any instant. Thus the discharge is calculated with the Glover Dumm Equation (Dumm, 1960) as follows.

If the polder water level is above the groundwater table at any instant, the discharge to the canal is as follows:

If h2(t) < D₂(t) then

$$G_{wo}(t) = D_2(t) + D_1(t) - h_2(t) + h_1(t) \qquad (6.156)$$

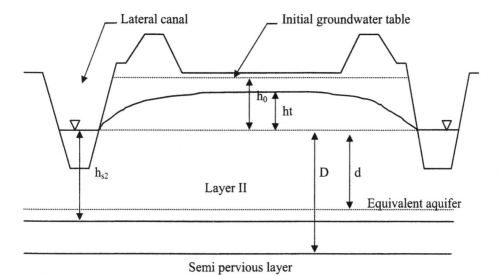

Figure 6.16 Drainage condition for the water level below the surface for a rice field

If h2(t) > D₂(t) then

$$G_{wo}(t) = D_2(t) - h_2(t) \qquad (6.157)$$

$$h_o(t) = G_{Wo}(t) - PL(t) \qquad (6.158)$$

If h₀(t) > 0 then

$$Q(t + \Delta t) = \frac{8 * K(t + \Delta t) * d(t + \Delta t)}{L^2} * h_0(t + \Delta t) * e^{-\alpha(t+\Delta t)*\Delta t} \qquad (6.159)$$

$$\alpha(t + \Delta t) = \frac{\pi * K(t + \Delta t) * d(t + \Delta t)}{\mu L^2} \qquad (6.160)$$

$$Q_g(\Delta t) = 0.5\big(Q(t) + Q(t + \Delta t)\big) \qquad (6.161)$$

$$Q_2(\Delta t) = Q_g(\Delta t) \qquad (6.162)$$

$$V_a(t + \Delta t)' = V_a(t + \Delta t) - Q_g(\Delta t) \qquad (6.163)$$

where:

$Q_2(\Delta t)$ = discharge due to different head in the field and the lateral canal (mm/time step)

$Q(t+\Delta t)$ = discharge to the lateral canal at any instant (mm/time step)

$Q_g(\Delta t)$ = discharge during each time step if the remaining volume is > 0 (mm)

$h_0(t)$ = initial height of the groundwater table at any instant (m)

$K(t+\Delta t)$ = saturated hydraulic conductivity of the soil (mm/time step)

$h_1(t)$ = groundwater depth above the bottom of layer I (m)

$h_2(t)$ = groundwater depth above the bottom of layer II (m)

h_{s2} = depth of the canal water level above the bottom of layer II (m)

$d(t+\Delta t)$ = equivalent thickness of the aquifer below the drain level (m)

$\propto$ = reaction factor (-)

L   = distance between the canals (m)

D   = depth of the aquifer below the water level in the lateral canal (m)

μ = drainable pore space which varies from 5% for clayey soil to 35% for coarse sand and gravely sand (Ritzema, 1994)

PL(t)   = polder water level (m-surface)

G_{w0} = groundwater table (m-surface)

If the water in the canal is higher than or equal to the initial groundwater table, there will be no flow to the drain. Reverse flow will occur which can be calculated as follows:

If $h_0(t) < 0$

$$Q(t + \Delta t) = \frac{8 * K(t + \Delta t) * d(t + \Delta t)}{L^2} * |h_0(t + \Delta t)| * e^{-\alpha(t+\Delta t)*(\Delta t)} \qquad (6.164)$$

$$Q_g(\Delta t) = 0.5\big(Q(t) - Q(t + \Delta t)\big) \qquad (6.165)$$

If $Q_g(\Delta t) < 0$ then $Q_2(\Delta t) = 0$ and

$$V_a(t + \Delta t)' = V_a(t + \Delta t) - Q_g(\Delta t) \tag{6.166}$$

If $Q_g(\Delta t) > 0$ then $Q_2(\Delta t) = Q_g(\Delta t)$ and

$$V_a(t + \Delta t)' = V_a(t + \Delta t) - Q_g(\Delta t) \tag{6.167}$$

Discharge to the lateral canal in case of ponding water and there is water in the canal

Surface drain discharge will be realized to the drain by pumping or through a culvert dependent on the water level in the lateral canal, the polder water level, and the maximum allowable water depth in the field and the drainage modulus as mentioned before. At the same time there is some water flow to the lateral canal as shown in Figure 6.17.

If $D_2(t) + D_1(t) < h_2(t) + h_1(t)$

In this situation, if the water level in the lateral canal is less than the water level in the field, the discharge will be calculated as unconfined groundwater flow by using Dupuit's formula (Forchheimer, 1930):

$$Q\ (t + \Delta t) = \frac{K(t) * (h_3(t + \Delta t)^2 - h_4(t + \Delta t)^2)}{L_b * L_S} * \Delta t \tag{6.168}$$

$$Q_2(\Delta t) = 0.5 * (Q(t) - Q(t + \Delta t)) \tag{6.169}$$

where:
$h_3(t+\Delta t)$ = height of the water level in the rice field above the equivalent of the aquifer level (m)
$h_4(t+\Delta t)$ = height of the canal water level above the equivalent of the aquifer level (m)
L_b = horizontal distance between the berm and the lateral canal (m)
L_s = distance between the lateral canals (m)

The equivalent thickness of the aquifer below the drain level is used in order to transform a combination of horizontal flow and radial flow into an equivalent horizontal flow. This thickness can be determined as follows:

$$d(t) = \frac{D}{\dfrac{8 * D}{\pi * L} * \ln\left(\dfrac{D}{u(t)}\right) + 1} \tag{6.170}$$

where:
$u(t)$ = wetted perimeter of the lateral canal (m)
D = thickness of the aquifer below the drain level (m)
$d(t)$ = equivalent thickness of the aquifer below drain level (m)

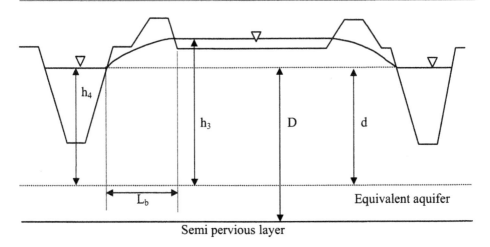

Figure 6.17 Drainage condition for the water level above the soil surface for a rice field

The wetted perimeter of the lateral canal with a bed width of 0.50 m is calculated as follows:

$$u(t) = 0.5 + 2 * \sqrt{1 + T_s^2} * (PL(t) - BL_L)$$ (6.171)

where:
PL(t)   = water level in the lateral canal (m-surface)
BL_L = bed level of the lateral canal (m-surface)
T_s = side slope for the lateral canal (-)

The saturated hydraulic conductivity of the cracks is assumed as function of the dynamic crack width and is derived from a slit model presented by Bouma and Anderson (1973) as:

$$K_{sc} = f_r * 10^{10} * \frac{w_c^3}{d_a}$$ (6.172)

where:
K_{sc} = saturated hydraulic conductivity of the cracks (m/day)
w_c = crack width (m)
d_a = diameter of aggregates (m)
f_r = empirical reduction factor accounting for tortuousity and necking of the cracks for the fact that not all the cracks are inter connected (-). Hendriks, et al., (1998) has found that this factor is about 0.001

The crack width can be calculated from the relative volume of cracks as given below (Hendriks, et al., 1998):

$$w_c = d_a * \left(1 - \sqrt{\frac{V_c}{D}}\right)$$ (6.173)

where:
V_c   = relative volume of cracks (m^3/m^2)
D   = thickness of the crack layer (m)

In this case for the composite permeability it is assumed that the flow is in vertical direction perpendicular to the layers I and II. The composite soil permeability is calculated as follows:

$$K_{avg} = \frac{D_1 + D_2}{\dfrac{D_1}{K_1} + \dfrac{D_2}{K_2}}$$ (6.174)

where:
D_1, D_2 = depth of layer I and layer II (m)
K_1, K_2 = permeability of layer I and layer II (m/day)
K_{avg} = composite permeability of the two layers (m/day)

Discharge to the lateral canal in case there is no water in the lateral canal

If there is no water in the lateral canal the discharge to the canal is calculated as seepage of water through a dike as shown in Figure 6.18.

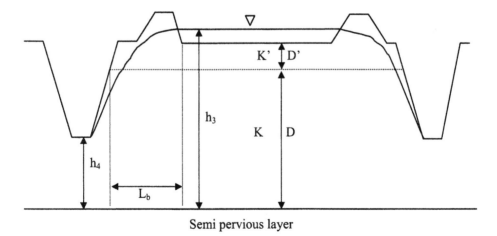

Semi pervious layer

Figure 6.18 Drainage condition for the water level above the soil surface for a rice field but no water in the lateral canal

$$Q\ (t+\Delta t) = \frac{K(t)/3,600 * D * (h_3(t+\Delta t)^2 - h_4(t+\Delta t)^2)}{L_s(2*L_b+2*L)} * \Delta t \qquad (6.175)$$

where:
K   = hydraulic conductivity of the aquifer (mm/day)
D   = aquifer thickness (m)
L   = leakage factor (m)

The leakage factor is calculated as follows:

$$L = \sqrt{KDc} \qquad (6.176)$$

where:
c   = hydraulic resistance of the confined layer (day)

The hydraulic resistance of the confined layer is calculated as follows:

$$c = \frac{D'}{K_a'/3,600} \qquad (6.177)$$

where:
D'   = thickness of the confined layer (m)
K_a' = hydraulic conductivity of the aquifer (mm/day)

Water management in a rice polder

Water management in a rice polder in the model may be divided into two rainfed and irrigated conditions.

Water management in rainfed conditions

In rainfed conditions the farmer will keep the water in the field as high as possible to compensate when there is a dry period. Hence the model calculates discharge from the plot when the water depth in the field is higher than the maximum water depth that crops can tolerate. In practice at first the farmer will drain by gravity, if the water layer depth continues rise until the water depth reaches a certain depth that is allowable to temporary store above the maximum water depth that crops can tolerate the farmer will start pumping. After the water depth in the field is equal to the maximum water depth that crops can tolerate the farmer will stop pumping. It is assumed that there will be no pumping of water from outside to the plot.

Water management in irrigated conditions

In irrigated conditions the farmer does not need to keep the water in the field as high as possible because the water is available. Hence the model calculates the discharge from the plot when the water depth in the field is higher than the maximum water depth that the crops can tolerate. In practice at first the farmer will drain by gravity, if the water depth continues rise until the water reaches a certain depth that is allowable to be temporary stored above the maximum water depth that crops can tolerate the farmer will start pumping. After the water depth in the field is equal the maximum water depth that crops can tolerate the farmer will stop pumping and drain by only gravity as long as possible. It is assumed that after the water layer depth reaches the maximum allowable water depth the farmer still continues to drain water out by gravity until the water depth is equal to the preferred water depth by gravity. If they cannot drain, pumping will be done.

Determination of the water level in rainfed conditions and irrigated conditions in the model is given in mathematical equations (see Appendix II).

Rainfed rice

Rainfed lowlands rice culture occupies 6.7 million ha of a total rice area of 9.6 million ha in Thailand where at 3.7 million ha of rainfed lowlands rice is grown in the Northeast region (Jongdee, et al., 1996). In general, the rainfed lowland rice system lacks control over the amount and timing of water, and hence both moisture insufficiency and moisture excess are encountered.

In rainfed conditions land preparation takes place after the beginning of the monsoon rains giving moisture to allow buffalo puddling. The starting of farming was formerly almost entirely dependent upon the uncertain pattern of monsoon precipitation. Although this problem has now been technically overcome by introduction of tractor and power tiller puddling, the sowing on the field has to be adjusted to monsoon rainfall. Due to this constraint the direct dry seeding method for sowing was developed for drought prone environments (Rice Research Institute, 1987). Rainfed lowlands are mostly planted with local photoperiod sensitive or improved traditional varieties (Poopakdi, 1999). Under rainfed conditions, rice can be grown only once a year in the rainy season (Kupkanchanakul, 1999).

The sowing time in rainfed lowlands is determined by one set of rainy season rainfall. Sowing early in the season, however, may be risky because there is a high probability of no standing water at the appropriate time of planting. On the other hand late-maturing cultivars may have a reduced grain yield due to late season drought (Rajatasereekul, et al., 1997). Most cultivars were strongly photoperiod sensitive, hence the variation in flowering date is rather small (Rajatasereekul, et al., 1997). An optimum flowering date was reported in late September to mid October when sown in June, whereas flowering in November produced lower yields for rainfed lowlands rice in the northern and northeastern part of Thailand (Rajatasereekul, et al., 1997).

From practice it is known that the time for land preparation is about 30 days before sowing. So the starting date of land preparation is in early May in order to be able to sowing in June. Normally in rainfed conditions, the soil in rice field is dry during the dry season and normally cracks will occur because most of the soils in rainfed rice fields are composed of a high percentage of clay. During the first stage of land preparation farmers are involved with soaking of soil. The amount of rainfall and the period of rain should be enough to soak the soil. The starting date of land preparation is calculated based on that amount of rainfall to fill in the crack volume (if any) and the depth of the water layer would at least be 10 mm above the surface for at least 10 days to saturate the puddled layer. Also the amount of rainfall has to compensate the volume lost by horizontal flow through cracks to the lateral canal.

The reduction in crop yield in rainfed conditions is computed in the model for certain boundary conditions based on the following parameters:

- soil moisture tension at the starting date of the land preparation pF <= 2.0;
- the period for land preparation under this condition will be considered in June and harvest in early November.

Irrigated rice

All rice in the Central Plain of Thailand is planted with modern high yield varieties (Poopakdi, 1999). Under irrigated conditions rice can be grown not only in the rainy season but also in the dry season. Farmers will let the field dry for about 30 days after harvesting and the land preparation can begin again. The farmer will prepare the land by puddling in the same way as rainfed rice.

In former times transplanting was the most popular planting method for the rice production, but the required labour for seedling preparation and transplanting, resulted in high cost of production. In order to reduce the cost of rice production and increase the yield, the wet seed rice technology or 'Nah Wan Namtom' has been developed. Wet rice technology is now widely adopted by the farmers, more than 90% of irrigated rice areas are currently using the wet seed rice broadcasting method (Kupkanchanakul, 1999). Farmers grow double rice crops or triple crops using high yielding non-photoperiod sensitive varieties in this area (Poopakdi, 1999).

The reduction of crop yield in irrigated conditions is computed in the model for certain boundary conditions based on the following parameters:
- soil moisture tension during the execution of tillage;
- water depth and duration of inundation.

Yield reduction in rice

Rice yield reduction due to drought

The yield reduction is calculated based on the assumption that the water supply does not meet the crop water requirement. The relative yield is the actual yield (Y_a) of the crop relative to the potential yield (Y_m). The potential yield will occur when there are ideal conditions during the plant growth stages such as sufficient radiation, no shortage of water, no shortage of nutrients, no plant diseases and no plant pests.

Due to insufficient moisture availability for crop transpiration, the relation between actual evapotranspiration and potential evapotranspiration can indicate the water deficit of the crops, which affects the crop growth. The reduction of the yield depends on the magnitude and stage of crop growth. The Food and Agriculture Organization of the United Nations (FAO) has published a relationship between relative yield (Y_a/Y_m) and relative evapotranspiration (ET_a/ET_m) for many crops. The relation between (Y_a/Y_m) and (ET_a/ET_m) for rice is shown in Figure 6.19.

From Figure 6.19, it can be seen that the yield is only influenced by the soil moisture status, which is expressed as actual evapotranspiration. So the relationship can be found as follows:

For the vegetative stage of rice:
$$\left(1 - \frac{ET_a}{ET_m}\right) * 1.49 = \left(1 - \frac{Y_a}{Y_m}\right) \qquad (6.178)$$

For the flowering stage of rice:
$$\left(1 - \frac{ET_a}{ET_m}\right) * 2.38 = \left(1 - \frac{Y_a}{Y_m}\right) \qquad (6.179)$$

For the yield formation stage of rice: $\left(1-\dfrac{ET_a}{ET_m}\right)*0.31=\left(1-\dfrac{Y_a}{Y_m}\right)$ (6.180)

where:

Y_a = actual yield of the crops (ton/ha)
Y_m = potential yield of the crops (ton/ha)
ET_a = actual evapotranspiration of crops (mm)
ET_m = potential evapotranspiration of crops (mm)

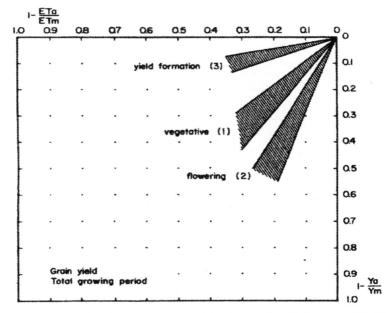

Figure 6.19 Relationship between relative yield reduction (1 - (Y_a/Y_m)) and (ET_a/ET_m) for rice (FAO, 1986)

Rice yield reduction due to inundation

The damage due to inundation of rice is dependent on the type of rice, age and growth stage at which inundation occurs (Palada and Vergara, 1972). The other environmental factors, which affect the damage to the rice, are the inundation period, the inundation depth, turbidity and temperature of the water (Palada and Vergara, 1972). Rice, which is inundated for a long time will die more than inundated for short time. Rice receives little light due to deep water or turbidity. The photosynthesis process is low; therefore the higher the depth and the more the turbidity the more rice will die.

A high temperature will cause high respiration. Due to the high respiration, the accumulation of carbohydrate of rice is low, therefore the higher the temperature the less recovery of rice after flooding. Fertilizing of nitrogen affects the tolerance of rice for inundation. A high fertilization will cause a low accumulation of carbohydrate, which makes the rice weak and less recoverable after flooding. Damage of rice due to flooding is different for the crop growth stages. For example, the older age rice can tolerate more than the younger age rice (Ghosh, et al., 1980).

Table 6.1 Yield reduction in % in different growth stages of Japonica rice and period of time of inundation experiments in Japan (ECAFE, 1969)

Growth stage	Clear water				Turbid water			
	Period of submergence in days				Period of submergence in days			
	1 - 2	3 - 4	5 - 7	>7	1 - 2	3 - 4	5 - 7	>7
20 days after planting	10	20	30	35				
Panicle initiation and partly submergence	10	30	65	90-100	20	50	85	95-100
Panicle initiation and fully submergence	25	45	80	80-100	70	80	85	90-100
Flowering	15	25	30	70	30	80	90	90-100
Harvesting	0	15	20	20	5	20	30	30

Note: partly submergence means the rice leaves were above the water level

Table 6.2 Yield reduction in % in different growth stages of Japonica rice and period of time of inundation experiments in Korea (ECAFE, 1969)

Growth stage	Clear water				Turbid water			
	Period of submergence in days				Period of submergence in days			
	1	3	5	7	1	3	5	7
Tillering	25	55	100	100	30	100	100	100
Panicle initiation	15	95	90	95	20	50	90	100
Panicle exertion	25	95	100	100	45	100	100	100
Flowering	15	90	50	50	45	85	85	85
Milk stage	5	5	10	10	15	35	40	65
Harvesting	5	20	20	30	10	20	30	30

Table 6.3 Yield reduction in % in different growth stages of rice and period of time of inundation experiments in the Philippines (Undan, 1977)

Growth stage	Period of submergence in days			
	1	3	5	7
Early tillering and partly submergence	7	4	10	12
Early tillering and fully submergence	25	61	64	84
Maximum tillering and partly submergence	0	8	9	5
Maximum tillering and fully submergence	25	38	82	95
Panicle initiation and partly submergence	10	8	10	11
Panicle initiation and fully submergence	74	94	96	100

Besides the age of rice the growth rate has also an effect to tolerance. It was reported that in 40 days rice, the amount of leaves has a relation with the percentage of the recovery of rice (Mallik, et al., 1988). The recovery of rice in the active tillering stage is better than in other stage (Rao and Murthy, 1986). An inundation at the early reproduction stage at which panicle is forming, causes more damage to rice than during other stages and yield reduction will decrease more than an inundation at tillering stage (Reddy, et al., 1986). There are many studies about yield reduction due to inundation at different growth stages of rice (Tables 6.1, 6.2 and 6.3). In Thailand, yield reduction due to flooding has been studied for Indica rice at the early growth stage with flood depth in Thailand as shown in Table 6.4 (Inthaiwong, 1996).

Table 6.4 Yield reduction in percentage for Indica rice at the age of 45 days in Thailand (Inthaiwong, 1996)

Depth in cm	Period of submergence in days			
	1	3	5	7
10	0	0	0	0
25	0	0	0	0
50	0	0	0	0
75 (full submergence)	0	0	19.12	40.57
100	0	0	22.61	41.48

Relative rice yield reduction due to a total submergence at different growth stages according to FAO is shown in Figure 6.20.

From flooding in the tillering stage rice will better recover compared to the other growth stages (Rao and Murthy, 1986). Flooding in the booting stage will cause more damage than in the tillering stages (Reddy, et al., 1986). Although rice in the tillering stage has the best recovery, in general rice in the tillering stage has the most probability of flooding (Van de Goor, 1973) and also in Thailand (Ratanaphol, 1994).

By studying of flooding in the Chao Phraya river basin, it has been found that inundation will generally occur between July and August, which is also in the early growth stage of rice (Ratanaphol, 1994 and Inthaiwong, 1996).

Damage of the rice model due to flooding

A water layer on the rice field is needed for weed control, pest control and also to meet the required temperature in the root zone to realize ideal conditions for root growth. The adequate water depth for maintaining the heat capacity to avoid harmful of rice when the air temperature is higher than 35 °C (Mao Zhi, 2001).

The water layer on the field, which produces optimal yields, is called 'preferred water depth'. The preferred water depth is determined empirically and practically. Its magnitude depends on the stage of crop growth and type of rice. The preferred water depth (Table 5.2) and damage are shown in Figure 6.20.

Although rice is a wet crop, which is grown under submergence conditions deep water and prolonged submergence may affect rice growth seriously and yield reduction may be expected (Tables 6.1, 6.2, 6.3 and 6.4). Based on this fact the model simulates yield reduction of rice when the water depth is not at the preferred level, the actual yield will be below the potential yield. The relative water layer at each crop stage is presented by the ratio of summation of the actual water layer depth and summation of the water layer depth. The assumption is that the rice yield will still be at the level of the potential yield when the maximum deviation from the preferred water level less or equal to 5%. If the deviation is larger than 5% the actual yield will decrease linearly (Figures 6.22 and 6.23). For the damage formulas due to flooding in the model computation see Appendix III.

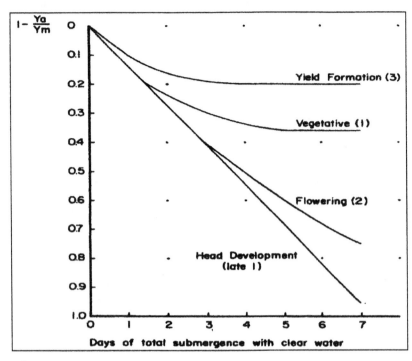

where:
Ya   = actual rice yield (ton/ha)
Ym   = potential rice yield (ton/ha)

Figure 6.20 Relative rice yield reduction due to a total submergence according to FAO
(Doorenbos and Kassem, 1986)

Table 6.5 Water layer depth and damage due to flooding

Stages of growth	Actual water layer as height of plant (Wa) in m	Preferred water layer (Wp) in m	$\Sigma Wa/\Sigma Wp$	$1-(Y_a/Y_m)$
1.Vegetative	0.25	0.05	5	0.36
2.Head development	0.50	0.044	11	0.96
3.Flowering	0.90	0.08	11	0.75
4.Yield formation	1.20	0.08	15	0.20

   The model simulates yield reduction and the relationship between water layer and
relative yield reduction at different rice growth stages based on Figure 6.20 by assuming
that the water height at full submergence is equal to the height of the rice at different
growth stages (Figure 6.21). In the calculation of the damage if the height of the water
level exceeds the maximum water depth (Table 6.3) that rice can tolerate, then yield
reduction will be calculated.

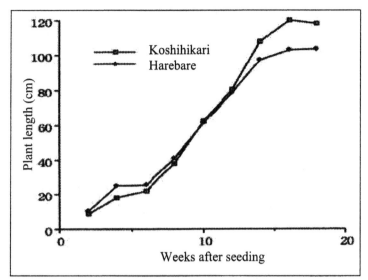

Figure 6.21 Change in rice height with time
(http://ss.jircas.affrc.go.jp/engpage/jarq/32-4/nagaya/fig10.htm)

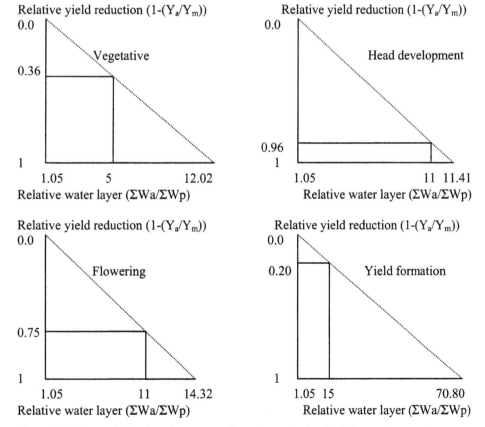

Figure 6.22 Extrapolation for relative water layer determination for different rice growth stages

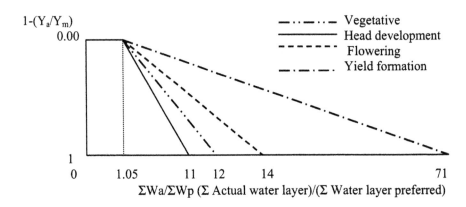

Figure 6.23 Relationship between relative water layer and relative yield reduction for rice in Thailand

Damage of rice model due to water shortage

Rice is susceptible to moisture at difference stages of the life cycle (Matsushima, 1962 and De Bie, 1992) see Tables 6.6 and 6.7. The time formation of sex cells is generally found to be sensitive to moisture stress (Salter and Goode, 1967). The most sensitive periods to water deficit are flowering and the second half of the vegetative period (head development) as also reported by FAO (Doorenbos and Kassem, 1986) (Figure 6.19). Tables 6.6 and 6.7 show the effect of water shortage in different stages in the growths of rice.

Yield reduction due to water shortage has been studied by data collection in paddy rice in the northern part of Thailand (De Bie, 1992). The relative yield reduction that has been found is shown in Table 6.7, but there are no water shortages at the yield formation stage occurring in this study.

It has been proved that rice can tolerate without yield reduction a shortage of water at 80%, 60% and 80% of the saturated moisture content at vegetative, head development and yield formation respectively (Mao Zhi, 2001).

Table 6.6 Effect of water shortage at different stages in the growth of rice (adapted from Matshushima, 1962)

Stage of rice growth	Yield in % of full controlled	Relative yield reduction $1-(Y_a/Y_m)$
Overall	82	0.18
Early tillering	74	0.26
Active tillering	86	0.14
Late tillering	88	0.12
Primary branch differentiation	75	0.25
Spikelet differentiation	79	0.21
Reduction division of sex cell (just prior to the heading)	29	0.71
Heading	37	0.63
Late heading	73	0.27
Ripening	90	0.10

Table 6.7 Impact of water shortage on rice yield in Thailand (adapted from De Bie, 1992)

Period of water shortage				Average yield in kg/ha	Relative yield reduction $1 - (Y_a/Y_m)$
Establishment	Vegetative	Heading + flowering	Yield formation		
yes	no	no	no	2,893	0.26
yes	no	yes	no	2,109	0.46
yes	yes	no	no	2,069	0.47
no	yes	no	no	2,605	0.34
no	yes	yes	no	2,019	0.49
no	no	yes	no	2,101	0.47
no	no	no	no	3,932	0.00

Drainage and crop yield effect for rice

One of the major problems affecting rainfed lowland rice is the poor drainability of fine-textured, puddled or compacted soils. Saturated hydraulic conductivity and percolation losses are reduced in puddled fields due to the close packing of soil particles in parallel orientation. A subsurface traffic pan further restricts the downward movement of water. This restricted percolation helps in maximizing nutrient efficiency and economizing on water use, but some percolation may be essential if phytotoxins accumulate in the root zone.

Gaseous diffusion of carbon dioxide and oxygen is 10,000 times lower in water than in air. In submergence-puddled soil, the exchange of gases, especially carbon dioxide, between the outer atmosphere and the soil is severely restricted. However, the rice crop does not suffer from an oxygen deficiency because most of the oxygen requirement is met through oxygen transported from the aerial parts of the rice plant to the roots through intercellular gaseous spaces.

Continued submergence without percolation increases the accumulation of CO_2, H_2S, organic acids and reduced iron, often in toxic quantities. Percolation increases the dissolved amount of O_2 content of the groundwater and will increase the ability to get rid of toxic substances. Low percolation rates reduce the aerated porosity in the soil. This causes that the uptake of nutrients by rice roots will become slower and the need for the application of chemical fertilizers will increase (Van den Eelaart, 1998).

The need for high percolation rates for poorly drained rice soils has been noted by Cheng (1984) in China. He recommends percolation rates of 7 - 20 mm/day, depending on the soil types and farming practices. Research in China showed that by improving the percolation rate and lowering the groundwater table in the off-season causes paddy yield increases from 2 - 3 ton/ha to 6 - 8 ton/ha. This was attributed, among others, to the changes in the types of organic matter by oxidation. By oxidation the organic matter became active in improving fertility. The C/N ratio will improve dramatically.

The required percolation for tidal lowland in Indonesia has been recorded from field data at about 5 - 6 mm/day during mechanized land preparation and early growth, while 2 - 4 mm/day for the remaining growth stages of the rice plant would be sufficient (Van den Eelaart, 1998) (http://www.eelaart.com/thesis.htm).

It is often impossible to drain rice fields during the rainy season, because impounding water in rice fields is an important way to conserve water and to control soil erosion.

Determination of the polder water level in a rice polder

In the rainy season the polder water is kept lower than in the dry season. Because in the rainy season the polder is more susceptible to flooding due to heavy rainfall caused by storms. On the other hand at the end of the rainy season the polder water level will be kept high in order to store water in the canal system for the irrigation.

In Thailand the polder water level is kept at 0.50 to 0.70 m-surface between June and October, which is the rainy season. But at the end of the rainy season from November the polder water level should be at 0.30 m-surface. Normally in December when there will be no rain at all the polder water level can be kept high at 0.10 m-surface in order to store water in polder.

In case of heavy rainfall due to a storm, draining of water in advance will be done and the polder water level can be kept lower to about 1.30 to 2.40 m-surface.

Open water in m³ from the polder water level up to the level when the pump is stopped is shown in Equation 6.64.

In the present computation several periods are used to compute the discharges and water levels for given rainfall and evapotranspiration in a similar way as computed in the module RICERUR.

Equation 6.69 calculates the rainfall surplus on the open water. The irrigation water that is taken from the canal is calculated as follows (see Appendix II):

$$Q_r(t) = N_{pt} * B_p * I_{rs}(\Delta t) * 100 \qquad (6.181)$$

where:
$Q_r(t)$ = total irrigation water supply taken from the canal system (m³/time step)
$I_{rs}(\Delta t)$ = irrigation water that is supplied to the plot (mm/time step)
B_p = size of the plot (ha)
N_{pt} = total number of plots (-)

The cumulative discharge from the five different land use areas into the open drainage system is computed by Equation 6.70. Equation 6.71 gives the discharge into the open water due to seepage. Therefore the total open water volume above the polder water level is given by the following formula:

$$V_o(t + \Delta t) = V_o(t) + P_n(t) + Q_s(t) - Q_r(t) \qquad (6.182)$$

For the computation of the discharge in the main system in the rural area, urban discharge is added to the total discharge.

Operation of pumps

The pump operation to pump out the excess water is handled in the program in the same manner as in a clay polder (Equations 6.73 to 6.75).

Fluctuation of the water level

If V_o is greater than 0 the rise of the water level in the open drains above the polder water level (Y_s) is computed by Equations 6.76 to 6.78. The new open water area in m² due to the rise of the water level is calculated by Equation 6.79. If V_o is equal to zero then the rise of the water level and open water volume is obtained by Equations 6.80 to 6.81.

Hydrological model in the rural area for irrigation and drainage for dry food crops in Thailand

In the model for the rural area the hydrological processes for dry food crops in tropical conditions involving surface and subsurface drainage are schematised into four layers where the water balance is done assuming each layer as a reservoir (Figure 6.24).

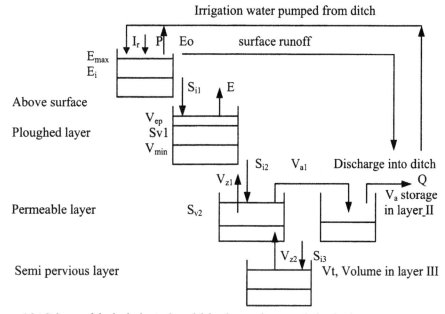

Figure 6.24 Scheme of the hydrological model for the rural area with dry food crops

The model components are as follows:

Above surface:
- precipitation (mm/day) (P)
- open water evaporation (mm/day) (E_0)
- maximum infiltration in layer I (mm/m) (S_{i1})
- interception (mm) (E_i)
- irrigation (mm/day) (I_r)

Layer I (ploughed layer):
- evapotranspiration (mm/day) (E)
- capillary rises from layer II (mm/day) (V_{z1})
- maximum infiltration in layer I (mm/m) (S_{i1})
- maximum percolation in layer II (mm/m) (S_{i2})
- storage above field capacity in layer I (mm/m) (S_{v1})
- soil moisture (mm/m) at which evapotranspiration is at potential rate (V_{ep})
- soil moisture (mm/m) at which evapotranspiration is zero (V_{min})

Layer II (permeable layer):
- capillary rise to layer I (mm/day) (V_{z1})
- capillary rise from layer III (mm/day) (V_{z2})
- percolation from layer I (mm/m) (S_{i2})

-   percolation of surplus water to layer III (S_{i3})
-   discharge of surplus water to the ditch (V_{a1})
-   storage above field capacity in layer II (mm/m) (S_{v2})

Layer III (semi pervious layer):
-   maximum percolation from layer II into layer III (mm/m) (S_{i3})
-   moisture at saturation in layer III (mm) (V_t)

Open water storage:
-   discharge from layer II through the subsurface drain (mm) (V_{a1})
-   change in the storage (mm) (V_a)
-   discharge through pumping (mm/day) (Q)

The above process components will be described in brief.

Interception

Interception is calculated in the same manner as described in Hydrological computation for the Netherlands as given by Equation 6.1.

In the polder for dry food crops in Thailand, it is assumed that surface runoff (Q_1) will go to the ditch because of the convex shape of the surface of the bed, after if the water level in the ditch is too high the farmer will drain the excess water into the lateral canal. Then it will discharge to the sub-main and main canal.

Evapotranspiration

Evapotranspiration is calculated in the same manner as described in Hydrological computation for the Netherlands as given by Equations 6.1 to 6.8.

Moistening of the layers

The amounts of moisture that will penetrate in layer I and in layer II is calculated in the same manner as described in Hydrological computation for the Netherlands as given by Equations 6.9 to 6.10.

Therefore maximum volume of water that can be absorbed at any instant is as follows:

In case no cracks occur at the soil surface ($\theta_1 > \theta$ at pF =2.53) the maximum volume of water that can be absorbed at any instant is given by Equation 6.11. When the interception at any instant (E_i) is greater or equal to V_{max}, there is maximum interception as well as maximum accommodation of water in the layers I and II. While if there is any surplus (see Equation 6.12) it will be add to the amount of water that is stored in the ditches and in the soil matrix. If the water level in the ditches is higher than the maximum allowable level it will discharge to the main drain.

In the model when there is a surplus in case the rainfall exceeds the maximum volume of water that can be absorbed at any instant (V_{max}) part of the rainfall excess will fill in the room in the soil matrix and part will go to the ditch.

$$V_{ald}(\Delta t) = C * V_{al}(\Delta t)$$ (6.183)

$$V_{als}(\Delta t) = (1 - C) * V_{al}(\Delta t)$$ (6.184)

where:
V_{ald} = volume of water that is added to the ditch at any instant (mm)
V_{als} = volume of water that is stored in the soil matrix at any instant (mm)
C   = runoff coefficient (-)

If however interception at any instant (E_i) is less than V_{max}, after meeting the maximum interception on the surface i.e. E_{imax}, the remaining goes to fill the room in the successive layer I and layer II in the following manner:

In case cracks occur at the soil surface ($\theta_1 < \theta$ at pF =2.53) due to drying or because farmers let the field dry before growing the next crop, when the precipitation is greater than maximum interception, this part of the water reaches the surface of the cracked soil, part of the water infiltrates into the soil matrix and the other part fall into the cracks. When rainfall exceeds the maximum infiltration rate of the soil matrix at the top layer (layer I), runoff water flows into the cracks. The calculation is in the same manner as described in Hydrological model for rice fields in Thailand as given by Equations 6.135 to 6.154.

When the interception at any instant (E_i) is greater or equal to V_{max}, there is maximum interception as well as maximum accommodation of water in layers I and II while if there is any surplus it will add to the amount of water that is stored in the ditches as given by Equation 6.12.where: V_{al} = surplus to be discharged to the ditch (mm).

If however interception at any instant (E_i) is less than V_{max}, after meeting the maximum interception on the surface i.e. E_{imax}, the remaining goes to fill the room in the successive layer I and layer II in the following manner as given by Equations 6.13 to 6.16.

Soil moisture tension and capillary rise

The soil moisture tension, capillary rise and percolation are calculated in the same manner as described in Hydrological computation for the Netherlands as given by Equations 6.17 to 6.42.

The net volume of moisture available for drains and to meet the deficit in layer III is given by Equation 6.185.

$$V_a(t + \Delta t) = V_a(t) + V_{a2} + V_{als}(\Delta t) - V_{z2}(\Delta t)$$ (6.185)

If the amount is less than 0 ($V_a < 0$) the net volume in layer III is updated with Equations 6.44 and 6.45. If the net volume available is less than the maximum moistening rate into layer III (S_{i3}) then Equations 6.46 to 6.49 are applicable. If the net available volume is greater than or equal to the maximum moistening capacity of layer III (S_{i3}) then Equations 6.50 to 6.53 are applicable. If there is no room in layer III to accommodate surplus water, then the net surplus goes as discharge to the ditches.

Irrigation schedule and drainage discharge

Determination of drainage capacity

The drainage will include surface drainage and subsurface drainage. The surface flow will occur when the rainfall exceeds the infiltration capacity and maximum interception in the field. The surface runoff is assumed to be temporary stored in the ditch. After that the excess water is discharged through the pipe culvert or pumped to the lateral canal dependent on the water level in the lateral canal. If the water level in the ditch is exceeding the allowable water level the farmer will pump the water out or let the water flow out by gravity dependent on the water level in the lateral canal (see Appendix V). The drainage capacity is limited by the storage capacity of the ditch, the lateral, the main canal and the pumping capacity from the plot.

Determination of irrigation capacity

The irrigation requirement is computed on time step basis based on the assumption that the farmer will irrigate the field until the moisture in the root zone is at field capacity. The irrigation water is calculated based on the assumption that the farmer will irrigate by movable pump when the moisture content of the soil in the plot is at -0.025 MPa or -0.25 bar (Sanders, 1997). The water, which is stored in the ditch, will be compensated by pumping or gravity flow until the water level in the ditch is equal to the required water level for moving the floating boat. The limit of the irrigation is the pumping capacity and the amount of water stored in the lateral and main canal (see Appendix V). It is assumed that irrigation by pumping can be done when the water level in the main canal is not lower than 0.30 m+bed level of the main canal.

The irrigation requirement for the crops is calculated as follows:

$$\text{If } \theta_2 > \theta_{2fc} \text{ and } \theta_1 < \theta_{1fc}, \ I_{rn}(\Delta t) = (\theta_{1fc} - \theta_1(t)) * D_1 / E_{ir} \qquad (6.186)$$

$$\text{If } \theta_2 < \theta_{2fc} \text{ and } \theta_1 < \theta_{1fc}$$

$$I_{rn}(t + \Delta t) = (\theta_{1fc} - \theta_1(t)) * D_1 / E_{ir} + (\theta_{2fc} - \theta_2(t)) * (D_r - D_1) / E_{ir} \ (6.187)$$

where:
$I_{rn}(t+\Delta t)$ = amount of water that needs to be supplied by spraying from the small boat (mm/time step)
D_r = root depth of the crops (m) (taken as 0.40 m for Chinese kale)
E_{ir} = irrigation efficiency (-)
θ_{1fc} = field capacity of soil layer I (%)
θ_{2fc} = field capacity of soil layer II (%)

Irrigation water is added to precipitation (if any) because irrigation water is sprayed to the crops by pumping from a movable boat. In vegetable farms the average water, which will be sprayed to the crops 1.6 hours/time by a pump on a boat at 6.45 litre per second for a plot of 1.2208 ha (Sutti, 1998). However, the time of spraying water is dependent on the size of the plot, the period of no water given and the capacity of the pump.

In the dry periods the irrigation capacity is calculated based on the available water in the lateral and main drain (see Appendix V). Based on the agricultural practices for dry food crops farmers will pump water from the field drain to irrigate the crops with a portable pump on a small boat. Normally irrigation water is given everyday. Hence the irrigation supply is calculated on a daily basis.

Discharge to subsurface drains and storage in the groundwater zone

The storage in the groundwater zone is conceptualised as a reservoir storage and discharge to the ditches as a function of actual storage in the reservoir. In reality flow to the ditches occurs mainly through the cracks or pores above the ditch level (Figure 6.25).

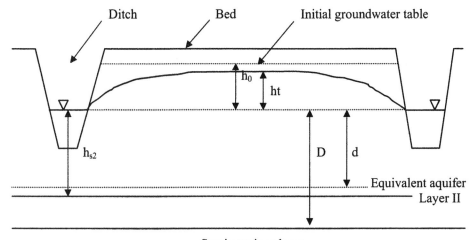

Semi pervious layer

Figure 6.25 Drainage condition for the water level below the soil surface for a dry food crops polder

The amount of water that will be discharged to the ditch is the average of the computation during the previous time step and the actual time step at any instant. Thus applied with the Glover Dumm Equation the discharge is calculated as follows.

If groundwater table is higher than the water level in the ditch at any instant, there is discharge to the ditch as given by Equations 6.156 to 6.163.

If the water level in the ditch is higher than or equal to the initial groundwater table. There will be no flow to the ditch. On the other hand there will be reverse flow from the ditch to the groundwater zone, which can be calculated with Equations 6.164 to 6.167.

The saturated hydraulic conductivity of the cracks is calculated in the same manner as described in Hydrological model for rice fields in Thailand as given by Equations 6.172 to 6.174.

Calculation of leakage to the lateral canal

In case there is water in the lateral canal the surface drain discharge will go to the lateral canal by pumping or through a culvert dependent on the water level in the lateral canal, polder water level, and maximum allowable water level in the ditch and drainage modulus as mentioned before. At the same time there is some water flow to the lateral canal as shown in Figure 6.26.

In this situation, if the water level in the lateral canal is lower than the water level in the ditch, discharge will be calculated as unconfined groundwater flow by using Dupuit's formula (Forchheimer, 1930); see Equations 6.168 to 6.169.

The equivalent thickness of the aquifer below the drain level (d) is used in order to transform a combination of horizontal flow and radial flow into an equivalent horizontal flow. Equation 6.170 can determine its thickness. Equation 6.171 calculates the wetted perimeter of the ditch with a bed width of 0.50 m. If the water level in the lateral canal is higher than the water level in the ditch the reverse flow has to be calculated.

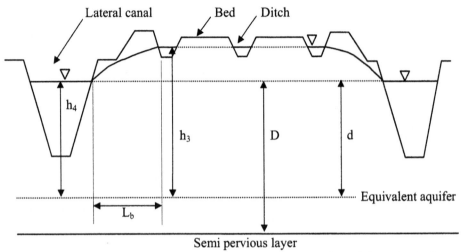

Figure 6.26 Leakage condition in case there is water in the lateral canal for a dry food crops polder

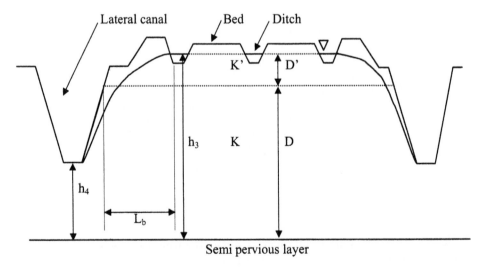

Figure 6.27 Leakage condition in case there is no water in the lateral canal for a dry food crops polder

In case there is no water in the lateral canal the discharge to the canal is calculated as seepage of water through a dike as shown in Figure 6.27 and Equations 6.175 to 6.177. The total discharge to the canal system is composed of discharge from the surface drain in the plot and leakage if the water level in the lateral canal is lower than in the plot.

Calculation of the groundwater table

The above computation is done for each time step (hours) and cumulated for presentation and averaged for each day. The computation of the groundwater table due to the surplus in layer II is based on the following schematisation of layer II and the rise in the water level in the cracks (Figure 6.7 and Equations 6.60 to 6.62).

Determination of the polder water level in a dry food crops polder

In the present computation several periods are used to compute the discharges and water levels for given rainfall and evapotranspiration in a similar way as computed in the module DRYRUR. Equation 6.69 calculates the rainfall surplus on the open water. The irrigation water supply to the plot is calculated as follows:

$$Q_r(t) = N_p * I_{rsd}(t) \qquad\qquad (6.188)$$

where:
$I_{rsd}(t)$ = irrigation water that is supplied to the dry food crop plot (m³/time step)
$Q_r(t)$ = total irrigation water supply taken from the canal systems (m³/time step)

The cumulative discharge from the five different land use areas into the open drainage system is computed by Equation 6.70. Equation 6.71 gives the discharge into the open water due to seepage. Therefore the total open water volume above the polder water level is given by Equation 6.182. For the computation of the discharge in the main system in the rural area, urban discharge is added to the total discharge.

Operation of pumps to pump water out of a plot

The pump operation to pump out excess water is handled in the program in the same way as in a clay polder (Equations 6.73 to 6.75).

Operation of pumps to supply water into a plot

Normally during the dry periods the farmers will try so keep water in the plot at the preferred water level. Therefore when the water level in the plot decreases from the preferred water level at certain level water will be pumped to the plot and pumping will be stopped when the water level in the plot increases to a certain level above the preferred water level. The pump operation in case water is needed for irrigated rice or dry food crops during the dry periods is handled in the program in the following manner.

When $Y_p(t) < Y_r - h_{is} : I_{wc} = 1$, start pumping

When $Y_p(t) > Y_r - h_{is} : I_{wc} = 1$, Gc (t) = capacity of pumping (m³/time step)

When $Y_p(t) > Yr + h_{ip} = 1$, $I_{wc} = 0$, stop pumping

where:

I_{wc} = indicator to start pumping into the plot (-)

G_c = pumping capacity (m³/time step)

Y_p = water level in the plot (m-surface)

Y_r = required water level in the plot (m-surface)

H_{is} = start level below the required water level in the plot (m)

h_p = stop level above the required water level in the plot (m)

Fluctuation of the water level

If V_o is greater than 0 the rise of the water level in the open drains above the polder water level (Y_s) is computed by Equations 6.76 to 6.78. The new open water area in m² due to the rise of the water level is calculated with Equation 6.79. If V_o is equal to zero then the rise of the water level and open water volume is obtained by Equation 6.80 to 6.81.

Model efficiency

The optimization model of Rosenbrock is used to determine the model efficiency as given by Equations 6.85 to 6.87.

Crop damage computations

The crop damage can be calculated by the relative yield, which can be defined as a percentage of yield reduction (Equation 6.190 to 6.192) between the actual yield (Ya) of the crop and the potential yield (Y_m). The damage to the crops for Thai conditions can occur in too wet and too dry conditions as follows.

Crop damage due to too wet conditions

The reduction of crop yield is computed in the model for certain boundary conditions based on the following parameters:
- soil moisture tension during the execution of tillage;
- date of sowing and harvesting;
- number of days and depth of the open water retained above the allowable depth.

Yield reduction due to air deficit

There may be reduction of yield due to insufficient air entry as a result of a bad soil structure in the upper layer. It was reported that crops (corn) tolerated soil submergence for 2 days without any detectable growth depression (Purvis and Williamson, 1972), probably because of internal diffusion of oxygen to their roots (Letey, et al., 1965). The relation between percentage of oxygen diffusion rate 3 days after rain in the soil layer and the average groundwater table (Figure 6.28) is obtained from field data (Chaudhary, et al., 1975), which can be evaluated with the following formula:

$$ODR = a + b * \ln h_g \qquad\qquad (6.189)$$

where:
ODR   = average oxygen diffusion rate during a rainy period ($g*10^{-8}/cm^2/min$)
a and b = constants equal to 22.234 and 32.858 respectively
h_g = average groundwater table (m-surface)

It was found that when the groundwater table is deeper the yield for corn decreases despite a larger amount of irrigation water supply (Chaudhary, et al., 1975). It can be ascribed to reduce the water availability or lower moisture at the upper part of the soil since water tension in the soil before each irrigation was higher under a deeper than under a shallower groundwater table. Follett, et al., 1974 and Williamson, 1964 reported a lower yield of soil bean, cabbage, corn, sugar beet, and alfalfa in treatments with a deeper groundwater table than shallower groundwater tables regardless the amount of irrigation. Chaudhary, et al., 1975 found that the main effect to the yield reduction is the value of ODR compared to the soil water depletion before irrigation and groundwater contribution.

Hence the relation for the relative yield with ODR after 3 days rain, which was obtained from field data, may be used to calculate the relative yield for a given ODR (Figures 6.29 to 6.32). From Figures 6.29 to 6.32 the relationship can be written as follows.

$$R_o = A * ODR^2 - B * ODR + C \qquad\qquad (6.190)$$

where:
R_o = relative yield reduction (%)
A, B, C = constant for yield reduction for tropical crops as shown in Table 6.8.

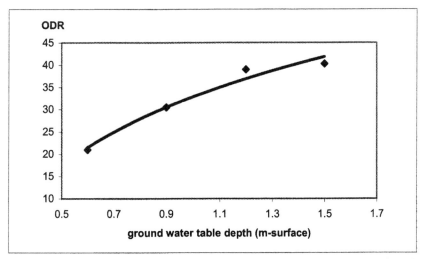

Figure 6.28 Oxygen diffusion rate after 3 days rain and groundwater table below the surface (adapted from Chaudhary, et al., 1975)

Table 6.8 Constants for crop yield reduction

Crop	Constants for percentage yield reduction		
	A	B	C
Corn	0.053	44.35	986.69
Cabbage	0.630	13.31	74.34
Soybeans	0.347	12.26	110.80
Sorghum	0.118	4.47	40.99

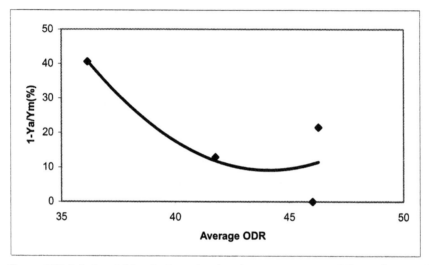

Figure 6.29 Relative yield reduction related to the average oxygen diffusion for corn (adapted from Chaudhary, et al., 1975)

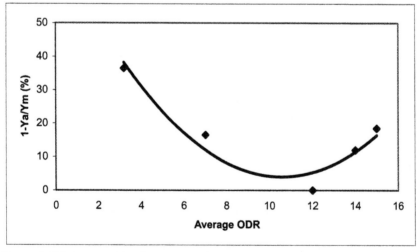

Figure 6.30 Relative yield reduction related to the average oxygen diffusion for cabbage (adapted from Williamson, 1964)

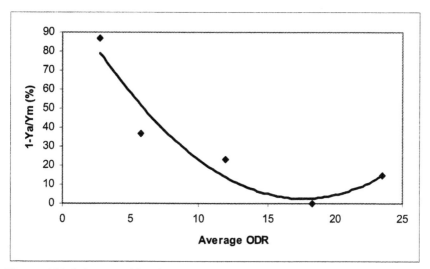

Figure 6.31 Relative yield reduction related to the average oxygen diffusion for soybeans (adapted from Williamson, 1964)

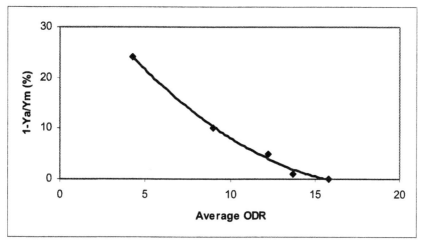

Figure 6.32 Relative yield reduction related to the average oxygen diffusion for sorghum (adapted from Williamson, 1964)

Crop damage due to too dry conditions

The yield reduction due to too dry conditions is calculated based on the assumption that the water supply does not meet the crop water requirement. The potential yield will occur when there are ideal conditions during the plant growth stages such as sufficient radiation, no shortage of water, no shortage of nutrients, no plant diseases and no plant pests.

Due to insufficient available moisture for crop transpiration, the relation between the actual evapotranspiration and the potential evapotranspiration can indicate the water deficit of the crops, which affects the crop growth. The reduction of the yield depends on the magnitude and stage of crop growth. FAO has published a relationship between the relative yield (Y_a/Y_m) and relative evapotranspiration (ET_a/ET_m) for various crops. The relation between (Y_a/Ym) and (ET_a/ET_m) for the Chinese kale (Brassica oleracea var. alboglabra) is assumed the same manner as cabbage (Brassica oleracea var capitata) shown in Figure 6.33 because the physical structure and family of breeding of these two vegetables is similar and there is no study for Chinese kale.

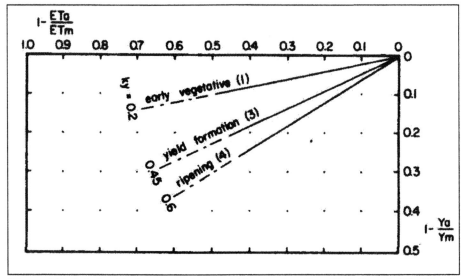

Figure 6.33 Relationship between relative yield reduction (1 - (Y_a/Y_m)) and (ET_a/ET_m) for cabbage (Doorenbos and Kassem, 1986)

From Figure 6.33 it can be assumed that the relative yield is only influenced by the soil moisture status, which is expressed by the actual evapotranspiration. So the relationship can be found as follows:

For the early vegetative stage of Chinese kale derived from Figure 6.33 (1):

$$\left(1-\frac{ET_a}{ET_m}\right)*0.20=\left(1-\frac{Y_a}{Y_m}\right) \tag{6.191}$$

For the yield formation stage of Chinese kale derived from Figure 6.33 (3):

$$\left(1-\frac{ET_a}{ET_m}\right)*0.45=\left(1-\frac{Y_a}{Y_m}\right) \tag{6.192}$$

where:
Y_a = actual yield of the crops (ton/ha)
Y_m = potential yield of the crops (ton/ha)
ET_a = actual evapotranspiration of the crops (mm)
ET_m= potential evapotranspiration of the crops (mm)

6.2.4 Economic computation for Thailand

Economic computation for the rural area of a rice polder

The open water area in the rural area is estimated with Equation 6.88, where a and b are constants (a = 271 and b = 2.08). The area of infrastructure is taken as a percentage of the net area by Equation 6.89. Equation 6.90 estimates the number of farms in the agricultural area. The area of the building facilities is calculated by using Equation 6.91. Equation 6.92 estimates the remaining area for agriculture and Equation 6.93 the areas covered by each crop.

Cost of pipe culverts

The surface drainage of the rice fields is calculated based on the drainage modulus and the requirement of the PVC pipe culvert, which are assumed to be used in the surface field drains. The maximum discharge is dependent on the hydraulic head and calculated based on the normal condition that the polder water level is lower than the ground level.

Maximum discharge from the pipe is calculated as follows (after Depeweg, 2000).

$$Q_d = C * \frac{\pi}{4} * \left(\frac{D}{1,000}\right)^2 \sqrt{2g(W_{LMR} - D/2)/1,000} \qquad (6.193)$$

where:
Q_d = design discharge of the pipe (m^3/s)
W_{LMR} = maximum allowable water depth in the rice field (mm)
C = discharge coefficient for a circular pipe = 0.90
D = internal diameter of the culvert (mm)

The drainage discharge requirement per ha, which is related to the drainage modulus is calculated as follows:

$$Q_{rd} = \frac{10 * D_m}{24 * 3,600} \qquad (6.194)$$

where:
Q_{rd} = drainage discharge requirement related to the drainage modulus (m^3/s/ha)
D_m = drainage modulus, about 46 mm/day (Jindasanguan, 1995)

The required number of pipe culverts is calculated as follows:

$$N_P = \frac{Q_{rd}}{Q_d} + 1 \qquad (6.195)$$

where:
N_p = number of pipe culverts per ha

The total costs for the pipe culverts is calculated under the assumption that one pipe is 1.0 metre long as follows:

$$\cos tds = N_p * P_p * Ar \tag{6.196}$$

where:
costds   = total costs for construction of the PVC pipe culvert (€)
P_p = cost for construction of the PVC pipe culvert (€/m)
Ar   = total area in the polder where rice is grown (ha)

The maintenance costs for a surface drainage pipe are estimated at 0.05 €/m/year.

$$\cos tdsm = 0.05 * N_p * Ar \tag{6.197}$$

where:
costdsm = maintenance costs for the PVC pipe culvert (€/year)

Costs for construction and maintenance of bunds

Costs for construction and maintenance of bunds are calculated as follows:

$$\cos tb = 0.3 * Cu * N_b * (P_L + P_W) \tag{6.198}$$

where:
costb   = total costs for bund construction (€/year)
C_u = earth compaction cost (€/m³)
P_L = length of the plot (m)
P_W = width of the plot (m)
N_b = total number of plots in the polder area (-)

The maintenance costs for the bunds are estimated at 0.20 €/m/year.

$$\cos tbm = 0.20 * N_b * (P_L + P_W) \tag{6.199}$$

where:
costbm   = total maintenance costs for the bund (€/year)

Costs for the canal system for a rice field

The length of the main drains in m is obtained from Equation 6.95, where: a = 0.271 and b = 0.00208 (data from Jindasanguan, 1995 and Suwannachit, 1996). Earthwork is the sum of the earth movement in the lateral canals, sub-main canals and main canals, the total costs for excavation and annual maintenance of the lateral canals, sub-main canals and main canals (Figure 6.10 and Equations 6.99 to 6.103 and 6.106).
The total costs for excavation of lateral canals, sub-main canals and main canals is:

$$K_{ot} = G_{ow} * K_{og} \tag{6.200}$$

Cost of pumping for the main pumping station

The required capacity of the pumping station in a rice polder can be calculated by Equation 6.107. The total investment costs for the installation of pumping station, which are calculated based on the report of the Master Plan for Flood Protection and Drainage in Samut Prakan East, 1988, the contract price for the construction the pumping station (Royal Irrigation Department, 2001) and the Thai price index (http://www.moc.go.th/thai/dbe/index/cal_k.html) is given in Equation 6.108 and maintenance is given in Equation 6.109.

But:

a   = constant taken as 0.111
b   = constant taken as 0.0046
hs   = pressure head which is taken as 1.5 m
h_l = headloss, taken as 1.5 m
B   = 138

The power required for the pumping station at least favourable circumstances is given by Equation 6.110. The operation power, based on the average annual rainfall and seepage, is given by Equation 6.111. The cost of energy for the pumping is determined by Equation 6.112, where: a and b are constants and taken as 40 and 0.0102 respectively.

Annual equivalent cost

The total costs for construction may be summarized as follows:

Patt   = pipe culvert + bunds + canal system + pumping station   (6.201)

Annual maintenance costs are the sum of the following:

Pott   = maintenance of the pipe culvert, bunds and canal system + operation and
           maintenance cost of pumping station   (6.202)

The annual recovery factor is expressed in Equation 6.115, where: r = annual interest rate, usually 5%. Therefore the annual equivalent costs for canal systems are determined by the construction and maintenance costs:

$$cdt = V_t * \cos td + \cos tdm + V_t * \cos tds + \cos tdsm \qquad (6.203)$$

The annual equivalent costs for the main canal system include excavation, and annual maintenance:

$$\cot = V_{t1} * G_{ow} * K_{og} + K_{at} \qquad (6.204)$$

where:
V_{t1} = annual distribution factor for excavation (-)

Annual equivalent costs for the pump are the costs for installation and operation as given by Equation 6.118. Equation 6.119 gives the total annual equivalent costs of the water management.

Cost of pumping for the field pumping

The costs for field pumping are calculated based on the quantity of water that was pumped for drainage and irrigation, dependent on the water level in the field (see Appendix IV).

Investment costs for crops, buildings and infrastructure

Maximum and minimum investments for each crop on a hectare of land that possibly can occur during each decade have been derived from a year list of investments for the crops. The value of buildings and other facilities is calculated with Equation 6.120. Equations 6.121 and 6.122 determine the total costs involved in the farm.

Crop damage computation for rice

Rice crop damage computation may be calculated separately for rainfed conditions and irrigated conditions because of different constrains.

Economic computation for the rural area for a dry food crops polder

In the economic computation for the rural area for a dry food crops polder the open water area is estimated with Equation 6.88. The area of infrastructure is taken as a percentage of the net area as given by Equation 6.89. Equation 6.90 calculates the number of farms and Equation 6.91 the area of the building facilities. Therefore the remaining area for agriculture and the areas covered by each crop are estimated by Equations 6.92 to 6.93.

Costs for surface drainage

Costs for pipe culverts

The surface drainage of the dry food crop fields is calculated based on the drainage modulus and the requirement of the PVC pipe drain culvert, which are assumed to be used in the surface field drain. The maximum discharge is dependent on the hydraulic head and calculated based on the normal condition that the polder water level is lower than the ground level.

Maximum discharge from the pipe is calculated with Equation 6.193. where W_{LM} in this Equation is the maximum allowable water depth in the plot in m above the bed level of the ditch.

The maximum allowable water depth in the plot in m above bed level of the ditch is calculated as follows:

$$W_{LMR} = D_r + P_d \tag{6.205}$$

where:
D_r = drain depth or minimum water level from the surface of the bed (m)
P_d = water depth in the ditch required for pumping by a removable boat (m)

The drainage discharge requirement per ha, which is related to the drainage modulus is calculated by Equation 6.194. Equation 6.195 calculates the required number of pipe culverts. The construction the PVC pipe is calculated under the assumption that 1 pipe is approximated at 2 m long as follows:

$$\cos tds = (2t_{dk} * H_d + W_{cr}) * N_P * P_p \tag{6.206}$$

where:
costds   = total costs for construction of the PVC pipe culvert (€)
P_p = cost for construction of the PVC pipe culvert (€/m)
H_d = height of the surrounding dike, 1 m+surface, (Sutti, 1998)
t_{dk} = side slope of the surrounding dike embankment (-)
W_{cr} = width of crest of the dike, 1 - 2 m (Sutti, 1998)

Costs for construction of the ditches

Costs for construction of the ditches are calculated from the volume of earth that has to be excavated dependent on the agricultural practice, geometry of the ditches, distance between the ditches and depth of the ditches.

$$\cos tfd = [(D_r + P_d)S_L + B_d](D_r + P_d)[(N_R P_L + 2P_W - 2N_R((D_r + P_D)S_L + B_d)]$$
$$* N_b E_c \tag{6.207}$$

where:
costfd   = total costs for construction of the ditches (€)
S_L = side slope of ditches (-)
B_d = bed width of ditches (m)
N_R = number of rows of the ditches in the plot (-)
E_C = cost for earth movement for the ditches (€/m^3)

The number of rows of the ditches in the plot is calculated as follows:

$$N_R = [P_W - 0.5(D_r + P_d)S_L]/D_S + 1 \tag{6.208}$$

where:
D_S = distance between the ditches (m)

The total costs for the surface ditch system is the summation of the cost of construction of the PVC pipe and the cost of excavation of the ditches system as follows.

$$\cos tdrf = \cos tds + \cos tfd \tag{6.209}$$

where:
costdrf = total costs for the surface ditches system for a dry food crops polder (€)

The maintenance costs for the pipe culvert are estimated at 0.05 €/m/year.

$$\cos tdsm = 0.05 * (2t_d * H_{dk} + W_{cr}) * N_P \tag{6.210}$$

where:
costdsm = maintenance costs for the pipe culvert (€/year)

The maintenance costs for the ditches per year are estimated at 10% of the construction costs.

$$\cos tfdm = 0.10 * \cos tfd \qquad (6.211)$$

where:
costfdm = maintenance costs for the ditches (€/year)

The total costs for maintenance of the surface ditches system are calculated as follows:

$$\cos tdrfm = \cos tdsm + \cos tfdm \qquad (6.212)$$

where:
costdrfm = total costs for maintenance of the surface ditches system for a dry food crops polder (€)

Costs for construction and maintenance of the surrounding dikes

The costs for construction and maintenance of the surrounding dikes are calculated as follows:

$$\cos tdk = (W_{cr} + t_{dk} * H_{dk}) * H_{dk} * C_u * N_b * (P_W + P_L) \qquad (6.213)$$

where:
costdk = total costs for the surrounding dike construction (€)

The maintenance costs for the surrounding dikes are estimated at 0.20 €/m/year.

$$\cos tdkm = 0.20 * N_b * (P_W + P_L) \qquad (6.214)$$

where:
costdkm = total maintenance costs for the surrounding dikes (€/year)

Costs for the canal system under conditions of a dry food crops polder

The length of the main drains in m is obtained from Equation 6.95, where: a = 0.271 and b = 0.00208 (data from Jindasanguan, 1995 and Suwannachit, 1996).

Earthwork is the sum of the earth movement in the lateral canals, sub-main canals and main canals, see Figure 6.10 and Equations 6.99 to 6.103. The total costs for excavation of the lateral canals, sub-main canals and main canals are calculated by Equation 6.105, but without timbering cost. Equation 6.106 calculates annual maintenance costs for the lateral canals, sub-main canals and main canals.

Cost of pumping for the main pumping station

The required capacity of the pumping station in a polder can be calculated in the same way as for the rice polder. The total investment costs for the installation of pumping station and operation and maintenance are also calculated in the same way as for the rice polder, because a dry food crops polder is usually modified from a rice polder.

Annual equivalent cost

The total costs for construction may be summarized as follows:

Patt   = surface ditch system + surrounding dikes + canal system + pumping station
                                                                      (6.215)

The annual maintenance costs are the sum of the following:

Pott   = maintenance of surface ditch system + surrounding dikes + canal system +
             pumping station   (6.216)

The annual recovery factor is expressed as shown in Equation 6.115. Therefore the annual equivalent costs for drainage are the summation of the construction and maintenance costs:

$$cdt = V_t * \cos tdrf + \cos tdrfm + V_t * \cos tdk + \cos tdkm \qquad (6.217)$$

The annual equivalent costs for the main drainage systems included excavation, and annual maintenance (Equation 6.117), but without timbering costs. The annual equivalent costs for the pump are the costs for installation and operation as given by Equation 6.118. Equation 6.119 calculates the total annual equivalent costs of the water management system.

Costs for the field pumping

The costs for pumping by spraying from a removable boat are calculated based on the moisture content of the soil in the bed and the water level in the ditches (see Appendix V). The costs for field pumping into or out of the plot are calculated based on the quantity of water that was pumped for drainage and irrigation, dependent on the water level in the ditches (see Appendix V).

The quantity of water to be pumped for irrigation depends much on the water level in the ditch, in the lateral canal and in the main canal and also on the ground level, the lateral canal bed level and the main canal bed level (see Appendix V).

If I_{WC2} (Δt) > 0, PPL(Δt) > = PL(Δt) then $Q_{IRi} = I_{Wc2} * \Delta t$ \qquad (6.218)

If I_{WC2} (Δt) > 0, PPL(Δt) > = PL(Δt) > = BL$_L$ + 0.10 then $Q_{IRi} = I_{Wc2} * \Delta t$ \qquad (6.219)

If I_{WC2} (Δt) > 0, PPL(Δt) <= BL$_L$(Δt) + 0.10 then $Q_{IRi} = 2 * I_{Wc2} * \Delta t$ \qquad (6.220)

If I_{WC2} (Δt) > 0, PPL(Δt) <= BL$_M$(Δt) + 0.30 then $Q_{IRi} = 0$ \qquad (6.221)

where:
$I_{wc2}(\Delta t)$ = discharge capacity of the irrigation water by pumping from the lateral canal
            (m^3/time step)
Q_{IRi} = amount of water to be pumped for irrigation during the time step (m^3)

Therefore the total amount of water to be pumped to the plot and costs for pumping by a removable pump on a small boat to spray the crops are calculated based on actual the quantity of water that was pumped per m^3.

Investment costs for crops, buildings and infrastructure

Maximum and minimum investments for each crop on a hectare of land that possibly can occur during each decade have been derived from a year list of investments for the crops. The value of buildings and other facilities is calculated with in Equation 6.120. The total costs involved in the farming are determined with Equations 6.121 and 6.122.

Set-up of the model

The model consists of several programs and associated subroutines and data files. With the main programs the following computations can be made:
- RAINRUR for the simulation of the rainfall-runoff process in the rural area in the Netherlands;
- RICERUF for the simulation of the rainfall-runoff process for rainfed rice in Thailand;
- RICERUI for the simulation of the rainfall-runoff process for irrigated rice in Thailand;
- DRYRUR for the simulation of the rainfall-runoff process for dry food crops in Thailand.
  The structure of the model is shown in Figures 6.34 to 6.37.

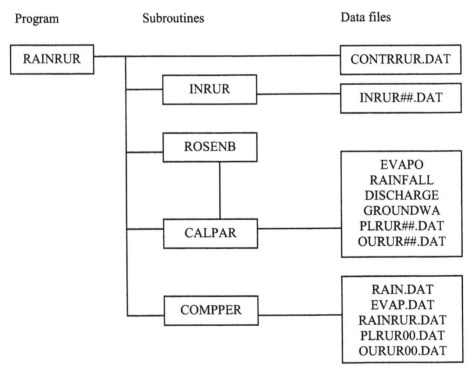

Figure 6.34 Scheme of the program RAINRUR for the simulation of the rainfall-runoff process in the rural area in the Netherlands

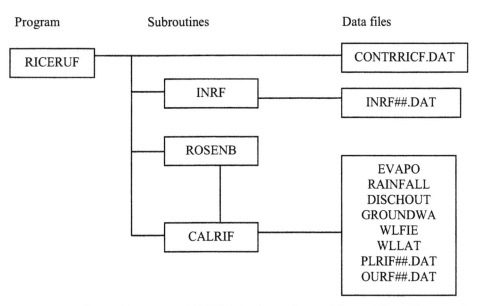

Figure 6.35 Scheme of the program RICERUF for the simulation of the rainfall-runoff process for rainfed rice in Thailand

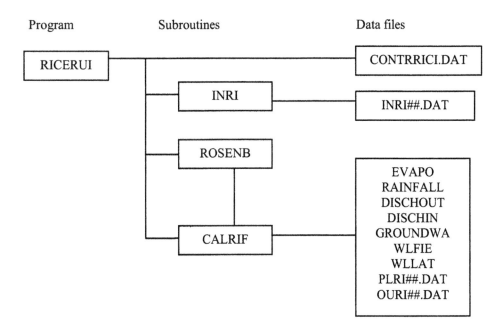

Figure 6.36 Scheme of the program RICERUI for the simulation of the rainfall-runoff process for irrigated rice in Thailand

Program   Subroutines   Data files

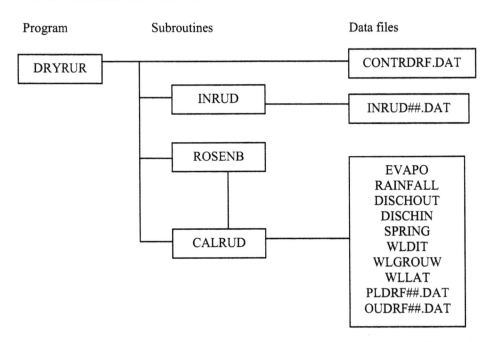

Figure 6.37 Scheme of the program DRYRUR for the simulation of the rainfall-runoff process for dry food crops in Thailand

6.2.5 Hydrological model for an urban area in the Netherlands and in Thailand

In the hydrological model for the urban area, the hydrological processes involved in the surface and subsurface part are schematised as shown in Figure 6.38.

In the urban environment excess rainfall is carried to the main drainage system through two paths, the sewer system and the subsurface drains. Both of them discharge through the urban canals and weir or pump to the main drainage system.

The inflow into the sewers is calculated by using a runoff coefficient (Van den Berg, et al., 1977) for the transfer of the total precipitation into the amount of runoff from the paved surface during a storm. Precipitation that is discharged through the sewer is only certain part of the total precipitation.

Surface runoff and sewer discharge

The model uses 1 hour measured data of rainfall, sewer inflow, sewer discharge and daily evaporation. During each time step, if the accumulated precipitation exceeds the open water evaporation, the difference contributes to the runoff from the paved area. Otherwise the precipitation will be transfer into evaporation.

$$V_{ia}(t + \Delta t) = V_{ia}(t) + P(t) \tag{6.222}$$

where:
$V_{ia}(t)$ = accumulated precipitation at instant t (mm)

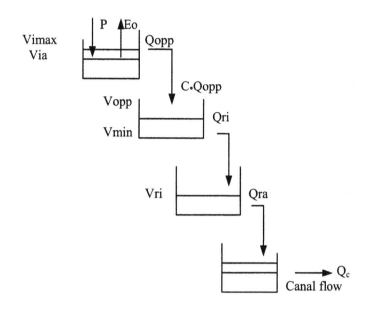

where:
Q_{opp} = discharge available for surface runoff (mm)
V_{imax} = maximum capacity of surface retention (mm)
V_{ia} = accumulate of precipitation (mm)
V_{ri} = maximum capacity of sewer retention (mm)
C   = runoff coefficient (-)
V_{opp} = maximum volume available for sewer discharge (m^3)
Q_{ri} = sewer discharge (mm/time step)
Q_{ra} = canal discharge (mm/time step)

Figure 6.38 Scheme of the model for the transformation of rainfall into runoff from a paved area

$$V_{ia}(t + \Delta t)' = V_{ia}(t + \Delta t) - E_0(t) \text{ , if } V_{ia} > E_0 \tag{6.223}$$

$$E(\Delta t) = E_0(\Delta t) \tag{6.224}$$

otherwise the following will be true:

$$E(\Delta t) = V_{ia}(t + \Delta t) \tag{6.225}$$

$$V_{ia}(t + \Delta t) = 0 \tag{6.226}$$

If the remaining runoff exceeds the maximum surface retention capacity of (V_{max}) then the surface runoff is calculated in the following way:

$$Q_{opp}(\Delta t) = V_{ia}(t + \Delta t) - V_{i\max}$$ (6.227)

where:
$Q_{opp}(\Delta t)$ = discharge available for surface runoff (mm)
$V_{i\max}$ = maximum capacity of surface retention (mm)

$$V_{ia}(t + \Delta t) = V_{i\max}$$ (6.228)

Therefore the surface runoff becomes:

$$V_{opp}(t + \Delta t) = V_{opp}(t) + C * Q_{opp}(t)$$ (6.229)

where:
C = runoff coefficient (-)
$V_{opp}(t+\Delta t)$ = maximum volume available for sewer discharge (m^3)

$$Q_{ri}(\Delta t) = \left(\frac{V_{opp}(t + \Delta t)}{k} \right)^{\frac{i}{n}}$$ (6.230)

where:
$Q_{ri}(\Delta t)$ = sewer discharge (mm/time step)
k, n = constants

If Q_{ri} is less than V_{opp} then:

$$V_{opp}(t + \Delta t)' = V_{opp}(t + \Delta t) - Q_{ri}(\Delta t)$$ (6.231)

$$Q_{ri}(t + \Delta t) = V_{0pp}(t + \Delta t)$$ (6.232)

$$V_{opp}(t + \Delta t) = 0$$ (6.233)

$$V_{ri}(t + \Delta t) = V_{ri}(t) + Q_{ri}(\Delta t)$$ (6.234)

$$Q_{ra}(\Delta t) = \left(\frac{V_{ri}(t + \Delta t)}{k} \right)^{\frac{i}{n}}$$ (6.235)

$$V_{ri}(t + \Delta t)' = V_{ri}(t + \Delta t) - Q_{ra}(\Delta t)$$ (6.236)

$$V_{ri}(t + \Delta t) = 0$$ (6.237)

The urban model consists RAINURB for the simulation of the rainfall-runoff process in the urban area of one program with subroutines and data files. The structure of the model is shown in Figure 6.39.

Program   Subroutines   Data files

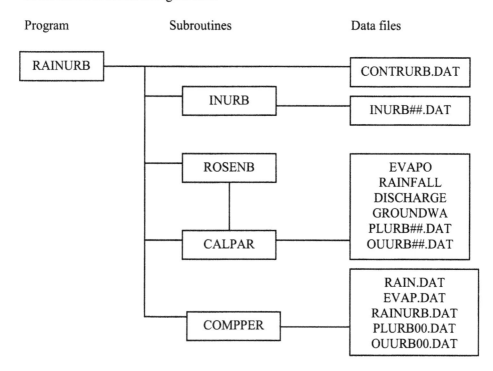

Figure 6.39 Scheme of the program RAINURB for the simulation of the rainfall-runoff process in the urban area

6.2.6 Economic computation for an urban area in the Netherlands and in Thailand

Economic computation

The economic computation has been done by the program OWAP (Figure 6.9). The economic computation is done separately for the rural and the urban area by the subroutines OWARUR and OWAURB. They are coupled with other modules for optimizing the components in the system.

Discharge into the urban canals is computed assuming 100 m strips of sub-areas i.e. housing, shops, industries along the canal divided into catchments at every 100 m perpendicular to the canal (Figure 6.40). Discharge is collected in the street gutters and conveyed by the secondary and main sewers into the urban canals. Pits are provided along the secondary sewers at each 50 m interval. Therefore the area for the rainfall volume is:

$$O_{af} = L_g * 100 \tag{6.238}$$

where:
O_{af} = urban area for the rainfall volume (m^2)
L_g = distance between canals or transport pipes (m)

Storage in the sewers is computed as follows:

Secondary $L_g * \pi * (0.5 * 0.3)^2$ $\tag{6.239}$

Main $L_g * \pi * (0.5 * D_{hr})^2$ $\tag{6.240}$

where:
D_{hr} = diameter of the main sewer (m)

Discharge out of urban polder is calculated using the formula for discharge over a weir or by a pumping station.
Costs in the urban area consist of:
-    construction and maintenance;
-    damage to different sub-areas.

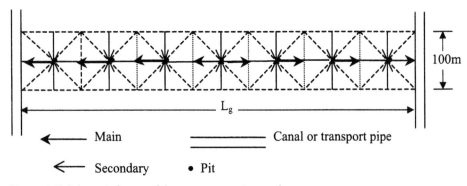

Figure 6.40 Schematic layout of the sewer system in an urban area

Cost of constructions

Cost of constructions in the area consists of the following:
-    sewers;
-    buildings;
-    canals or transport pipes;
-    weirs or pumping stations.
    Based on the schematisation in Figure 6.40 the length of the secondary and main sewers is computed as follows:

$$L_{wr} = \frac{L_g + 50}{100} * \frac{O_{si}}{L_g} * 10,000$$ (6.241)

where:
O_{si} = sub-areas like housing, shops, etc. (ha)
L_{wr} = length of secondary sewers (0.3 m diameter)

$$L_{hr} = \left(\frac{O_{si} * 100}{L_g} + 0.5 \right) * L_g$$ (6.242)

where:
L_{hr} = length of the main sewer (variable diameter) (m)

Number of manholes (pits) along the main sewer at an interval of 50 m is computed as follows:

$$N_p = \left(\frac{L_g + 25}{100} - 1 \right) * \left(\frac{O_{si} * 100}{L_g} + 0.5 \right)$$ (6.243)

The costs for urban canals are based on canal width, surface level (Figure 6.42) as follows:

$$(SP + y) * B + T_{gb} * (SP - 0.5 * H_b)^2 - T_{go} * (y - 0.5 * H_b)^2$$ (6.244)

where:
B   = bed width (m)
H_b = height of timbering (m)
SP   = depth of the water level (m-surface)
T_{gb} = horizontal component of the side slope above the water level (-)
T_{go} = horizontal component of the side slope below the water level (-)
y   = water depth in canal (m)

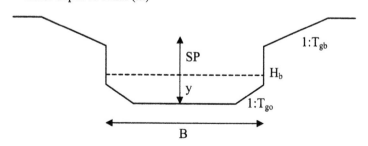

Figure 6.41 Canal section for the cost computation

Cost of damage

Damage is calculated separately for different sub-areas as follows:

- housing;
- shops;
- office;
- industries, etc.
  Based on the following criteria:
- calculated volume of water at the streets;
- groundwater table;
- exceedence of critical level in the canals.
  The following classes of damages are distinguished:
- direct;
- indirect;
- secondary;
- not qualified;
- certain.

In the model only direct and indirect costs of damage are considered. Direct damage is originating from the contact of the water with goods that can be affected and the quantification is based on the costs for repair and replacement. Infrastructure is affected from inundation but not so serious as damages in industries and housing. Indirect damages consist of the value of lost services, working hours, industrial production, inconveniences in traffic facilities, etc.

In the model direct damages are computed by using an empirical relation with the inundation or the depth of the groundwater table, while the indirect damage costs are taken as a fixed percentage of the direct costs as follows:

- housing   15%;
- shops   35%;
- industries 45%;
- facilities 10%;
- others   34%.

6.2.7 Optimization technique

The general problem in the optimization is to optimize the goal function.

$$F(X_1, X_2, X_3, \ldots\ldots\ldots, X_n) \tag{6.245}$$

Subject to constraints:

$$Gk(X_1, X_2, X_3, \ldots\ldots\ldots, X_n) \le 0 \tag{6.246}$$

$$k = 1, 2, 3, \ldots\ldots, M \tag{6.247}$$

The decision variables or the independent variables may be a finite or infinite set. Many models involving infinite variables by numerical difference techniques transferring an infinite independent set of variables into a finite set, which is suitable for the optimization technique. There are several methods to get solutions. In general they may be classified into indirect methods, mathematical programming and search methods, which require a computer for the solutions.

In the present study the optimizing routine used in the optimizing model is based on the direct search method as proposed by Rosenbrock, 1960 (Kuester and Mize, 1973).

The algorithm finds the minimum value of a non-linear goal function with multi variables and without constraints. The procedure assumes a unimodel function. Therefore several sets of starting values for the independent variables should be used if it is known that more than one minimum exists or if the shape of the surface is unknown. The algorithm proceeds as follows:

-   a starting point and initial step sizes, Si, I = 1, 2, 3,...., N are picked and the goal function is evaluated;
-   the first variable X1 is stepped a distance S1 parallel to the axis, and the function is evaluated. If the value of the goal function (F) decreases, the move is termed a success and S1 is increased by a factor α, $\alpha >= 1.0$. If the value of F increases, the move is termed a failure and S1 is decreased by a factor β, $0 < \beta <= 1.0$ and the direction of the movement is reversed;
-   the next turn variable, X2, is stepped a distance S2 parallel to the axis. The same acceleration or deceleration and reversal procedure are followed for all variables in consecutive sequences until a success (decrease in F) and failure (increase in F) have been encountered in all directions.

The axes are then rotated by the following equations. Each rotation of the axes is termed as stage:

$$M_{i,j}^{(k+1)} = \frac{D_{i,j}^{(k)}}{\left[\sum_{i=1}^{N} \left(D_{i,j}^{(k)}\right)^2\right]^{\frac{1}{2}}} \tag{6.248}$$

$$D_{i,j}^{(k)} = A_{i,j}^{(k)} \tag{6.249}$$

where:

$$D_{i,j}^{(k)} = A_{i,j}^{(k)} - \sum_{i=1}^{j-1}\left[\left(\sum_{n=1}^{j} M_{n=1}^{(k+1)} * A_{n,j}^{(k)}\right) * M_{i,1}^{(k=1)}\right] \quad j = 2, 3,...., N \tag{6.250}$$

$$A_{i,j}^{(k)} = \sum_{i=j}^{(k)} d_1^{(k)} * M_{i,1}^{(k)} \tag{6.251}$$

where:
i   = variable index = 1, 2, 3,............., N (-)
j   = direction index = 1, 2, 3,..........., N (-)
k   = stage index (-)
d$_i$ = sum of distances move in the i direction since last rotation of axes (-)
M$_{i,j}$ = direction vector component (normalized)

A search is made in each of the X directions using the new coordinate axes. The procedure terminates when the convergence criterion is satisfied.

6.3 Optimization with a GIS

6.3.1 General

GIS is a system of hardware and software used for storage, retrieval, mapping, and analysis of geographic data. Spatial features are stored in a coordinate system (latitude/longitude, state plane, UTM, etc.), which references a particular place on the earth. Descriptive attributes in tabular form are associated with spatial features. Spatial data and associated attributes in the same coordinate system can then be layered together for mapping and analysis. GIS can be used for scientific investigations, resource management, and development planning.

GIS differs from other graphical computer applications in that all spatial data are geographically referenced to a map projection in an earth coordinate system. For the most part, spatial data can be 're-projected' from one coordinate system into another, thus data from various sources can be brought together into a common database and integrated using GIS software. Boundaries of spatial features should 'register' or align properly when re-projected into the same coordinate system. Another property of a GIS database is that it has 'topology' which defines the spatial relationships between features. The fundamental components of spatial data in a GIS are points, lines (arcs), and polygons. When topological relationships exist, a user can perform an analysis, such as modelling the flow through connecting lines in a network, combining adjacent polygons that have similar characteristics, and overlaying geographic features.

6.3.2 Spatial data

Spatial data are the ones that drive a GIS. Every functionality that makes a GIS different from another analytical environment is rooted in the spatially explicit nature of the data.

Spatial data are often referred to as layers, coverage, or themes. The term of themes will be used from this point on, since this is the recognized term used in Arc View 3.x (http://www.esri.com). Themes represent, in a special digital storage format, features on, above, or below the surface of the earth. Depending on the type of features they represent, and the purpose to which the data will be applied, themes will be one of 2 major types, vector data and raster data.

Vector data

Vector data represent features as discrete points, lines, and polygons.

Points

Points represent discrete locations on the ground. Either these are true points, such as the point marked by a brass cap, such as a section corner, or they may be virtual points, based on the scale of representation. For example, a city's location on a driving map of the United States is represented by a point, even though in reality a city has an area. As the map's scale increases, the city will soon appear as a polygon. Beyond a certain scale of zoom (i.e., when the map's extent is completely within the city), there will be no representation of the city at all; it will simply be the background of the map.

Lines

Lines represent linear features, such as rivers, roads and transmission cables. Each line is composed of a number of different coordinates, which make up the shape of the line, as well as the tabular record for the line vector feature.

Polygons

Polygons form bounded areas. In the point and line data sets shown above, the landmasses, islands, and water features are represented as polygons. Polygons are formed by bounding arcs, which keep track of the location of each polygon.

The vector data structure is as follows: ArcInfo Coverages, ArcView 3.x Shape Files, CAD (AutoCAD DXF & DWG, or Micro Station DGN files), and ASCII coordinate data.

Raster data

Raster data represent the landscape as a rectangular matrix of square cells. Raster data sets are composed of rectangular arrays of regularly spaced square grid cells. Each cell has a value, representing a property or attribute of interest. While any type of geographic data can be stored in raster format, raster data sets are especially suited to the representation of continuous rather than discrete data. The data structure is as follows: ArcInfo Grids Images, Digital Elevation Models (DEMs), and generic raster data sets.

6.3.3 Application of GIS to optimization

The GIS has been used for the investigation and interpretation of data on land use, soil distribution, elevation, polder water level, drainage systems in the clay polder, the peat polder, the urban polders in the Netherlands and in Thailand and the rural polder in Thailand. To obtain the real situation of water management and it was used as input for the optimizing model.

7 Case studies

In order to analyse the method and the results of it compared to the present conditions the method has been applied to several case studies in the Netherlands and in Thailand. In the Netherlands data from polders in the Principal Water-Board of Delfland were used and in Thailand data from experimental plots and data that were collected in areas in the Central Plain of Thailand and in Bangkok Metropolitan.

Water management system of Delfland

The main open water system in Delfland is the collection and transport system for superfluous polder water, a system of interconnected watercourses and lakes between various polders, which is used to store water temporarily, with a target water level of 0.40 m-NAP. Three types of areas discharge into the collection and transport system: low-lying polders (72%), high-lying polders (6%), areas along the collection and transport system (22%), (Lobbrecht, et al., 1999). The lowest open water level in the polders that discharge into the collection and transport system is 6.30 m-NAP, whereas the highest level is 1.4 m+NAP. Excess water is pumped from the low-lying polders into the collection and transport system by 110 polder-pumping stations with a total capacity of 48 m³/s. High-lying polders discharge via fixed weirs into the collection and transport system. Water from the collection and transport system is discharged to the North Sea, the Nieuwe Maas and the Nieuwe Waterweg by six 'main pumping stations' with a total capacity of 54 m³/s.

The water level in the collection and transport system has to be kept below 0.25 m-NAP. When it rises to a higher level, there is a chance of flooding and dike bursts. This level is called 'stop pumping level'. During excessive precipitation, the discharge from the polders and areas along the collection and transport system is stored in the collection and transport system. Consequently, the open water may rise above the stop pumping level at some locations in the collection and transport system. If the water level exceeds this limit, the principal water-board can impose a pumping stop at the location where problems are foreseen. In this case certain polder pumping stations have to stop pumping, resulting in possible flooding of polder land.

In Delfland it is common practice that urban areas primarily drain to combined sewer systems. The flows from these sewer systems are treated in sewage treatment plants, most of which discharge the effluent outside the polder water system. Via these routes, part of the precipitation is discharged from the polder water system into the outside water through the bypassing channels.

All polder pumping stations in Delfland are automated and operate on the basis of water level set points, which represent target open water levels in the polders. There is no facility to control all these pumping stations from a central place. The water management system in Delfland is presented in Figure 7.1.

Selected case-study areas in Delfland

For this study three polders in Delfland were selected: a clay polder, polder Schieveen, a peat polder, Duifpolder and an urban polder, Hoge and Lage Abtswoudse polder.

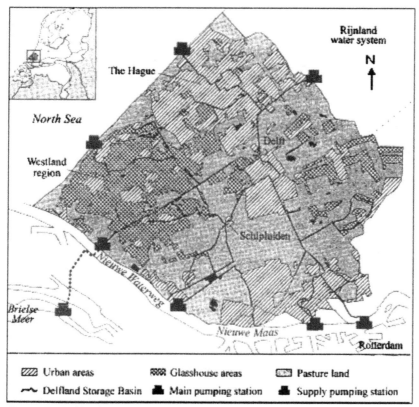

Figure 7.1 Delfland water management system (Lobbrecht, et al., 1999)

7.1 The clay polder Schieveen

7.1.1 General

The polder Schieveen lies within the municipality of Rotterdam. The polder Schieveen is limited by:
- Berkelsche Zweth;
- Oude Bovendijk;
- Doenkade;
- Doenpad;
- Delftweg.

7.1.2 Physical conditions

Area and open water

The area of polder Schieveen is 584 ha. The total length of the main drains is 16,760 m, the drains along dike are 5,447 m and there is 114,507 m of other drains. The area of the main drains is 114,486 m^2 and the water area of the other drains and main drains, based on the width of the water in winter, is about 212,554 m^2. Therefore the total area of the open water is about 327,040 m^2, or 5.6% of the total area.

Land use

Most of the land use in the area is grassland, with a small area of greenhouses. The land use in the area is shown in Table 7.1.

Table 7.1 Land use in polder Schieveen

Type of land use	Area in ha	Percentage of the total area
Greenhouse	8.0	1.3
Grassland	487.2	81.3
Cemetery	24.0	4.1
Urban and other	32.1	7.7
Water area	32.7	5.6
Total	584.0	100.0

Source: Topografische Dienst, 2003

The total area of houses is 56,603 m², which is composed of: houses 55,293 m², storage tanks (open) 723 m² and storage tanks (closed) 587m² (Topografische Dienst, 2003).

Soil

Most of the soil in the polder is sandy clay, clay and peat. The soil is classified into four main groups as follows: sea clay (46 ha), peat (3 ha), peat with non-decomposed plants (338 ha) and mixed soil (104 ha).

Sea clay soil (46 ha)

Tochteerdgronden (10 ha)

The upper soil layer exists from 0.20 - 0.30 m depth of humus rich to very humus lime false light clay. Under the upper soil 0.50 m is moderately to very poor humus lime rich heavy sandy clay, which proceeds down in light sandy clay. This sandy clay has not entirely ripened. Below 0.80 m there is very humus, lime rich clay and extremely fine sand.

Drachtvaaggrond (24 ha)

The upper soil layer exists of 0.40 - 0.70 m thick moderate humus to rich humus lime false heavy clay. This layer proceeds gradually up to approximately 0.80 m depth. Below there is rich clay peat up to a depth of 1.20 m, then starts very rich humus lime heavy clay, which is almost unripe.

Lime poor vague polder grounds (12 ha)

The topsoil exists from 0.10 - 0.15 m of rich humus lime false light clay. Below the topsoil there is 0.20 - 0.30 m thick moderate humus lime false heavy clay, which gradually proceeds in moderate humus poor lime rich heavy clay. Between 0.50 and 0.80 m the subsoil starts of moderately heavy to very humus poor lime rich clay, which is almost ripe.

Peat soil (3 ha)

The topsoil exists of lime false rich humus to a low clay content soil of 0.15 to 0.20 m. Below the topsoil there is clay to a depth of 0.30 to 0.40 m lime false, moderately to very humus. The sub soil exists of peat, which is loose to very loose.

Peat with the non-decomposed plants (338 ha)

The topsoil exists of low clay to rich clay peat at a depth of 0.20 to 0.30 m. Sometimes directly under the topsoil thin low rest peat is present. The subsoil exists of moderately heavy to very humus poor lime false clay. Sometimes this proceeds between 0.70 and 1.20 m in lime false or rich lime light clay or sandy clay. Above 0.50 m the layers are almost ripe or half matured, below they consist of half ripe to unripe soil.

Mixed soil (104 ha)

Moerige eerdgrond with elder tree peat ground (100 ha)

The topsoil exists of rich clay peat. In the swampy parts the topsoil is thinner, about 0.40 m, while at the peat grounds the depth of the topsoil varies between 0.40 and 0.80 m. The subsoil exists of lime false loose clay.

Warmoezerijgronden (4 ha)

The grounds are associated with different soil types. It is a heterogeneous ground, which strongly varies in humus, lime and clay quality. The topsoil exists mainly of rich humus to humus sandy clay or clay. The subsoil, which is 0.80 m under the surface, exists of half ripe sandy clay or clay.

Subsidence

Subsidence has been caused by the following components:
- oxidation;
- shrinkage;
- settlement.

For the peat grounds oxidation makes the largest contribution to the surface subsidence, the settlement and shrinkage are small. For peat with non-decomposed plants grounds no oxidation takes place but there is a certain subsidence due to settlement and shrinkage. In Table 7.2 the subsidence is indicated for a period of 30 years, according to the Instituut voor Culturrtechniek en Waterhuishouding (Delfland, 1990).

The C value of the soil is between 5,000 and 10,000 days (Staringcentrum, 1976). The C value is defined by:

$$C = d / k \qquad (7.1)$$

where:
C   = hydraulic resistance (days)
d   = thickness (m)
k   = permeability (m/day)

Table 7.2 Subsidence for a period of 30 years in polder Schieveen

Peat ground	Field drain depth		Subsidence		
	(m-surface)	settlement in cm	shrinkage in cm	oxydation in cm	total in cm
	0.40	1	1	9	11
	0.60	2	1	13	16
	0.80	3	2	17	22
Peat with non decomposed plants ground	0.40	1	1	0	2
	0.60	2	1	0	3
	0.80	3	2	0	5

Source: Principal Water-board of Delfland, 1990

Hydro topography

The phreatic groundwater table is shown in Figure 7.2. Seepage in this area is between 0.00 to 0.25 mm/day (Staringcentrum, 1976).

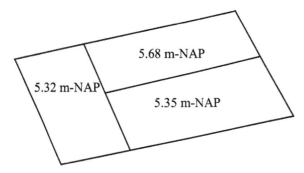

Figure 7.2 Phreatic groundwater table in polder Schieveen (Staringcentrum, 1976).

7.1.3 Present situation of operation and maintenance

Water management

The water management can be divided into three categories as follows: field drainage, main drainage and water supply.

The field drainage of the grassland plots takes place by means of open field drains or subsurface pipe drains; the drainage of the greenhouse plots is established by subsurface pipe drains. Generally the subsurface drainage is done by means of corrugate pipes and the excess water is stored in a pit. The superfluous water is pumped out by small pumps. With these conditions drainage of the greenhouses is more or less independent of the polder water level. At the greenhouse plots without pumping, the drains discharge directly into the open water.

The pumping station Hofweg (pumping station B, see Figure 7.3) collects the water, which is coming from three main watercourses via regulating collector culverts. The ways of carrying water to the pumping station Hofweg are as follows.

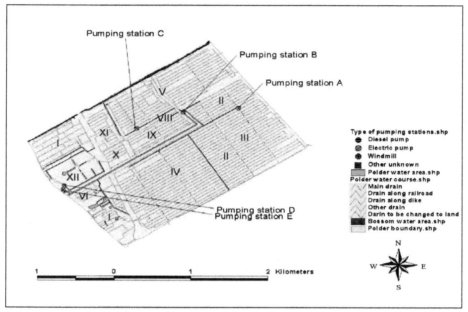

Figure 7.3 Layout of polder Schieveen

The drainage area of the West block-system (XI) drains its water by means of a movable weir along the Hofweg to the new part of the polder (V). North of the pumping station Hofweg (C), there is the regulating collector drain culvert for discharge of the new part of the polder (V). At Bergboezem West (X) the water is discharged by means of a movable weir on Bergboezem West (IX). Halfway the main watercourse of area IX a weir is present that has a small opening at 5.20 m-NAP and at a higher level the area IX drains by means of a fixed weir on area VIII. West of pumping station Hofweg (B) the drain culvert collects the discharge of Bergboezem Oost (VIII) to the Hofweg pumping station. The drainage area north of the Tempel (VI) and south of the Tempel (VII) discharges the water by two-fixed weirs through RW13 to the old area West (IV). The old part of the polder (III) discharges the water out via a pumping station (A) to the eastern part of the old part of the polder (II). The pumping station (A) has an automatic electrical pump with a capacity of 0.085 m³/s.

The eastern part of the old part of the polder (II) is drained by means of a collector drain culvert at the east of the pumping station (A). The western part of the old part of the polder (IV) drains the water to the eastern part of the old part of the polder (II). The pumping station Hofweg drains the low part of the polder on the higher part of the polder (I). The pumping station Hofweg (B) has an electrical centrifugal pump with a capacity of 0.667 m³/s. The higher part of the polder (I) is drained by a pump (E) along the Delftweg on the Schie. This is an electrical centrifugal pump with a capacity of 0.867 m³/s, dependent on the intensity of the discharge. The northern part of the high part of the polder at the cemetery has an evacuation culvert at the east side of the cemetery to the main watercourse along the Tempelweg. The cemetery (XII) has its own pumping station (D) near the main canal (Schie). This is an automatic electrical centrifugal pump with the capacity of 0.1 m³/s.

Water supply is arranged through several inlets from the main canal (Schie), the Berkelsche Zweth and the inside collection and transport system of the polder Berkel along the present Oude Bovendijk. At the Bovenvaart (I), there is inlet of water to the lower part of the polder.

The average other drains width at 0.25 m-NAP is 1.96 m and the average drain width in the urban area at 0.00 m-NAP is 2.98 m. The water level during winter in most of the other drains is kept at 0.25 m-NAP, 0.40 to 0.50 m-NAP for the main drain (average main drain width is 4.58 m) and 0.60 - 1.00 m-NAP for the open water area in the rural area (average open water width is 8.67 m). The total length of the main drain is 16,760 m, the drain along the dike is 5,447 m and the other drains are in total 114,507 m. The geometry of the drains in the polder is shown in Figure 7.4.

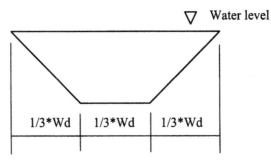

Wd   = width of the drain at polder water level

Figure 7.4 Geometry of a drain in the polder area in Delfland

There are five pumping stations in the polder as follows (Figure 7.3):

-   pumping station A at Oude Bovendijk, there is one pump and the maximum capacity is 0.085 m³/s. The function of this pump is to pump excess water from the main drain, which receives the water from the southeastern part of the polder area;
-   pumping station B at Hofweg 150, consists of one pump with a capacity of 0.667 m³/s. The function of this pump is to lift the collector water from pump A and C from the low part of the transport canal to the high part of the transport canal;
-   pumping station C at Hofweg has one pump with a capacity of 0.022 m³/s. The function of this pump is to pump excess water from the main drain, which receives the water from the northwestern part of the polder area;
-   pumping station D at cemetery Hofwijk has also one pump with a capacity of 0.1 m³/s. The function of this pump is to pump excess water from the drain, which receives excess water from the graveyards;
-   pumping station E at Schieweg-Delftweg has two pumps with a capacity of 0.867 m³/s (main pump capacity 0.70 m³/s and extra pump capacity 0.167 m³/s). The function of this pump is to drain out excess water from pumping station B and water received from the adjacent area of this pumping station.

7.1.4 Hydrological and water management data

For this case study the following hydrological and water management data have been used in the analysis:

-   the data for the model calibration were as follows:
    -   hourly pumping discharge of pumping station E in mm over the drained area for 2002;
    -   hourly rainfall in mm at Holierh & Zouteveen for 2002;

- reference evapotranspiration according to Makkink in mm/day, De Bilt, for 2002;
- the data for validation and computation:
  - daily rainfall in mm at Hoofddorp for 1960 to 2001;
  - reference evapotranspiration according to Makkink in mm/day, De Bilt, for 1960 to 2001.

While the graveyard has its own drainage system (pumping station D), the simulation and calibration were done for only 560 ha.

Characteristics of the clay polder

Based on average conditions the soil storage is assumed to exist in the soil above the winter polder water level, which consists of 0.10 m of ploughed layer (layer I) and 0.61 m of layer II. The crop is grass. The estimated number of houses is about 100 (data obtained from Delfland, 2004). Therefore the farm size is averaged to 5.6 ha. The field drain is applied in most of the area, which is subsurface pipe drains with a distance between this drains of approximately 20.00 m and a depth of 0.50 m. The dimension of one parcel is 252 * 225 m as shown in Figure 7.5. The side slopes of the watercourses are:

Collector drains   = 1:1.5;
Sub-main drains   = 1:2;
Main drains   = 1:2.

The soil type has been regrouped according to the soil classification in the Netherlands as selected from Wosten, et al. (2001) for the interpretation of the soil moisture content and other properties. The topsoil is composed of light clay (B10) 26 ha or 5.3%, heavy clay (B12) 24 ha or 4.9% and clayey peat with non-decomposed plants (B18) 441 ha or 89.8%. The subsoil composes of clayey peat with non-decomposed plants (O18) 27 ha or 5.5%, sandy clay (O11) 26 ha or 5.3%, heavy clay (O13) 338 ha 68.8% and moderate clay (O12) 100 ha or 20.4% of the total area. Based on the groups of soil the soil properties have been averaged as shown in Tables 7.3 and 7.4.

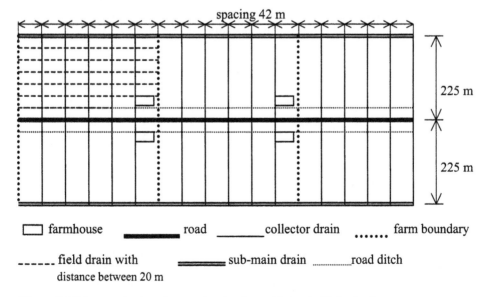

Figure 7.5 Schematic layout of a parcel in the clay polder in the Netherlands

Table 7.3 Average organic matters content in the soil profile in the clay polder (Wosten, et al., 2001)

Depth m-surface	Description	Clay content in %	Organic matter content in %
0.00 - 0.30	Mostly clayey peat	45.2	43.0
0.30 - 1.00	Mostly heavy clay	53.8	2.7
> 1.00	Mostly heavy clay	53.8	2.7

The soil moisture tension computation in the model is based on an empirical relation as follows:

$$\log p = a + b * \ln((\frac{\theta}{n})^c - 1)$$
(7.2)

where:

p	= moisture tension (cm)
θ	= moisture (%)
a, b and c	= constants for soil type (-)
n	= porosity (%)

Table 7.4 Average soil moisture tension in pF and soil moisture content in percentage for the clay polder (Wosten, et al., 2001)

Layer \ pF	1.0	1.3	1.7	2.0	2.4	3.0	3.4	4.2
I	74.1	71.3	67.8	63.8	57.2	47.1	41.8	30.9
II	56.0	54.6	52.9	50.7	47.1	41.0	37.2	30.6
III	56.0	54.6	52.9	50.7	47.1	41.0	37.2	30.6

The constants a, b and c (Table 7.5) for the different soil profiles have been calculated in a subroutine of OPOL called PFGRAPH using the typical soil properties. Detailed output of PFGRAPH is given in Appendix VI.

Table 7.5 Constants and moisture contents for the clay polder

Layer	Constant			Moisture content in %		
	a	b	c	Sat	FC	WP
I	5.085	0.934	-0.261	73.8	65.3	21.3
II	5.433	0.950	-0.261	56.3	51.5	22.6
III	5.433	0.950	-0.261	56.3	51.5	22.6

Note: Sat = saturation
FC = field capacity
WP = wilting point

Hydraulic conductivity

The saturated hydraulic conductivity has been taken as 0.063 m/day for layer I and 0.058 m/day for layers II and III (Wosten, et al., 2001). The groundwater table does not influence the hydraulic conductivity, while the flow mainly takes place through the pore space in the soil. The unsaturated hydraulic conductivity above the phreatic line is dependent on the clay content, porosity, and instantaneous moisture status. The model computes the unsaturated hydraulic conductivity and capillary flow using the following empirical formula:

$$k_{ot}(t) = A * p(t)^{-B} \tag{7.3}$$

$$B = 2.07 - \left[\frac{(L_1 + L_2)}{69.6} \right] \tag{7.4}$$

$$A = 10^{(3,21*B-4.78)} \tag{7.5}$$

where:
$k_{ot}(t)$ = unsaturated hydraulic conductivity at instant t (mm/day)
$p(t)$ = moisture tension (m)
L_1 = clay content in layer I (%)
L_2 = clay content in layer II (%)

Evapotranspiration and crop factor

The potential evapotranspiration is based on the reference crop evapotranspiration according to Makkink in mm/day. The crop factor is equal to 1.0 for the whole year for grass multiplied with the reference crop evapotranspiration as based on Commissie voor Hydrologisch Onderzoek (1988).

7.1.5 Economical data

The economic computation consists of the following:
- costs for the water management system;
- costs for buildings and infrastructure;
- investment costs for crops;
- cost of damage to crops, buildings and infrastructure.

These costs are based on the prices in 2003 (Van den Bosch, 2003, Rijkswaterstaat, 2004 and Principal Water-board of Delfland, 2004) for construction and maintenance work with Value Added Tax included in it, which is 19.5% of the total costs. The costs are given in €.

Costs for the water management system

The costs for the water management system include the following:
- costs for the field drainage system;
- costs for the main drainage system;
- costs for pumping.

The costs for the field drainage systems are based on function of the distance between the field drains (subsurface pipe drains) and their depth according to Equation 6.94. a, b, c and d are estimated according to the Tables 7.6 and 7.7, which in this case are equal to 7.463, 0.038, 0.209 and 1.1 respectively. The field drain depth is taken at 1.0 m-surface, which is generally the value for clay polders. Table 7.6 gives a list of costs involved in subsurface field drain for various distances between the drains.

Cost of maintenance is estimated around 0.39 €/m and is based on the assumption that the maintenance is required once in four years. Cost of maintenance per metre comes to about 0.10 €/m annually.

Table 7.6 Costs for subsurface drainage at a field drain depth of 1.10 m-surface in relation to the distance between the field drains for the clay polder

Distance between the field drains in m	Material cost in €/ha	Construction cost in €/ha	Total costs in €/ha	Total costs in €/m
12	3,306	3,130	6,436	7.96
24	1,653	1,724	3,377	8.36
36	1,102	1,272	2,374	8.81
48	827	1,043	1,870	9.25
60	661	909	1,570	9.72
72	551	815	1,366	10.14
84	472	748	1,220	10.57
96	413	715	1,128	11.17

Note: estimated from Van den Bosch, 2003

Table 7.7 Relation between construction cost of field drains and field drain depth for the clay polder (distance between the field drains 20 m)

Field drain depth in m-surface	Construction cost in €/m
0.50 or shallower	6.87
0.50 - 0.95	7.99
1.00 - 1.95	9.29
2.00 - 2.95	10.71

The formulas used for the estimation of the costs for construction of the main drains, sub-main drains and collector drains have been explained in Chapter 6. The estimation of costs for the main drainage system is based on the overall area of the polder, parcel length and distance between the collector drains. Basic values for the estimation of the main components of the water management system are given in Table 7.8.

Table 7.8 Unit cost for the determination of the costs for the main components of the water management system for the clay polder

Item	Construction cost in €	Annual maintenance cost in €	Lifetime in years
Subsurface field drain/m	9.72	0.09	30
Earth movement/m^3	2.63	0.15	50
Timbering sub-main drain/m	19.73	1.32	25
Timbering main drain/m	30.36	2.20	25

The interest rate is estimated at 5%, which is normally used in cost-benefit analyses for hydraulic works in the Netherlands.

The pumping head is based on the water levels before and behind the pumping station which is for Schieveen Delftweg (pump E) 2.55 m and for Schieveen Hofweg (pump B) 2.85 m. The annual rainfall surplus and seepage are estimated to be around 394 mm and 91 mm respectively for the estimation of the operation cost for pumping. The installation and operation and maintenance cost are given in Table 7.9.

Table 7.9 Pumping costs for the clay polder

Installation in million €	Operation in million €/year	Maintenance in million €/year	Lifetime in years
0.341	0.0196	0.0051	50

Note: Upward seepage is estimated at 0.25 mm/day.
Drainage modulus is estimated at 13.4 mm/day.
The pumping costs are only the cost of the main pump at Schieveen Delftweg.

Costs for buildings, infrastructure and crops

The average farm size has been estimated at 5.6 ha. The value of the buildings is estimated at 600,000 €/farm. The area of greenhouses is about 7.96 ha and the value is estimated at 900,000 €/ha. The area of buildings and yards is estimated at 1,130 m^2/farm. Average farm size included greenhouse area is 1,926 m^2/farm and the value of the buildings is 662,000 €/farm. For facilities such as shops, factories, schools, etc., 20% of the value of the farms has been taken.

The infrastructure in the rural area is mainly composed of roads. In order to compute the costs for infrastructure the density of paved roads are estimated from the topographic map at 330 m^2/ha. The estimated cost of roads is 24.26 €/m^2.

The crop in the clay polder is only grass. The investment costs and time schedule are shown in Figure 7.6 and Table 7.10.

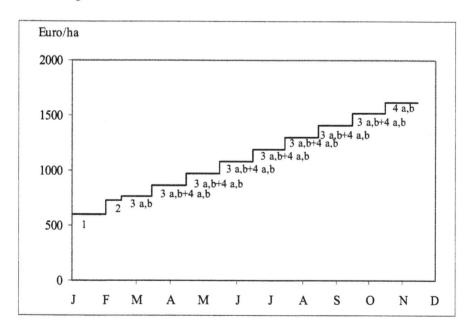

where:

1   = lease/rent
2   = inject manure slurry
3 a,b = fertilizers and spreading respectively
4 a,b = harvesting, drying grass and transportation to stock respectively

Figure 7.6 Investment costs and time schedule for grass

Table 7.10 Agricultural activities, investment costs and scheduling for grass

Activity	Investment cost in €/ha	Starting date
1 lease/rent	600	1 January
2 inject manure slurry	125	1 February
3a fertilizers (nitrogen 300 kg/ha)	275	15 Feb. - 15 Oct.
b spreading	35	15 Feb. - 15 Oct.
4a harvesting and drying grass	460	15 Mar. - 15 Nov.
b transportation to stock	200	15 Mar. - 15 Nov.
total	1,695	
5 gross profit	2,130	15 Mar. - 15 Nov.

Note: Grass cost, which not includes transportation, is estimated at 0.15 € per kg dry matter (de Laat, 2003).

Cost of damage to crops, buildings and infrastructure for the clay polder

Crop damage is supposed to occur when there are too wet or too dry conditions. The method of calculation and the parameters used for estimating crop damage have been described in Chapter 6.

Damage to buildings and infrastructure is estimated as a percentage of value of buildings and infrastructure in relation with the groundwater table as shown in Table 7.11.

Table 7.11 Relation of the groundwater table and damage in percentage of the value of buildings and infrastructure (US Army Corps of engineers, 1996\2000)

Groundwater table	Buildings	Infrastructure
at 0.60 m-surface	0	0
equal to surface	10.0	10.0

7.1.6 Calibration of the parameters

Calibration of the parameters of model in the rural area for the clay polder involved determination of the hydrological parameters, which have been explained in Chapter 6 as follows:
- maximum interception in mm E_{max};
- potential evapotranspiration from layer I in mm/m V_{ep};
- parameters k and n in the discharge formula;
- maximum moistening rate in layer I in mm/m/hour V_{max1};
- maximum moistening rate in the columns of layer II in mm/m/hour V_{max2}.

Sensitivity of the parameters for the clay polder

The interception is a function of canopy coverage, type of vegetation, the crop stage and the period of the year. The canopy coverage of grass is assumed to be constant all over the year. Any excess beyond E_{max} and the absorption capacity of the underneath layer will be recharged to the groundwater and stored on the soil surface. Therefore the higher the interception the lower the values of recharge, the lower the value of the groundwater table and water depth above the ground surface. The optimal value of the maximum interception has been found at 2.9 mm. This value is high, which may be due to some depression storage inside the area. In reality the interception may change throughout the year due to differences in the grass cover.

The upper limit for potential evapotranspiration has been influenced by the water losses from the soil through evapotranspiration. The calibrated parameter value for this case was the value from Table 7.12, which is 28.2 mm.

Table 7.12 Calibrated parameters of the hydrological model for the clay polder

Year	Maximum interception in mm	Upper limit for potential evapotranspiration in mm	Parameter discharge model		Maximum moistening rate in mm/m/hour		Goal function	Year discharge computed in mm
			k	n	Layer I	Layer II		
2002	2.9	28.2	0.0649	2.48	2.2	1.4	650	502

In the model the non-linear model is used as given by Equation 6.54. The value of the goal function is significantly related with the discharge parameters. The discharge is more sensitive to the parameter k than to the parameter n. As a result of the calibration for 2002, the values for k and n are taken as 0.0649 and 2.48 respectively for further calculations.

The maximum moistening rates in layer I and layer II affect the amount of moisture that will penetrate in layer I and layer II. The other factors are interception, moisture at the previous time step and thickness of the layer. The rate of increasing storage during a time step is higher when the maximum moistening rate increases.

The value of the maximum moistening rate gives an effect to the discharge to the drain and the soil storage, but there is little effect on the yearly discharge (the lower the moistening rate the higher the discharge). The higher the maximum moistening rate, the lower the time that is required to recharge the subsoil. Hence the maximum moistening rate also affects the soil moisture, potential evapotranspiration and the groundwater table in the soil (the lower of moistening rate the higher the groundwater table).

However, in this study the maximum moistening rate in layer I is less sensitive than in layer II, because the thickness of layer I is smaller than layer II.

7.1.7 Model validation with fixed calibrated parameters

The model has been validated with the data from the calibrated parameters in Table 7.12 for 2002 as shown in Figures 7.7 to 7.9.

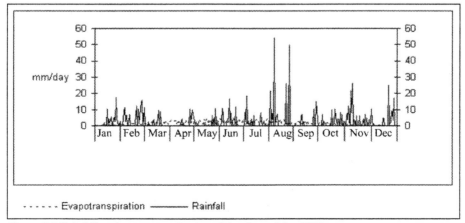

Figure 7.7 Rainfall and evapotranspiration of the clay polder, 2002

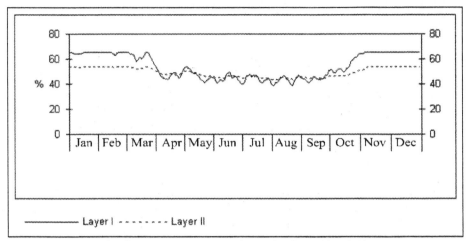

Figure 7.8 Computed soil moisture in layers I and II in the clay polder, 2002

In Figure 7.8 the soil moisture during winter is more or less at field capacity. The soil moisture is gradually decreasing during spring. The lowest soil moisture content is occurring at late June due to low rainfall and high evapotranspiration (Figures 7.7 and 7.8). In the summer period the soil moisture content in layer I can drop below the soil moisture content in layer II due to high evapotranspiration. The capillary rise to layer I cannot compensate this evapotranspiration, because the clay soil has a high tension to hold the moisture in the soil.

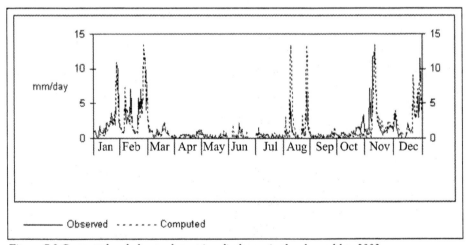

Figure 7.9 Computed and observed pumping discharge in the clay polder, 2002

There is a short lagtime between rainfall and discharge during the winter, while the soil is more or less at field capacity. On the contrary in the summer when the soil has a lower moisture content there is a larger lagtime between rainfall and discharge, which can be described by the fact that the rainfall has to refill the soil storage before the excess water discharges to the drain (Figures 7.7, 7.8 and 7.9).

7.1.8 Analysis with the optimization model

Hydrological analysis

The hydrological analysis with the validated model for the clay polder has resulted in the data as given in Table 7.13.

Table 7.13 Review of annual data in mm for the clay polder

Year	Rainfall in mm	Evaporation in mm	Pumping discharge in mm		Capillary rise in mm	
			computed	observed	II → I	III → II
2002	969	558	502	481	169	105

Economic computation

The results and review computations for optimal conditions are shown in Tables 7.14 to 7.17.

Table 7.14 Review of the total values of the crops, buildings and infrastructure in the clay polder at optimal water management

Item	Area in ha	Highest value		Lowest value	
		in € per ha	in € * 10^6	in € per ha	in € * 10^6
Buildings	19	4,122,000	79.08	4,122,000	79.08
Infrastructure	16	243,000	3.88	243,000	3.88
Grass	450	2,130	0.96	600	0.27
Total	485		83.91		83.23

Table 7.14 shows that the value of buildings is about 94% of the total value in the clay polder. Therefore the damage to buildings may be the most important in the clay polder. In fact the buildings and infrastructure in this area were constructed in a high part of the area or at some landfill.

Table 7.15 Optimization analysis of the main components of the water management system in the clay polder

Polder water level in m-s.	Open water in %	Pumping capacity in mm/day	Distance between field drains in m	Field drain depth in m-s.	Costs in € * 10^6	Damage in € * 10^6	Value of the goal function
1.51	1.9	16.3	23.20	1.30	0.317	0.028	345,000

In Table 7.15 the simulated optimal main components of the water management system are shown. The optimal field drain depth is 1.30 m-surface. At this level the subsidence can be high due to oxidation of the peat layer in the topsoil. The field drain depth and distance between the field drains are important factors for damage in the field. The percentage of open water at optimal conditions is calculated as the least required area according to existing conditions such as parcel width and length, dimension and slope of the drains, distance between sub-main drains and length of the main drains and geometry.

Table 7.16 Total costs for the water management system for the clay polder at optimal water management

Item	Construction costs in € * 10^6	Annual maintenance costs in € * 10^6	Annual equivalent costs in € * 10^6
Field drainage	1.68	0.019	0.13
Open water	0.97	0.061	0.12
Pumping	0.60	0.028	0.06
Total	3.26	0.108	0.31

Table 7.16 shows that the overall annual maintenance costs in the clay polder are about 3.3% of the construction costs, while the overall annual equivalent costs are about 9.6%. The field drainage composes the highest amount of the total construction costs for water management, which is about 51.7%. While the pumping costs comprise the lowest part, which are about 18.5%.

Table 7.17 Specification of average annual damage and yield reduction for the clay polder at optimal water management

Item	High groundwater tables		Too late sowing		Too late harvesting	
	in €	in % of the value	in €	in % of the value	in €	in % of the value
Buildings	42,600	0.05	0	0.0	0	0.0
Infrastructure	2,090	0.05	0	0.0	0	0.0
Grass	0	0.00	8,360	0.9	20,760	2.2
Total	44,690	0.05	8,360	0.9	20,760	2.2

In Table 7.17 the simulated average annual damage costs and average annual damage in percentage of the total value are shown. The main damage to the crops may be caused by too late harvesting, and there is no damage due to too high groundwater tables. The latter is caused by the deep field drain depth.

Economical analysis for the water management system in the present situation

In practice the control criteria for the polder water level may be different throughout the year. In this simulation the polder water level in winter was used, because damage will occur in this period. The general values of the main components of the water management system for the clay polder may be approximated as follows:
- field drain depth   = 0.50 m-surface;
- distance between the field drains   = 20.0 m;
- polder water level   = 0.71 m-surface;
- open water area   = 5.60%;
- pumping capacity   = 13.4 mm/day.

The results and review computations for conditions in practice are shown in Tables 7.18 to 7.21.

Table 7.18 Review of the total values of the grass, buildings and infrastructure in the clay polder
at present water management

Item	Area in ha	Highest value		Lowest value	
		in € per ha	in € * 10^6	in € per ha	in € * 10^6
Buildings	20	4,122,000	80.98	4,122,000	80.98
Infrastructure	16	243,000	3.98	242,600	3.98
Grass	459	2,130	0.98	600	0.28
Total	495		85.94		85.23

Table 7.18 shows that the value of buildings is about 94% of the total value in the clay polder at present conditions. Therefore the damage to buildings may be the most important in the clay polder.

Table 7.19 Optimization analysis of the main components of the water management system in the clay polder at present water management

Polder water level in m-s.	Open water in %	Pumping capacity in mm/day	Distance between field drains in m	Field drain depth in m-s.	Costs in € * 10^6	Damage in € * 10^6	Value of the goal function
0.71	5.6	13.4	20.0	0.50	0.403	6.166	6,569,000

Table 7.19 shows that damage is larger than in Table 7.15 and also the costs are higher than in Table 7.15. Damage is high due to too high groundwater tables because of the shallow field drain depth and costs are higher due to the high pumping capacity.

Table 7.20 shows that the overall annual maintenance costs in the clay polder at present are about 3.9% of the construction costs, while the overall annual equivalent costs are about 10.0%. The open water composes the highest amount of the total construction costs, which is about 46.3%, while the pumping costs comprise the lowest amount, which are about 10.3%. The pumping costs and open water costs are higher than at optimal conditions (Table 7.16) but the field drainage costs are lower because of the deeper field drains at optimal conditions.

Table 7.20 Total costs for the water management system in the clay polder at present water management

Item	Construction costs in € * 10^6	Annual maintenance costs in € * 10^6	Annual equivalent costs in € * 10^6
Field drainage	1.74	0.02	0.14
Open water	1.85	0.11	0.22
Pumping	0.41	0.02	0.04
Total	4.00	0.15	0.40

In Table 7.21 the damage due to too high groundwater tables is higher compared to optimal conditions (Table 7.17) because of the higher probability to have a too high groundwater table. As the field drain depth decreases damage due to too late sowing and harvesting increases significantly compared to the optimal conditions.

Table 7.21 Specification of average annual damage and yield reduction in the clay polder at present water management

Item	High groundwater tables		Too late sowing		Too late harvesting	
	in €	in % of the value	in €	in % of the value	in €	in % of the value
Buildings	5,085,000	6.3	0	0.0	0	0.0
Infrastructure	250,000	6.3	0	0.0	0	0.0
Grass	301,000	30.7	1,650	0.2	30,860	3.2
Total	5,635,000	6.6	1,650	0.2	30,860	3.2

Table 7.22 Comparison between the optimal and present values of the main components of the water management system in the clay polder

Item	Polder water level in m-s.	Open water in %	Pumping capacity in mm/day	Distance between field drains in m	Field drain depth in m-s.	Costs in € * 10^6	Damage in € * 10^6	Value of the goal function
Optimal	1.51	1.9	16.3	23.20	1.30	0.317	0.028	345,000
Present	0.71	5.6	13.4	20.00	0.50	0.592	6.166	6,569,000
% difference	+113	-66	+22	+16	+222	-46	-100	-95
Fixed depth	1.90	2.3	32.30	3.70	0.50	0.911	2.268	3,179,000

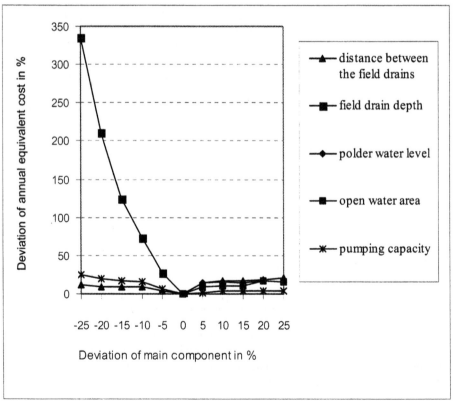

Figure 7.10 Deviation of the main components and of the annual equivalent costs in percentage in the clay polder

In Figure 7.10 the influence of deviations from the simulated optimal values is shown. It shows that the field drain depth has most influence on the annual equivalent costs in the system. If the field drain depth is 25% shallower compared to the optimal value the annual equivalent costs will increase approximately 336%, because of the increased damage due to a too high groundwater table. A 25% lower the pumping capacity than the optimal value increases the annual equivalent costs with approximately 25%. The lower the pumping capacity, the higher the probability to get a high polder water level. Therefore the higher the chance that the area will be subject to a too high groundwater table and consequently the higher the chance to get too wet conditions for sowing and harvesting. A deeper polder water level with 25% results in an increase of the annual equivalent costs with 15% because the construction costs for open water increase. A decrease in the distance between the drains of 25% results in an increase of the annual equivalent costs with 12%.

Discussion

Based on the results as shown above the following points can be discussed.

From Table 7.22 it can be concluded that the field drain depth and the polder water level may be deeper but further study on the effect on subsidence is recommended.

In Figure 7.10 it is shown that a reduction in the field drain depth has most influence on the annual equivalent costs.

The polder water level during the winter period can be kept lower to reduce the probability of submerge of the soil in the area during this period. Therefore the chance of too wet conditions for sowing and harvesting can be reduced. The polder water level during the summer period can be kept higher to get more capillary rise in the grassland, which can compensate the soil moisture deficit during this period.

In the simulation with fixed depth (Table 7.22) it was found that the distance between the field drains is small enough to obtain a fast discharge in order to get the least damage. While the pumping capacity will have to be increased because of the fast discharge from the drainage system, more open water area will be required and the polder water level needs to be deeper. In this simulation it was found that the polder water level could be lower than at present, which is indicated in Table 7.22. This will result in lower damage, but damage to the bank slope of the main drain may occur. This depends on the soil properties. However, a lower water level before heavy rainfall with no harmful effect to the stability of the bank slope of the main drain can be applied.

7.2 The peat polder Duifpolder

7.2.1 General

The Duifpolder is part of the region Midden-Delfland and lies in the municipalities Schipluiden and Maasland.

7.2.2 Physical conditions

The area of the Duifpolder is 370.38 ha. The water area of collector drains plus drains along the dike and main drains, based on the drain width at polder water level in winter is about 154,400 m^2 and the water area of the other drains is 7,376 m^2 (Figure 7.11). Therefore the total area of the open water is about 161,775 m^2 or 4.4% of the total area.

The average of the other drains and collector drains at winter water level is 1.85 m and for the main drains it is 5.02 m. The bottom of the other drains and the collector drains is 0.35 m below the water level and the bottom depth of the main drain is between 0.65 and 1.00 m-water level.

The total length of the main drains is 6,738 m, of the drains along the dike 3,816 m and of the other drains and collector drains 61,359 m. The cross section of the drains in this polder is the same as the clay polder, which is shown in Figure 7.4.

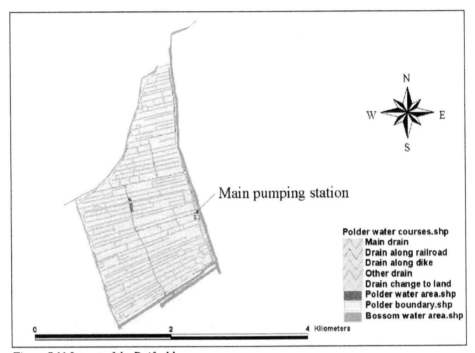

Figure 7.11 Layout of the Duifpolder

Soil

The soil type of the Duifpolder is mostly peat. The soil map is shown in Figure 7.12. The following soil groups are found in the polder:

- *pMn85c composed of 100% homogeneous sea clay.* The topsoil has humus and/or rich humus sandy clay 0.25 up to 0.35 m thick. Underneath the soil becomes gradually barge and from 0.50 m it consists of lime rich and laminated matured sandy clay and grey spots within 0.50 m depth;
- *Wo composed of 50% clays, 50% peat.* This soil is characterized by peat with non-decomposed plants. The topsoil is generally well mature and exists of a rich clay peat or rich humus clay proceeding down into peat. Generally the organic-substance is spread above the mineral subsoil at a thickness of 0.10 to 0.20 m. At many places in the upper layer 0.10 to 0.20 m moderately fine sand is found, because it has been brought in to stabilise the soil surface;

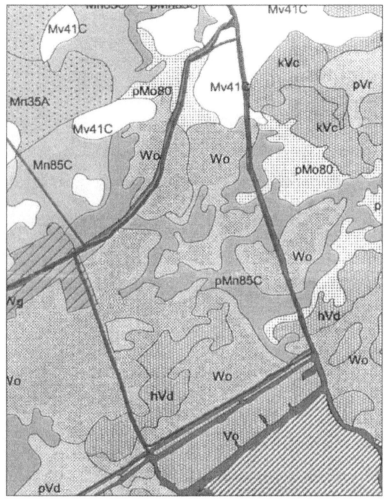

Figure 7.12 The soil map of the Duifpolder (Jeurink, et al., 2000)

- *pVr composed of 100% peat.* This surface is described as meadow peat soils. These exist from 0.15 up to 0.25 m and are composed of humus rich, light or heavy clay. By means of 0.10 m thick, very humus or humus rich, heavy clay layer proceeds the upper soil in strongly oxidized peat. The subsoil is mostly peat;
- *hVd; composed of 100% peat.* Peat soil on mud is the general description. The topsoil is 0.15 to 0.20 m thick and consists of mostly of rich clay peat. Under this topsoil generally lies strong peat that between 0.30 and 0.60 m proceeds down in mud or in a thin clay layer. The subsoil, the mud, exists of very finely divided organic substance and thin clay layers. Under the mud (from 1.20 to 1.50 m) is generally peat;
- *pMo80; composed of 100% sea clay.* These soils have a 0.20 up to 0.30 m thick topsoil, which consists of very humus or rich humus heavy clay. Below mostly 0.15 up to 0.30 m thick is low lime heavy clay with cat clay spot. The thin clay layer is in the subsoil from 0.70 m. Between 0.80 and 1.20 m turf can occur locally;

- *Mv41C; composed of 100% heavy clay.* The heavy clay soils are lime arm. The mineral part of the soils exists of lime false heavy clay. The humus topsoil is thin, the pH low. The upper 0.15 to 0.30 m consists of peat. The other soil profile has humus topsoil, which exists up to approximately 0.10 to 0.15 m of very humus, sometimes rich humus, lime false light or moderately heavy clay. Below the topsoil the humus content decreases. Within 0.15 to 0.30 m depth mostly lime false moderately heavy clay starts, which passes through to the peat underground. The upper part of the peat has been strongly oxidized. Locally, generally at a depth of 0.80 up to 1.00 m, there is a very low permeable mud. In a number of cases the peat underground reaches 1.20 m depth; below this is slack clay.

Most of the soil groups in the polder area are Wo, which covers about 60.6% of the total area. The soil group pMn85c is the second large soil group, which covers about 18.4% of the total area. The percentage area of the different soil groups is shown in Figure 7.13.

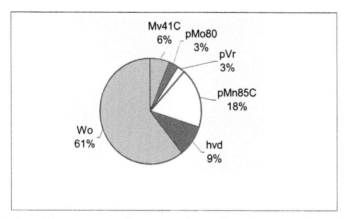

Figure 7.13 Percentage of area according to the soil groups in the Duifpolder

Peat soil means that from the surface at least 0.40 - 0.80 m of peat present. Peat soils are generally used for meadow. The C value of the soil varies from 5,000 to 10,000 days (Staring Centrum, 1976).

Surface level

The area and the surface level are shown in Figure 7.14 and Table 7.23. The average ground surface is 2.65 m-NAP.

Hydro topography

The phreatic groundwater table is 2.92 m-NAP (Staring Centrum, 1976). The average highest groundwater table is up to 0.40 m-surface and the average lowest groundwater table is 1.20 m-surface. In the relatively high area D7 (on the spot of an old transverse creek) the average groundwater table is 0.80 m-surface and the lowest groundwater table is more than 1.20 m-surface (Jeurink, et al., 2000).

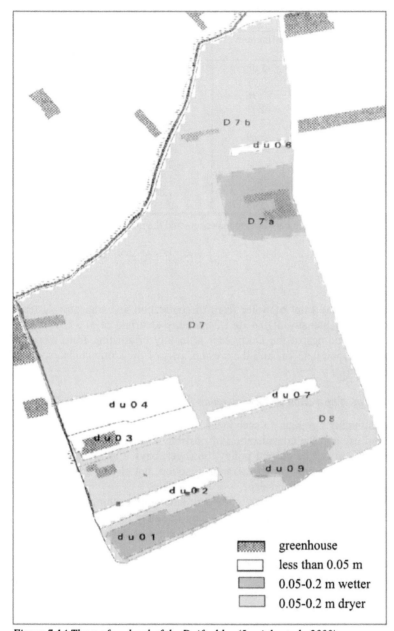

Figure 7.14 The surface level of the Duifpolder (Jeurink, et al., 2000)

Land use

Most of the area is pasture (88.8%) (Figure 7.15). The greenhouse area is about 1.5% of the total area, while houses occupy 0.7%. Other areas, such as roads, bicycle paths and green areas cover 9.1% of the total area.

The total area of houses is 23,944 m^2, which is composed of: houses 23,238 m^2, and storage tanks (open) 706 m^2 (Topografische Dienst, 2003).

Table 7.23 Surface level in the Duifpolder (Jeurink, et al., 2000)

Code level area/category level	Surface level in m+NAP			
	Average	Minimum	Maximum	Large differrence[1]
D7	-2.40	-3.06	-0.08	yes
D8	-2.58	-3.02	-0.40	yes
do01	-2.75	-2.96	-2.43	yes
du02	-2.75	-3.01	-2.08	yes
du03	-2.87	-3.01	-2.42	yes
du04	-2.86	-3.29	-2.57	yes
du05	-2.89	-3.29	-2.56	yes
du06	-2.55	-2.66	-2.40	no
du07	-2.41	-3.02	-1.69	yes
do08	-2.59	-2.93	-1.86	yes
du09	-2.56	-2.93	-0.45	yes

Note: [1] It is noticed that the difference between the maximum and the minimum surface altitude is larger than 0.40 m.

Seepage

In addition to rainwater and inlet of water from the collection and transport system, there is seepage with an intensity of 0.4 to 1.0 mm/day (Jeurink, et al., 2000). The seepage from the surroundings of the Duifpolder is mainly originating from the area north of it (level difference 0.45 m) and the western part of the Commandeurspolder (level difference 0.25 m).

7.2.3 Present situation of operation and maintenance

Midden-Delfland is mainly agrarian. In such areas the polder water level is dominated by the agrarian land use and the consideration of possible disadvantageous impacts on buildings and infrastructure. The current policy, however, pays a lot of attention to a sustainable environment, not only for the agrarian function, but also to a balance with other interests, such as nature and landscape.

Moreover, important item with respect to this are, among other things, enlarging of the water area, measures against drought, diminishing of saline seepage, improvement of the water quality, possibilities for durable agriculture and conservation of cultural heritage.

The water level during winter in the main drain is 3.15 m-NAP. The water level during winter in the other drains is kept from 3.40 m-NAP to 3.00 m-NAP dependent on the ground surface. The main pumping station has a maximum capacity of 0.63 m³/s. The water level in the drains varies from 0.37 to 0.93 m-surface, dependent on the height of the ground surface. On the average it is about 0.67 m-surface.

7.2.4 Hydrological and water management data

For this case study the following hydrological and water management data have been used in the analysis:
- the data for the model calibration were as follows:
  • hourly pumping discharge of the pumping station in mm for 2002;
  • hourly rainfall in mm at Holierh & Zouteveen for 2002;
  • hourly polder water level at pumping station for 2002;

- reference evapotranspiration according to Makkink in mm/day, De Bilt, for 2002;
- the data for validation and computation:
  - daily rainfall in mm at Hoofddorp for 1960 to 2001;
  - reference evapotranspiration according to Makkink in mm/day, De Bilt, for 1960 to 2001.

residentail area	paved road	public area
social area	unpaved road	sport areas
service sector area		camping area
area reserved for public	water area for recreation	recreation area
area plan to build office	water width wider than 6m	green house
factory area	forest area	other agriculture
area for factory office	park area	nautral area
house	cemetry area	other land use

Figure 7.15 Land use in the Duifpolder and adjacent area (Jeurink, et al., 2000)

Characteristics of the peat polder

Grass is grown in this polder area for diary farming. Based on the average conditions the soil storage is assumed to exist in the upper 0.67 m, which consists of 0.10 m of ploughed layer (layer I) and 0.57 m of layer II. The estimated number of houses is 70 (data obtained from Delfland, 2004). Therefore the farm size is averaged to 5.06 ha. The dimension of a parcel is about 66 * 235 m as shown in Figure 7.16. The water level in the open field drains in this area is kept as high as possible to avoid over subsidence due to oxidation of the peat soil. For grass the water level should be at least 0.20 m-surface. The area is drained by open field drains with a distance between the drains of approximately 10.00 m. The depth is 0.30 to 0.40 m. The side slopes of the watercourses are:

Open field drains = 1:0.75;
Collector drains   = 1:1.5;
Sub-main drains = 1:2.5;
Main drains   = 1:2.5.

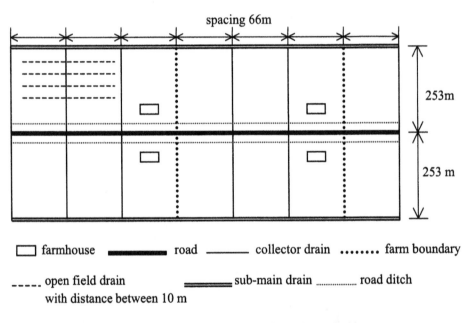

Figure 7.16 Schematic layout of parcels in the peat polder in the Duifpolder

As in the clay polder the soil types have been regrouped in a topsoil composed of clay peat (B18) 72.5%, sandy clay (B11) 18.4% and heavy clay (B12) 9.2%. The subsoil is composed of peat with non-decomposed plants (O18) 78.7%, sandy clay 18.4% (O11) and heavy clay (O13) 2.9%. Based on the groups of soil, the soil properties have been averaged as shown in Tables 7.24 to 7.26.

The soil moisture tension computation in the model is based on the empirical relation as given in section 7.1. The constants a, b and c are shown in Table 7.26 and Appendix VI.

Table 7.24 Organic matter content in the soil profile in the peat polder (Wosten, et al., 2001)

Depth in m-surface	Description	Clay content in %	Organic matter content in %
0.00 - 0.20	Clayey peat	45.0	47.5
0.20 - 1.00	Peat with non-decomposed plants	7.5	22.5
> 1.00	Peat with non-decomposed plants	7.5	22.5

Table 7.25 Soil moisture tension in pF and soil moisture content in percentage for the peat polder (Wosten, et al., 2001)

Layer \ pF	1.0	1.3	1.7	2.0	2.4	3.0	3.4	4.2
I	71.6	68.9	65.7	62.0	56.0	46.7	41.1	31.7
II	54.2	52.2	48.6	44.1	36.4	25.7	20.3	12.7
III	54.2	52.2	48.6	44.1	36.4	25.7	20.3	12.7

Table 7.26 Constants and moisture contents for the peat polder

Layer	Constants			Moisture content in %		
	a	b	c	Sat	FC	WP
I	5.117	0.908	-0.251	71.0	63.4	20.9
II	4.718	0.775	-0.168	54.3	46.1	4.7
III	4.718	0.775	-0.168	54.3	46.1	4.7

Hydraulic conductivity

The saturated hydraulic conductivity for the peat soil is taken as 0.028 m/day (Wosten, et al., 2001) for layers II and III. For layer I the saturated hydraulic conductivity is taken as 0.062 m/day (Wosten, et al., 2001). The unsaturated hydraulic conductivity above the phreatic level is determined dependent on the clay content, porosity, and instantaneous moisture status. The model computes the unsaturated permeability and capillary flow using the same empirical formulas as for the clay polder (sub section 7.1.4).

Evapotranspiration and crop factor

The potential evapotranspiration for grass is based on the Makkink formula. The crop factors for grass multiplied with the reference evapotranspiration are taken from the Commissie voor Hydrologisch Onderzoek (1988) and are the same as for the clay polder.

7.2.5 Economical data

The economic computation consists of the following:
- costs for the water management system;
- costs for buildings and infrastructure;
- investment costs for grass;
- damage to grass, buildings and infrastructure for the peat polder.

These costs are based on the prices in 2003 (Van den Bosch, 2003, Rijkswaterstaat, 2004 and Principal Water-board of Delfland, 2004) for construction and maintenance work with Value Added Tax included in it, which is 19.5% of the total costs. The costs are given in €.

Costs for the water management system

The costs for the water management system include the following:
- costs for the open field drainage system;
- costs for the main drainage system;
- costs for pumping.

The costs for the open field drainage systems are computed according to Equation 6.94. a, b, c and d are estimated according to the Tables 7.27 and 7.28, which results in 0.96, 0.0057, 2.62 and 0.50 respectively. The open field drain depth is taken at 0.50 m-surface. The typical open field drain for the cost estimation is shown in Figure 7.17. Table 7.27 gives a list of costs involved in the open field drains for various distances between the drains. Cost of maintenance is estimated at around 0.18 €/m, based on the assumption that maintenance is required once per year.

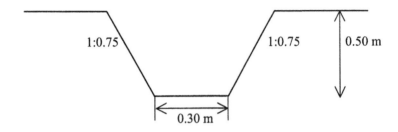

1:0.75   1:0.75   0.50 m

0.30 m

Figure 7.17 A typical open field drain cross-section

Table 7.27 Cost of open field drains in relation to the distance between the drains for the peat polder at a drain depth of 0.50 m

Distance between the open field drains in m	Canal length in m/ha	Excavation volume in m³/ha	Unit cost in €/m³	Total costs in €/ha	Total costs in €/m
10	1,100	371	3.03	1,126	1.02
20	600	202	3.18	645	1.07
30	433	146	3.36	491	1.13
40	350	118	3.53	416	1.19

Table 7.28 Relation between construction cost of open field drains and drain depth in the peat polder at a distance between the drains of 20 m

Open field drain depth in m-surface	Construction cost in €/m
0.30 to 0.50	0.65
0.50 to 0.70	1.12
0.70 to 1.00	1.88

The formulas used for the estimation of the costs for construction of the main drains, sub-main drains and collector drains have been explained in Chapter 6. The estimation of costs for the drainage system is based on the overall area of the polder, parcel length and the distance between the collector drains. Normally costs for construction and maintenance in a peat polder are about 1.2 - 1.5 times of the costs for a clay polder due to less workability (De Bakker and Van den Berg, 1982). Basic values for the estimation of the main components of the water management system are given in Table 7.29.

Table 7.29 Unit cost for the determination of the costs for the main components of the water management system in the peat polder

Item	Construction cost in €	Annual maintenance cost in €	Lifetime in years
Open field drain /m	1.07	0.18	30
Earth movement /m^3	3.55	0.20	50
Timbering sub-main drain/m	19.73	1.32	25
Timbering main drain/m	36.42	2.20	25

The pumping head is based on the average water level before and behind the pumping station, which is for the Duifpolder 2.70 m (field data from Delfland, 2003). The drainage modulus for the peat polder is at 14.9 mm/day. The annual rainfall surplus is estimated to be around 394 mm and annual seepage is 250 mm. The installation cost and maintenance cost are shown in Table 7.30.

Table 7.30 Pumping costs for the peat polder

Installation in million €	Operation in million €/year	Maintenance in million €/year	Lifetime in years
0.248	0.0155	0.0037	50

Costs for buildings, infrastructure and crops

The average farm size has been estimated at 5.06 ha. The value of the buildings is estimated at 797,000 €/farm. The area of buildings and yards is estimated at 1,210 m^2/farm. The area of greenhouses is about 5.40 ha and the value is estimated at 1,215,000 €/ha. The area of buildings including the greenhouse area is 1,981 m^2/farm and the value of buildings is 890,000 €/farm. For facilities such as shops, factories, schools 20% of the value of the farms has been taken.

The infrastructure in the rural area is mainly composed of roads. In order to compute the cost of infrastructure, the density of paved roads is, based on a width of 5 m, determined at 121.5 m^2/ha (Figure 7.16). The cost of roads is estimated at 32.75 €/m^2.

Investment costs for grass and time schedule have been given in Figure 7.6 and Table 7.10.

Cost of damage to crops, buildings and infrastructure for the peat polder

The method of calculation and the parameters used have been described in Chapter 6 and investment costs for grass and time schedule are the same as for the clay polder. Damage to buildings and infrastructure is assumed to be a percentage of the value in relation to the groundwater tables as given in Table 7.11.

7.2.6 Calibration of the parameters

Calibration of the parameters of model for the rural area of the peat polder involved the determination of the hydrological parameters, which have been explained in Chapter 6 as follows:

- maximum interception in mm E_{max};
- potential evapotranspiration from layer I in mm/m V_{ep};
- parameters k and n in the discharge formula;
- maximum moistening rate in layer I in mm/m/hour V_{max1};
- maximum moistening rate in the columns of layer II in mm/m/hour V_{max2}.

Sensitivity of the parameters for the peat polder

The optimal value of the maximum interception has been found to be 2.64 mm. In reality the interception may change throughout the year due to differences in the grass cover.

The upper limit for potential evapotranspiration has been influenced by the water losses from the soil through evapotranspiration. The calibrated parameter value for this case was the average value from Table 7.31, which is 65.4 mm.

As a result of the calibration for 2002, the values for k and n are taken as 0.0646 and 2.26 respectively for further calculation.

The calibrated values for the maximum moistening rate in layer I and layer II were found to be 1.44 and 1.27 respectively.

Table 7.31 Calibrated parameters of the hydrological model for the peat polder

Year	Maximum interception in mm	Upper limit for potential evapotranspiration in mm	Parameter discharge model		Maximum moistening rate in mm/m/hour		Goal function	Year discharge computed in mm
			k	n	Layer I	Layer II		
2002	2.64	65.4	0.0646	2.26	1.44	1.27	623	552

7.2.7 Model validation with fixed calibrated parameters

The model has been validated with the data from the calibrated parameters in Table 7.31 for 2002 as shown in Figures 7.18 to 7.20.

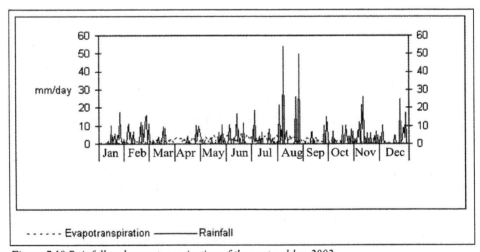

Figure 7.18 Rainfall and evapotranspiration of the peat polder, 2002

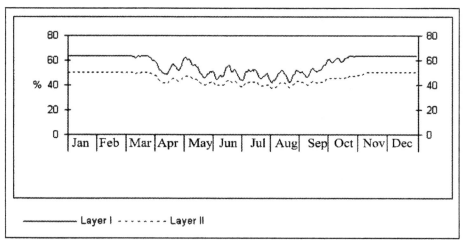

Figure 7.19 Computed soil moisture in layers I and II in the peat polder, 2002

In Figure 7.19 the soil moisture during winter is more or less at field capacity. The soil moisture content is gradually decreasing during spring. The lowest soil moisture content occurs at late June due to low rainfall and high evapotranspiration (Figures 7.18 and 7.19).

There is a short lagtime between rainfall and discharge during the winter, while the soil is more or less at field capacity. On the contrary in the summer when the soil has less moisture there is a certain lagtime between rainfall and discharge, which can be described by the fact that the rainfall has to refill the soil storage before the excess water discharges to the drain (Figures 7.18, 7.19 and 7.20).

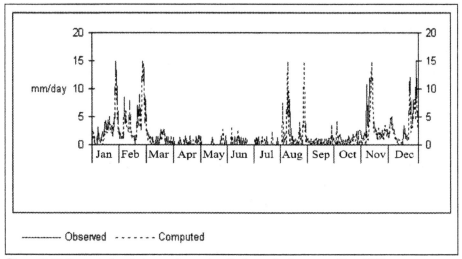

Figure 7.20 Computed and observed pumping discharge in the peat polder, 2002

The total computed annual discharge out of the area is 552 mm, while the total measured annual discharge is 644 mm for 2002 (Table 7.32). The model efficiency, which has been calculated by the observed and computed pumping discharge, is 0.719.

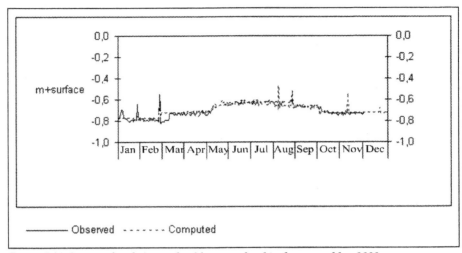

Figure 7.21 Computed and observed polder water level in the peat polder, 2002

In Figure 7.21 the polder water level during the summer period is higher than during the winter period about 0.10 - 0.15 m, because the chance of damage due to a too high water level during winter is higher and also because during summer a shallow polder water level can be of benefit to the crops due to the higher capillary rise.

7.2.8 Analysis with the optimization model

Hydrological analysis

The hydrological analysis with the validated model for the peat polder has resulted in the data as given in Table 7.32.

Table 7.32 Review of annual data in mm for the peat polder

Year	Rainfall in mm	Evaporation in mm	Pumping discharge in mm		Capillary rise in mm	
			computed	observed	II → I	III → II
2002	969	582	552	644	415	568

Economic computation

The results and review computations for optimal conditions are shown in Tables 7.33 to 7.36.

Table 7.33 shows that the value of buildings is about 95.8% of the total value in the peat polder. Therefore the damage to buildings may be the most important in the peat polder. In fact the buildings and infrastructure in this area were constructed in a high part of the area or at some landfill.

Table 7.33 Review of the total values of the grass, buildings and infrastructure in the peat polder at optimal water management

Item	Area in ha	Highest value		Lowest value	
		in € per ha	in € * 10^6	in € per ha	in € * 10^6
Buildings	15	5,394,000	83.34	5,394,000	83.34
Infrastructure	9	328,000	3.02	328,000	3.02
Grass	312	2,130	0.67	600	0.19
Total	336		87.03		86.55

Table 7.34 Optimization analysis of the main components of the water management system in the peat polder

Polder water level in m-s.	Open water in %	Pumping capacity in mm/day	Distance between open field drains in m	Open field drain depth in m-s.	Costs in € * 10^6	Damage in € * 10^6	Value of the goal function
1.35	1.9	4.9	26.50	1.15	0.166	0.008	174,000

In Table 7.34 the simulated the optimal main components of the water management system are shown. The optimal open field drain depth is 1.15 m-surface. However, at this level the subsidence can be high due to oxidation of peat. Therefore the relation of the open field drain depth and damage due to subsidence will also have been taken into account. However, during the summer period the polder water level can be kept higher to prevent drying out of soil and subsidence. In a peat polder open field drains can be used to supply water during a dry period when the water level is kept high. The open field drain depth and the distance between the open field drains are important factors for damage in the field. The optimal percentage of open water is calculated based on the least required area according to the existing conditions, such as parcel width and length, dimension and slope of the drains, distance between sub-main drains and length of the main drains and geometry.

Table 7.35 Total costs for the water management system for the peat polder at optimal water management

Item	Construction costs in € * 10^6	Annual maintenance costs in € * 10^6	Annual equivalent costs in € * 10^6
Field drainage	0.36	0.020	0.04
Open water	0.90	0.054	0.11
Pumping	0.08	0.007	0.01
Total	1.34	0.081	0.16

Table 7.35 shows that the overall annual maintenance costs in the peat polder are about 6.1% of the construction costs, while the overall annual equivalent costs are about 12.4%. The costs for open water constitute the highest part of the total construction costs, which are about 67.2%. Pumping costs concern the smallest part, which are about 5.8%. The low head of pumping may cause this.

Table 7.36 Specification of the average annual damage and yield reduction for the peat polder at optimal water management

Item	High groundwater tables		Too late sowing		Too late harvesting	
	in €	in % of the value	in €	in % of the value	in €	in % of the value
Buildings	0	0.00	0	0.0	0	0.0
Infrastructure	340	0.01	0	0.0	0	0.0
Grass	0	0.00	5,710	0.9	1,450	0.2
Total	340	0.00	5,710	0.9	1,450	0.2

In Table 7.36 the simulated average annual damage costs and average annual damage in percentage of the total value are shown. The main damage may be caused by too late sowing, and a very small damage is due to a too high groundwater table. The damage due to a too high groundwater table is low because of the deep open field drain depth.

Economical analysis for the water management system at the present situation

In practice the control criteria for the polder water level are different throughout the year. In this simulation the polder water level in winter was used because damage may occur in this period. The general values of the main components of the water management system of the peat polder may be approximated as follows:

- polder water level   = 0.78 m-surface;
- open water area   = 4.4%;
- open field drain depth   = 0.35 m-surface;
- distance between the open field drains   = 10.0 m;
- pumping capacity   = 14.9 mm/day.

The results and review computations for practical conditions are shown in Tables 7.37 to 7.40.

Table 7.37 Review of the total values of the grass, buildings and infrastructure in the peat polder at present water management

Item	Area in ha	Highest value		Lowest value	
		in € per ha	in € * 10^6	in € per ha	in € * 10^6
Buildings	15	5,394,000	83.34	5,394,000	83.34
Infrastructure	9	328,000	3.04	328,000	3.04
Grass	315	2,130	0.67	600	0.19
Total	339		87.05		86.57

Table 7.37 shows that the value of buildings is about 95.8% of the total value in the peat polder. Therefore the damage to buildings may be the most important in the peat polder.

Table 7.38 shows that the damage is larger than in Table 7.34 and also that the costs are higher than in Table 7.34. Damage is higher due to a too high groundwater table because of the shallower open field drain depth and the costs are higher due to a larger open water area, a smaller distance between the open field drains and a higher pumping capacity.

Table 7.38 Optimization analysis of the main components of the water management system in the peat polder at present water management

Polder water level in m-s.	Open water in %	Pumping capacity in mm/day	Distance between open field drains in m	Open field drain depth in m-s.	Costs in € * 10^6	Damage in € * 10^6	Value of the goal function
0.78	4.4	14.9	10.00	0.35	0.273	3.840	4,114,000

Table 7.39 Total costs for the water management system in the peat polder at present water management

Item	Construction costs in € * 10^6	Annual maintenance costs in € * 10^6	Annual equivalent costs in € * 10^6
Field drainage	0.223	0.059	0.074
Open water	1.448	0.085	0.171
Pumping	0.303	0.016	0.032
Total	1.974	0.160	0.277

Table 7.39 shows that the overall annual maintenance costs at present are about 8.1% of the construction costs, while the overall annual equivalent costs are about 14.0%. The costs for open water constitute the highest amount of the total construction costs, which is about 73.3%. The pumping costs concern the smallest part, which are about 15.3%. The costs for the pumping station and for open water are higher than under optimal conditions (Table 7.35), but the costs for field drainage are lower because of the deeper open field drains at optimal conditions.

Table 7.40 Specification of average annual damage and yield reduction in the peat polder at present water management

Item	High groundwater tables		Too late sowing		Too late harvesting	
	in €	in % of the value	in €	in % of the value	in €	in % of the value
Buildings	2,888,000	3.5	0	0.0	0.00	0.0
Infrastructure	131,000	4.3	0	0.0	0.00	0.0
Grass	301,000	44.9	1,550	0.2	29,160	4.3
Total	3,320,000	3.8	1,550	0.2	29,160	4.3

In Table 7.40 the damage due to a too high groundwater table is higher compared to the optimal conditions (Table 7.36), because of the higher probability of a too high groundwater table due to a shallower open field drain depth. Too late harvesting more often occurs than too late sowing in the present day practice, but when the open field drain depth increases the damage due to too late harvesting decreases significantly (Table 7.36).

Table 7.41 Comparison between the optimal and present values of the main components of the water management system in the peat polder

Item	Polder water level in m-s.	Open water in %	Pumping capacity in mm/day	Distance between open field drains in m	Open field drain depth in m-s.	Costs in € * 10⁶	Damage in € * 10⁶	Value of the goal function
Optimal	1.35	1.9	4.9	26.50	1.15	0.166	0.008	174,000
Present	0.78	4.4	14.9	10.00	0.35	0.273	3.840	4,114,000
% difference	+73	-56	-67	+165	+229	-64	-100	-96
Fixed depth	1.90	3.3	18.0	2.62	0.35	0.445	1.952	2,398,000

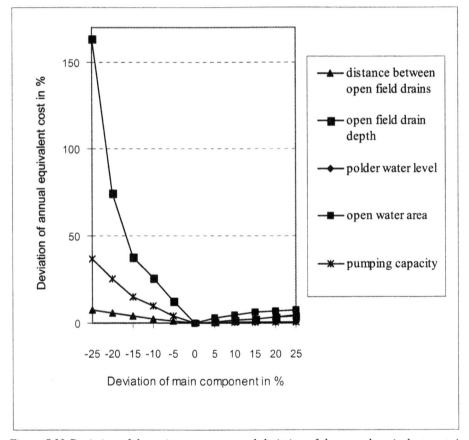

Figure 7.22 Deviation of the main components and deviation of the annual equivalent costs in percentage in the peat polder

In Figure 7.22 the influence of deviations from the simulated optimal values is shown. It shows that the open field drain depth has most influence on the annual equivalent costs. If the open field drain depth is 25% shallower compared to the optimal value the annual equivalent costs will increase approximately 164%, because of the increased damage due to a too high groundwater table. While a deeper open field drain depth will not result in a substantial increase of the annual equivalent costs, because it only result in an increase in the construction costs. A reduction of 25% in the pumping capacity results in an increase of the annual equivalent costs with approximately 37%.

The lower the pumping capacity, the higher the probability to have a high water level and therefore the higher the chance that the area will be subject to a too high groundwater table. Consequently, the higher is the chance to have too wet conditions for fertilizing and harvesting. A 25% smaller the distance between the open field drains results in an increase of the annual equivalent costs with 7% because the construction costs for the open field drains increase. A 25% increase in the open water area results in an increase of the annual equivalent costs with 7% because the construction costs for open water increase. A deeper polder water level compared to the optimal conditions has not so much influence on the annual equivalent costs, may be while at optimal conditions the depth is already enough.

Discussion

Based on the results as shown above the following points can be discussed.

From Table 7.41 it can be concluded that the open field drain depth may be increased but further study is recommended, the pumping capacity at present conditions may be more than enough.

In Figure 7.22 the open field drain depth has most influence on the annual equivalent costs, hence a lower open field drain depth is suggested to reduce the chance of damage to the grass. However, it is essential to keep a high groundwater table in the peat to prevent an increase in subsidence.

The polder water level during the winter period can be kept lower to reduce the probability of submergence of the soil during this period. Thereafter the chance of too wet conditions for fertilizing and harvesting is reduced. During the summer period the polder water level can be kept high to create more capillary rise in the grassland, which can compensate the soil moisture deficit during this period.

In the simulation with fixed depth (Table 7.41) it was found that the distance between the open field drains is very small to obtain a fast drain discharge in order to get the least damage. While the pumping capacity will be increased because of the fast discharge from the drainage system also a larger open water area will be required and the polder water level needs to be deeper than under optimal conditions. In this simulation it was found that the polder water level could be lower than at present, which is indicated in Table 7.41. This will result in a lower damage, but damage to bank slopes of the main canal may occur. Therefore the depth depends on soil properties. However, a lower water level before heavy rainfall to a level, which will not be harmful to the bank slope stability of the main drain, can be applied.

From the maintenance point of view the level and the cross section of the open field drains in a peat polder should be maintained in a good condition because they have a large influence on the damage to the crops. The ground level would have to be checked regularly to follow the subsidence. This may result in changes and redesign of the main components of the water management system.

7.3 The urban polder Hoge and Lage Abtswoudse polder

The Hoge and Lage Abtswoudse polder belongs to Delft municipality.

7.3.1 Physical conditions

Hoge Abtswoudse polder

The area of the Hoge Abtswoudse polder is 216.4 ha (Figure 7.23). The area of open water is 8.37 ha. Therefore the open water covers about 3.8% of the total area. Total length of the drain along the dike is 790 m.

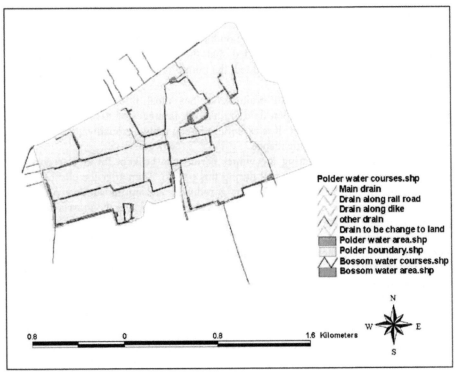

Figure 7.23 Layout of the Hoge Abtswoudse polder

During the winter the water level is kept at 0.00 m-NAP in the western part and at 0.30 m-NAP in the southern part of the polder, while the water level in the main drain is kept at 0.35 m-NAP.

Percolation is between 0.00 and 0.25 mm/day. The C value of the soil is varying from 5,000 to 10,000 day (Staring Centrum, 1976).

The total area of houses is 24.43 ha, which is composed of buildings and houses 21.80 ha, high buildings 2.62 ha and storage tanks (closed) 0.01 ha (Topografische Dienst, 2003).

Lage Abtswoudse polder

The total area of the Lage Abtswoudse polder is about 828.3 ha which is composed of 496.7 ha urban area and 331.6 ha rural area (Figure 7.24).

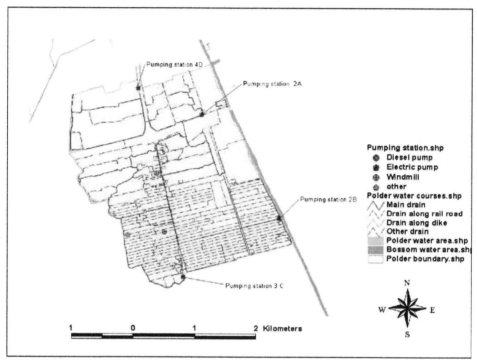

Figure 7.24 Layout of the Lage Abtswoudse polder

There are 3 pumping stations in this area. Pumping station 2A is situated at the Voorhof and there are 2 electrical pumps with a maximum capacity of 0.663 m^3/s. Pumping station 2B is situated at Schieweg 160 and there is one electrical pump with a maximum capacity of 0.533 m^3/s. The location of the two pumping stations is at the eastern side of the polder to pump excess water to the collection and transport system. Pumping station 4D is situated inside the polder area in the northern part. The function of this pumping station is to pump water from the Hoge Abtswoudse polder into the main drain of the polder when the water level in the drain is low. There are no data available for the pumping capacity of the pumping stations 3C and 4D.

The average field drain width at 0.30 m-NAP is 1.54 m. The water level during winter in most of the field drains is kept at 0.75 to 0.85 m-NAP and 0.85 to 1.00 m-NAP for the open water in the rural area.

The area of open water in the urban area, including the main drain is 18.93 ha and the total water area, based on the drain width at the water level in winter is about 25.33 ha or 5.1% of the urban area. Total length of main drains in the urban area is 24,555 m, of the drains along the dike 2,641 m, of the drains along the railway 2,117 m and of the other drains 51,056 m. This study will focus on the urban area, which is drained by the pumping station Voorhof (2A). This area consists of the total area of the Hoge Abtswoudse polder and the urban part of the Lage Abtswoudse polder.

The total area of houses in the Lage Abtswoudse polder that is drained by the pumping station Voorhof is 62.35 ha, which is composed of: buildings and houses 60.56 ha and high buildings 1.79 ha (Topografische Dienst, 2003).

The percentages of water, paved area and unpaved area for the Hoge and Lage Abtswoudse polder as far as drained by the pumping station Voorhof are shown in Table 7.42.

Table 7.42 Percentage of paved and unpaved area that is drained by the pumping station Voorhof (De Ron and Van der Werf, 2003)

Item	Hoge Abtswoudse polder	Lage Abtswoudse polder
Total area in ha	216.4	496.7
% unpaved	49	59
% paved	47	36
% water	4	5

7.3.2 Present situation of operation and maintenance

In the Hoge Abtswoudse polder the water level is controlled by weirs. The excess water discharges to the water management system in the Lage Abtswoudse polder. The total width of these weirs is 3.52 m and their depth is 1.5 m. The present situation of operation of the Hoge and Lage Abtswoudse polder is shown in Tables 7.43 and 7.44.

Table 7.43 Present situation of the operation in the urban polder (De Ron and Van der Werf, 2003)

Name	Polder water level in m+NAP	Lowest ground surface in m+NAP	Lowest sill level in m+NAP	Average ground surface in m+NAP	Allowable water level rise in m
Hoge Abtswoudse polder	-1.5	-1.35	-1.12	-1.35	0.15
Lage Abtswoudse polder	-2.7	-2.15	-	-2.15	0.55

Table 7.44 Present situation of the polder water level related to the ground surface in the urban polder

Name	Polder water level in m-surface	Lowest sill level in m-surface	Allowable water level rise in m
Hoge Abtswoudse polder	0.15	0.23	0.15
Lage Abtswoudse polder	0.55	-	0.55

In the Hoge Abtswoudse polder the allowable level increase is according to the current situation only 0.15 m, as a result of which too little storage is available (De Ron and Van der Werf, 2003)

The drainage works standards for the urban area have been determined by the Principal Water-board of Delfland (2000). The mentioned standards for paved and unpaved area have been given in the Table 7.45 and evaluation according to this standard is shown in Table 7.46. The pumping capacity in this area is as shown in Table 7.47.

Table 7.45 Drainage work requirement standards of the Principal Water-board of Delfland (2000)

	Paved area	Unpaved area
Requirement	20 m³/min/100 ha (28.8 mm/d)	10 m³/min/100 ha (14.4 mm/d)

Table 7.46 Evaluation of pumping capacity according to the standards of the Principal Water-board of Delfland (2000)

Pumping station	Drained area	Pumping capacity in m³/min	Total paved area in ha	Total unpaved area in ha	Maximum capacity according to standard in m³/min
Voorhof	Hoge and Lage Abtswoudse polder	41.7	280	398	95.8 (20.2 mm/d)

Table 7.47 Pump operation in the Lage Abtswoudse polder for modelling of the water system

Location		Capacity in m³/min	Level to start in m+NAP	Level to stop in m+NAP
Voorhof	pump 1	21	-2.68	-2.74
	pump 2	21	-2.63	-2.70

7.3.3 Hydrological and water management data

The wastewater disposal systems, which discharge into wastewater treatment plants, have very little influence on the discharge into the urban canals. Therefore this amount of water has been neglected.

Hydrological data

The hydrological data used in the model were taken from Delfland, Hoofddorp and De Bilt as follows:
- the data for the model calibration:
  - daily rainfall in mm at pumping station Voorhof for 2003;
  - daily pumping discharge in mm of the pumping station Voorhof for 2003;
  - daily average water level at the pumping station Voorhof for 2003;
  - daily open water evaporation according to Makkink in mm, at De Bilt for 2003;
- the data for validation and computation:
  - daily rainfall in mm at Hoofddorp for 1960 to 2001;
  - daily open water evaporation according to Makkink in mm, De Bilt, for 1960 to 2001.

Parameter values for the simulation of the unpaved area according to the formula of Horton:

$$f_t = f_c + (f_0 - f_c) * e^{(-K_f * t)} \tag{7.6}$$

where:
f_t, f_0 and f_c = infiltration capacity of time step t, 0 and infinite
f_0 = 14.0 mm/hour
f_c = 2.0 mm/hour
K_f = 1.7 hour^{-1}

The storage in depressions is set at 3.0 mm, which is a normal value for paved areas in flat urban areas.

The unpaved area is provided with a subsurface drainage system as in the rural area. It is assumed that the discharges from this area behave in the same manner as in the rural area.

As said the study concerns the total area of the Hoge Abtswoudse polder and the urban part of the Lage Abtswoudse polder. The area drained by the Voorhof pumping station for the Hoge Abtswoudse polder is 216.4 ha, which has an open water area of 8.4 ha. For the Lage Abtswoudse polder it is 496.7 ha, which has an open water of 25.2 ha at polder water level. The total area is 713.1 ha, which contains an open water area of 33.6 ha or 4.7%.

The total area of houses in the Hoge and Lage Abtswoudse polder that is drained by pumping station Voorhof is 86.78 ha, which is composed of buildings and houses 82.36 ha, high buildings 4.41 ha and storage tanks (closed) 0.01 ha.

The discharge from the paved area is directly conveyed by the sewer system while the unpaved area usually discharges into the subsurface drainpipes.

The levels simulation based on average conditions are as follows. The level of houses is set at 0.50 m+surface. The level of square and path is set at 0.40 m+surface. The quarter road is at 0.30 m+surface and main road is at 0.00 m+surface.

The water depth in the canal is 1.20 m. Canal profile:

Under water slope   = 1:2.0;
Bank slope   = 1:2.0;
Timbering   = 0.30 m.

The paved area may comprise the following:
-    housing area;
-    parking lot;
-    commercial centre;
-    industrial area;
-    roads.

The paved area is composed of flat or slope roof houses, asphalt roads, brick roads, etc. The composition of the paved area, which is used in the hydrological analysis, is given in Table 7.48.

Squares, paths and quarter roads are classified according to their capacity to infiltrate the rainfall. The composition of the different paved surfaces is described in Table 7.49.

Table 7.48 Estimated the composition of the paved areas in the Hoge Abtswoudse polder and in the Lage Abtswoudse polder that is drained by the pumping station Voorhof (Delft municipality, 2002)

Paved area	Hoge Abtswoudse polder	Lage Abtswoudse polder
	Area in ha	Area in ha
Roofs:		
- sloping	6	18
- flat	18	58
Squares and paths	58	60
Quarter roads	24	43
Main roads	-	11
Green areas and gardens	102	293
Water	8	25
Total	216	497

Table 7.49 Percentage of paved surface areas for different types of roads in the urban polder

Type of road	Tiles in %	Bricks in %	Asphalt in %
Squares and paths	50.0	50.0	0.0
Quarter roads	0.0	20.0	80.0
Main roads	0.0	10.0	90.0

Note: approximated by field inspection

7.3.4 Economical data

Data of paved areas for the economic computation

For the economic computation for urban areas the percentage of different paved areas are taken into account as shown in Table 7.49.

Data for construction and maintenance cost of the water management system

The cost of the water management system may be divided into the following:
- subsurface drainage;
- sewer system;
- open canal system.

Data of construction, operation and maintenance cost of the water management system are given in Tables 7.50 to 7.55.

Table 7.50 Costs for subsurface drainage in an urban polder (Rijkswaterstaat, 2004)

Item	Cost in €/m
Construction	4.49
Maintenance	0.39

Note: Lifetime for the subsurface drains is taken as 30 years for the economic analysis.

Table 7.51 Construction costs for sewers in an urban polder (Rijkswaterstaat, 2004)

Diameter in m	Cost in €/m
0.30	147
0.40	190
0.50	261
0.60	334
0.70	417
0.80	512

Construction cost of gutters is taken as 1,519 €/piece (Delft municipality, 2003). The cleaning of sewers consists of flushing, which is estimated at one time in two years. Annual cost is taken as 0.87 €/year/m (Delft municipality, 2003). The lifetime for the sewers is taken as 30 years for the economic analysis. The canal profile is as follows: under water slope is 1:2.0, bank slope is 1:2.0 and the height of timbering is 0.50 m.

Table 7.52 Construction costs for urban canals (Rijkswaterstaat, 2004)

Item	Cost in €/m
Digging	4.81 €/m^3
Timbering	36.30 €/m
Fence	2.95 €/m

Table 7.53 Annual maintenance costs for the canals in an urban polder (Rijkswaterstaat, 2004)

Item	Unit cost
Cleaning	1.22 €/m
Mowing	0.64 €/m
Timbering	2.20 €/m

Table 7.54 Lifetime for the economic analysis of the urban polder

Item	Lifetime in years
Canals	100
Timbering	25
Weir	50

The interest rate has been set at 5.0% for the determination of the present or capital value in the urban polder.

The pumping head is about 3.8 m based on the water level inside which is 2.70 m-NAP and the outside water level in the collection and transport system, which is 0.40 m-NAP and head loss, which is taken as 1.50 m. The pumping capacity is 8.5 mm/day. The costs for pumping are shown in Table 7.55.

Table 7.55 Pumping costs for pumping station Voorhof

Installation in million €	Operation in million €/year	Maintenance in million €/year	Lifetime in years
0.400	0.012	0.003	50

Data for the damage computation are given in Tables 7.56 to 7.61.

Table 7.56 Data on values of buildings for the urban polder (data obtained from a real estate company in Delft, 2003)

Item	Number/ha	Average value in €/number	Furniture value in %
Houses with one floor	11	158,000	50
Houses with two floors	18	290,000	50
Shops and offices	5	1,280,000	80
Industrial buildings	0.05	2,130,000	70

Table 7.57 Data on the value of infrastructure in an urban polder (Rijkswaterstaat, 2004)

Type of Infrastructure	Value of infrastructure	Unit
Squares and paths	19.42	€/m^2
Quarter roads	24.26	€/m^2
Main roads	32.22	€/m^2
Public facilities	11,372.00	€/ha
Utilities	11,372.00	€/ha

Table 7.58 Damage functions for buildings in percentage of the value in an urban polder (adapted from US Army Corps of engineers, 1996\2000)

Level of water in m+surface	Houses one floor	Houses two floors	Shops and offices	Industrial buildings
-0.40	1.25	2.50	1.25	2.50
-0.20	7.00	5.50	2.75	5.50
0.00	13.40	9.30	4.65	9.30
0.30	23.30	15.20	7.60	15.20
0.60	32.10	20.90	10.45	20.90

Table 7.59 Damage functions for infrastructure in percentage of the value in the urban polder (adapted from US Army Corps of engineers, 1996\2000)

Level of water in m+surface	Infrastructure			Public facilities	Utilities
	Squares and paths	Quarter roads	Main roads		
-0.60	0.0	1.6	1.6	0.0	1.6
-0.40	0.0	3.2	3.2	0.0	3.2
-0.20	1.6	4.0	4.8	1.6	4.8
-0.00	4.8	4.8	8.0	4.8	8.0
0.20	8.0	8.0	16.0	8.0	16.0

Table 7.60 Damage functions due to exceedence of the water level in the urban canals in percentage of the construction cost of canals and main roads (Schultz, 1992)

Water level	Canals	Main roads
Top of protection	0.5	0.0
0.30 m-surface of main roads	1.0	1.0
Surface of main roads	2.0	3.0
0.30m -land surface	3.0	5.0

Table 7.61 Values for the indirect damages in percentage of the direct damage in the urban polder (Schultz, 1992)

Type of infrastructure	Indirect damage in % of direct damage
Living areas	15
Shops and office centre	35
Industrial areas	45
Utilities	34
Public facilities	10

7.3.5 Calibration of the parameters

The calibration of the parameters of model for the urban area involved the following parameters:
- storage in depressions;
- runoff coefficient;
- parameters k_1 and n_1 for the transformation of rainfall into sewer inflow;
- parameters k_2 and n_2 for the transformation of sewer inflow into sewer discharge.

The calibration of these parameters was done with the Rosenbrock method as described before. The calibration in this area was done for the year 2003. The calibrated parameters, which were obtained for 2003, are shown in Table 7.62.

Table 7.62 Calibrated parameters of the hydrological model for the Hoge and Lage Abtswoudse
         polder

Storage in mm	Runoff coefficient	Rainfall > sewer inflow		Sewer inflow > sewer discharge		Pumping discharge in mm		Goal function
		k_1	n_1	k_2	n_2	obs.	comp.	
0.82	0.59	2.34	0.175	2.36	0.220	208	211	139.23

Discharge to the canal in the urban area comes from both storm sewers and subsurface drains. It has been approximated that there is no lagtime between rainfall and storm sewer discharge and that the subsurface drains behave in the same manner as in the rural area. Due to the fact that the discharge from the Hoge Abtswoudse polder to the Lage Abtswoudse polder takes place over the weir, the model simulation for calibration and optimization considers the discharge of the Hoge Abtswoudse polder over the weir as given input to the Lage Abtswoudse polder.

Sensitivity in the paved area

In flat urban areas, there is a very small storage in depressions. The main depression storage is only on flat roofs, at the streets and the static storage in the sewer. Therefore the storage in depressions in the urban area has little effect on the optimization. During the calibration a value of 0.82 mm has been found for storage in depressions.

The runoff coefficient is the factor, which transforms the rainfall into runoff. It means that this coefficient includes all losses from rainfall to runoff. The losses depend on the kind and the conditions of the surface. In the model the runoff coefficient is applied only for the paved surface in the urban area. The main three losses in the urban area are as follows:

- *evaporation loss.* For a warm and wet surface condition with water stored on the surface the evaporation loss can be considered as open water evaporation;
- *infiltration loss.* The infiltration may be large through brick and tile roads while asphalt and concrete are almost impervious;
- *initial losses,* Initial losses are composed of wetting loss and depression loss. The depression loss is large on a flat roof.

All kinds of losses above are summarized by the runoff coefficient. The higher the coefficient the smaller is the loss. The value of 0.59 from the calibration process has been used as a representative runoff coefficient to simulate the transformation of rainfall into runoff.

Transformation of runoff into sewer inflow and from inflow to sewer discharge in the model is being assumed to behave as a non-linear function as shown below.

$$Q_{ri}(\Delta t) = \left(\frac{V_{opp}(t + \Delta t)}{k_1} \right)^{\frac{1}{n_1}} \text{ Transformation of runoff into sewer inflow} \qquad (7.7)$$

And

$$Q_{ra}(\Delta t) = \left(\frac{V_{ri}(t + \Delta t)}{k_2} \right)^{\frac{1}{n_2}}$$ Transformation of sewer inflow into sewer discharge

(7.8)

The values of the transformation parameters as found during the calibration are shown in Table 7.62.

The model was run for time steps of 1 hour by distributing daily rainfall data equally over the time step. Owing to the fact that during the computation of sewer inflow and sewer discharge no lagtime is taken into account and the input of hydrological data is 1 hour, there may be differences when the observed values and computed values are compared. However, the model efficiency at this time step is still 0.881.

7.3.6 Model validation with fixed calibrated parameters

The model has been validated with the data from the calibrated parameters in Table 7.62 for 2003 as shown in Figures 7.25 to 7.28.

In Figures 7.25 and 7.26 there is small lagtime between rainfall and pumping discharge because the time step of 1 hour is large for the concerned process. During the summer period pumping is also required due to seepage to the polder area, especially for the Lage Abtswoudse polder. Around August there was no rainfall but still a high pumping discharge due to seepage to the Lage Abtswoudse polder and also due to percolation from the Hoge Abtswoudse polder. It is estimated through the calibration that during winter and spring (November to May) seepage and percolation are between 0.2 to 0.3 mm/day and during summer and autumn (June to October) at about 0.6 mm/day. The maximum used pumping capacity for this year was around 5 mm/day or only 59% of the maximum capacity.

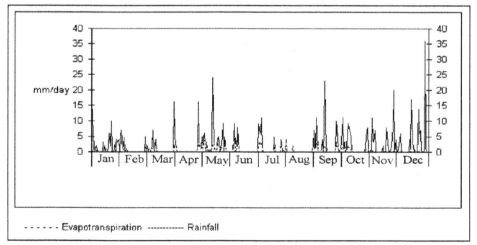

Figure 7.25 Rainfall and evaporation, 2003

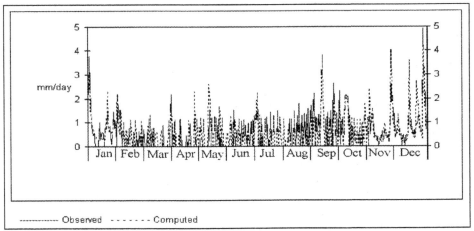

Figure 7.26 Computed and observed pumping discharge at pumping station Voorhof, 2003

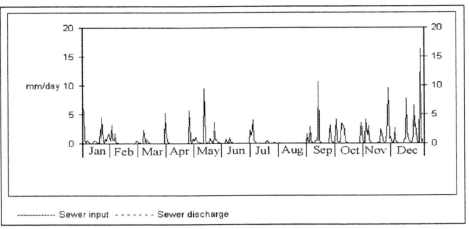

Figure 7.27 Sewer inflow and sewer discharge in the Hoge and Lage Abtswoudse polder, 2003

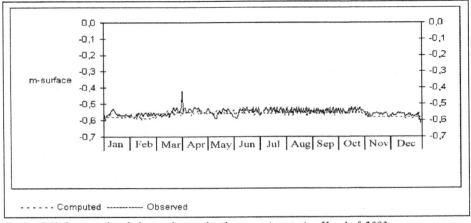

Figure 7.28 Computed and observed water level at pumping station Voorhof, 2003

In Figure 7.27 there is a small lag time between sewer inflow and sewer discharge because there is a small volume of water stored in the sewer pipe before discharge into the canal occurs.

In Figure 7.28 the water level in the polder during summer period is about 0.05 m higher than during winter, because the chance of a too high water level during winter is higher.

7.3.7 Analysis with the optimization model

Hydrological analysis

The hydrological analysis with the validated model for the urban polder has resulted in the data as given in Table 7.63.

Table 7.63 Review of annual data for the urban polder

Year	Rainfall in mm	Evaporation in mm	Infiltration in mm	Sewer discharge in mm	Pumping discharge in mm
2003	650	174	195	246	211

In Table 7.63 it can be seen that there is some difference between sewer discharge and pumping discharge due to the fact that discharge from the sewer is temporary stored in the canal system and evaporation from open water takes place there.

Economic computation

The economic computations are based on the values and functions as described above. The main components of the water management system in the urban area are shown in Table 7.64 and results are shown in Tables 7.65 to 7.67.

Table 7.64 Optimal values of the main components of the water management system in the urban area

Diameter sewer in m	Distance between canals in m	Canal water level in m-s.	Open water area in %	Pumping capacity in mm/day	Costs in € * 10⁶	Damage in € * 10⁶	Goal function
0.30	1,870	0.87	0.23	5.0	0.970	0.004	973,000

In Table 7.64 it can be seen that at optimal conditions the required area of open water is only 0.23%. The computed open water area may be so small because it is based on daily rainfall data, which were equally distributed over the time step. Hence, the flash floods were spread over at least one day. Moreover, constraints of layout and topography, which can affect this component, were not included in this study.

Table 7.65 shows that the cost of the sewers is the highest component in the urban area, which is about 89% of the total construction costs at optimal conditions. Operation and maintenance costs for canals are 54% of the total operation and maintenance costs due to the required dredging, bank slope mowing and the maintenance of the timbering.

Table 7.65 Costs for the drainage system in the urban area at optimal water management
         conditions

Item	Living areas $€ * 10^6$	Shopping- and office centres $€ * 10^6$	Industrial areas $€ * 10^6$	Total $€ * 10^6$
Drains				
- construction costs	0.596	0.260	0.347	1.230
- maintenance/year	0.052	0.023	0.030	0.107
- annual equiv. costs	0.090	0.040	0.053	0.187
Sewers				
- construction costs	6.195	2.789	3.671	12.656
- maintenance/year	0.036	0.016	0.021	0.073
- annual equiv. costs	0.363	0.163	0.215	0.742
Canals				
- construction costs				0.198
- maintenance/year				0.011
- annual equiv. costs				0.021
Pumping station				
- construction costs				0.165
- maintenance/year				0.010
- annual equiv. costs				0.019

Table 7.66 shows that the value of infrastructure and facilities is relatively small,
which is about 0.76% of the total value in the urban area. Therefore, damage may be
expected especially to the houses, shops, offices and the industrial area.

Table 7.66 Review of the total value of buildings and infrastructure in the urban area at optimal
         water management conditions

Item	Area in ha	Value including furniture in €/ha * 10^6	in € * 10^6
Houses with one floor	76	4.961	377
Houses with two floors	123	14.790	1,819
Shops and offices	87	23.040	2,004
Industrial buildings	116	0.797	92
Infrastructure	103	0.223	23
Public facilities	412	0.011	5
Public utilities	412	0.011	5
Total	412		4,325

Table 7.67 Specification of average annual damage in € in the urban area at optimal water
         management conditions

Item	High groundwater table	Water at the street	Exceedence of canal water level
Living areas			
- Houses with one floor	690	250	
- Houses with two floors	690	250	
Shops and offices	780	230	
Industrial areas	1,120	540	
Green areas	0		
Urban area			0
Total			4,550

Table 7.67 shows that at optimal conditions the damage due to water at the street was less than damage due to too high groundwater tables. The rainfall intensity is not high and there is no strong water level fluctuation in the canal due to flash floods in this simulation. The damage to houses with one floor and with two floors is equal because harmful effects occur only at the street, not at the houses while the level of houses is higher than the level of the streets. The damage due to too high groundwater tables and water at the street was high in the industrial area due to more indirect damage and less green area.

Economical analysis for a water management system at the present situation

The values of the main components of the water management system of the Hoge and Lage Abtswoudse polder at present conditions are as follows:
- diameter of the sewer   = 0.30 m;
- distance between canals   = 175 to 350 m;
- canal water level during the winter period = 0.55 m-surface;
- open water area   = 5.0%;
- pumping capacity   = 8.5 mm/day.

The results and review computations for present conditions are shown in Tables 7.68 to 7.71.

Table 7.68 Analysis with optimization of the values of the main components of the water management system at the present conditions in the urban area

Diameter sewer in m	Distance between canals in m	Canal water level in m-s.	Open water area in %	Pumping capacity in mm/day	Costs in € * 10^6	Damage in € * 10^6	Goal function
0.30	262.5	0.55	5.0	8.5	1.186	0.024	1,210,000

Table 7.69 Costs for the drainage system in the urban area at present water management

Item	Living areas € * 10^6	Shopping- and office centres € * 10^6	Industrial areas € * 10^6	Total € * 10^6
Drains				
- construction costs	0.596	0.260	0.347	1.230
- maintenance/year	0.517	0.023	0.030	0.107
- annual equiv. costs	0.090	0.040	0.053	0.187
Sewers				
- construction costs	6.461	2.836	3.775	13.072
- maintenance/year	0.038	0.017	0.022	0.077
- annual equiv. costs	0.379	0.166	0.222	0.767
Canals				
- construction costs				2.296
- maintenance/year				0.074
- annual equiv. costs				0.195
Pumping station				
- construction costs				0.289
- maintenance/year				0.015
- annual equiv. costs				0.031

Table 7.69 shows that the cost of the sewers is the highest, which is about 77.5% of the total construction costs at present conditions. Operation and maintenance costs for the sewers are 28.4% of the total operation and maintenance costs, which are almost the same amount as for the operation and maintenance costs for canals.

Table 7.70 Review of the total value of buildings and infrastructure in the urban area at present water management conditions

Item	Area in ha	Value including furniture	
		in €/ha * 10^6	in € * 10^6
Houses with one floor	76	4.961	377
Houses with two floors	123	14.790	1,819
Shops and offices	87	23.040	2,004
Industrial buildings	116	0.798	93
Infrastructure	103	0.223	23
Public facilities	412	0.011	5
Public utilities	412	0.011	5
Total	412		4,325

Table 7.70 shows that the value of infrastructure and facilities is relatively small, which is about 0.75% of the total value in the urban area. Therefore main damage may be expected especially to the houses, shops, offices and the industrial area.

Table 7.71 shows that too high groundwater tables are much more harmful to the urban area than damage due to water at the street because the water level in the urban canal is rather high. The damage to houses with one floor and with two floors is equal because there is only a harmful effect of too high groundwater and water at the street only influences squares and paths, quarter roads and main roads in this area due to the lower level than of the houses.

In Table 7.72 a comparison is shown between the simulated optimal water management conditions and at the present situation. From this table it can be derived that the sewer diameter in the present situation is equal to optimal conditions. The canal water level would have to be kept lower compared to the present situation, but it needs further study. The pumping capacity at present is enough to evacuate excess rainfall and seepage out of the area. It can be seen that the area of open water may be more than enough.

Table 7.71 Specification of the average annual damage in € in the urban area at present water management conditions

Item	High groundwater table	Water at the street	Exceedence of canal water level
Living areas			
- Houses with one floor	2,000	240	
- Houses with two floors	2,000	240	
Shops and offices	2,030	220	
Industrial areas	2,950	520	
Green areas	0		
Urban area			0
Total			10,200

Table 7.72 Comparison between the optimal and the present values of the main components of the water management system in the urban polder

Item	Diameter sewer in m	Distance between canals in m	Canal water level in m-s.	Open water area in %	Pumping capacity in mm/day	Costs in € * 10⁶	Damage in € * 10⁶	Goal function
Optimal	0.30	1,869	0.87	0.23	5.00	0.969	0.004	973,000
Present	0.30	263	0.55	5.00	8.5	1.186	0.024	1,210,000
% difference	-	+612	+58	-94	-41	-18	-83	-20

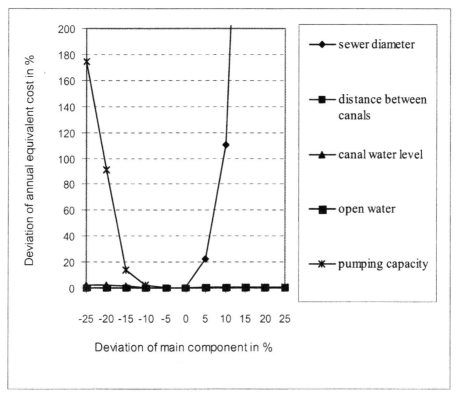

Figure 7.29 Deviation of the main components and deviation of annual equivalent costs in percentage in the urban polder

In Figure 7.29 the influence of deviations from the simulated optimal values is shown. It shows that the sewer diameter and the pumping capacity have most influence on the annual equivalent costs. If the sewer diameter is 25% increased in comparison with the optimal value the annual equivalent costs will increase approximately 600%, because the damage will increase due to a more rapid discharge to the canal systems and therefore a higher water level rise in the urban canals and also the cost of the sewers is high. A decrease in pumping capacity by 25% from the optimal value will increase the annual equivalent costs by approximately 175%, because damage due to too high canal water levels increases. Whereas an increase or decrease in the open water area and distance between the canals has no significant influence on the annual equivalent costs, because of construction costs at optimal conditions are not a large portion of the water management costs. The canal water level has not much influence on the annual equivalent costs, may be due to the fact that the fluctuation of the water level due to flash floods may have been underestimated in this simulation. Moreover, the pumping capacity at optimal conditions is high enough to control the water level fluctuation in the canal systems.

Discussion

Based on the results as shown above the following points can be discussed.

A shallow canal water level results in a higher chance to get more damage to the urban area due to a too high groundwater table.

At present the area of open water is higher than the optimal area and the pumping capacity is about 1.7 times the pumping capacity at optimal conditions. This may be due to the two pumps in this polder and that the pumps will work alternatively, for different parts of the polder. However, flash floods due to heavy rainfall were not included. The required pumping capacity according to the standards of Delfland is about 4 times the pumping capacity at optimal conditions. This value is high compared to optimal conditions, because this pumping capacity may also include a safety margin.

The canal water level under the present condition can be lower in order to get lower damage due to too high groundwater tables, but the side effects of lowering the water level, such as settlement, slope stability and water quality have to be further studied.

The distance between canals can be increased from the present situation and this parameter is related to the open water area and the storage volume in the sewer system before discharge occurs to the urban canals.

7.4 A rice polder under rainfed conditions

7.4.1 General

The experiments with respect to the rice polder under rainfed conditions and dry food crop conditions were done in 2003 and for the rice polder under irrigated conditions in 2004 at Samchuk Irrigation Water Use Research Station in Suphanburi, Thailand. Suphanburi province is located in the northwestern part of the Central Plain (Figure 7.30). The terrain of Suphanburi province consists mostly of low river plains, with small mountain ranges in the north and the west of the province. The southeastern part of the province is the very low plain of the Tha Chin river, where the paddy rice farming area is located. The experimental area was located at a distance of about 208 km to the northwest of Bangkok. In this station the dry food crops, rainfed rice and irrigated rice experiments were done in such a way that they represented more or less the polder enviroment. This area was chosen because it is near Bangkok where most polders are located in the central part of Thailand and there were the Royal Irrigation Department officers who could perform the experiments, collect data and take care of the operation and maintenance of the experiment instruments.

Aim of the experiment

The aim of the experiment was to find the response of ground/open water and discharge related to the cropping pattern, soil moisture conditions, hydrological conditions, drainage system and water management conditions in a rainfed rice polder.

Set-up of the instruments

The instruments were composed of the following:

- *rain gauge.* The Tipping bucket rain gauge with a freestanding receptacle was used for measuring precipitation;
- *evaporation pan (Class A).* The evaporation of water from this evaporation pan was recorded on a daily basis. The relationship between the pan evaporation and crop evapotranspiration has been used;
- *water level by float sensor.* The pulley and counterweight sensor was installed on a perforated PVC pipe with a diameter of 0.20 m. This pipe was covered with geo textile to prevent small particles to settle inside of the pipe. The pulley and counterweight sensor was in the box above the PVC pipe and protected by a cover box to against sun or rain;

Figure 7.30 Location of Suphanburi province (http://en.wikipedia.org/wiki/Suphanburi_province)

- *data logger.* A data logger is an instrument for recording and storing data, for which it has a data processor. So it can read and write data according to time, which can be manipulated by the user. It can be connected with more than one sensor at the same time. The data logger can interface to a microcomputer directly on site and also be connected to a telephone by a modem. The data can be downloaded to a computer by using the program 'Data Logger manager';
- *microcomputer.* A microcomputer CPU Pentium 3 was used for running the program and downloading data from the data logger;
- *pumping station.* The pumping station was constructed to pump the water out or in the field of the rainfed rice, irrigated rice and dry food crops area;
- *instrument box.* The instrument box was constructed to keep the data logger and back up power supply e.g. batteries for the data logger;

-    *small cottage.* A small cottage was used for staying of the security guard and the
     caretaker for the instruments and crops, including for the water management
     operation, such as operation of the pump and agricultural activities.

7.4.2 Physical conditions

The rainfed rice area in this experiment was located at a high part of the area (Figure
7.32), because in the rainfed conditions deep groundwater was required. In the area
previously vegetables were grown. The soil was almost dry throughout the year. The
groundwater table in the lower part is 1.0 to 2.0 m-surface. In the lower part irrigated
rice was cultivated and there was a lined canal for transporting water to the irrigated
rice. The land in the high part was prepared by puddling, which is a normal practice for
rice, in order to reduce the macro pores in the upper layer and consequently reduce the
permeability in the upper layer of the soil. In this way the land was suitable for growing
rice. The soil in this area is clay and cracks occur with a diameter of the columns of
about 0.50 m, width of the cracks during the dry period is 2.0 to 3.0 cm and the depth is
approximately 0.40 to 0.60 m. The groundwater table was about 1.30 m-surface.

7.4.3 Set-up of the experimental plot

Layout of the experimental area

Layout and location of the water level sensors in the experimental area are shown in
Figures 7.31 to 7.33.
    The total experimental area for rainfed rice was 213.85 m^2. The area where the crops
were grown was 153.3 m^2. In rainfed conditions it is assumed that only a drainage
system is available. The water management in this area was done on the principle that
the excess water, which was related to the maximum water level that rice can tolerate,
will be drained by pumping to the lateral canal (Figure 7.33). The KDML105 rice,
which is normally grown in rainfed areas, was used in this pilot area.

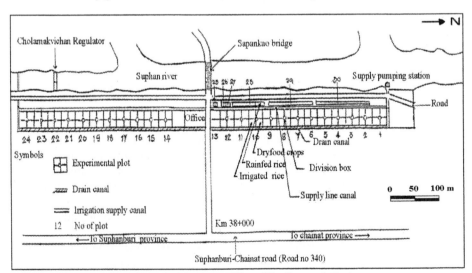

Figure 7.31 Layout plan of Samchuk Irrigation Water Use Research Station

Figure 7.32 Rainfed rice experimental area at Samchuk Irrigation Water Use Research Station, 2003

Monitoring and data recording for the rainfed rice experimental plot

The monitoring and data recording for rainfed rice was done from 1 January 2003 to 31 December 2003. The data were recorded as follows:
- hourly rainfall in mm;
- daily pan evaporation in mm;
- pumping discharge to the lateral canal;
- surface water level or groundwater table in the field;
- surface water level or groundwater table in the lateral canal;
- crops and agricultural practice schedule;
- crop growth and growth stage;
- investment costs and time schedule for agricultural practices;
- production.

The data mentioned above were used for calibration and verification of the model.

Cropping pattern and harvesting method

The cropping pattern for rainfed rice depends on the moisture conditions and growing method. For this experiment the rice was planted on 1st August 2003 and harvested on 3rd December 2003. The dry seed method (Table 7.76) was used for growing rice in this experiment.

Soil properties

The soil properties were as follows:
- soil is clay loam, which is approximately composed of sand 38%, silt 26% and clay 36%;
- the clay mineral analysis has shown that it is composed of Kaolinite 42.4%, Illite 48.3% and Montmorillonite 9.3%;
- average saturated permeability at 0.30 m-surface is 7.6 $*10^{-6}$ m/day and at 0.60 m-surface is 5.9 $*10^{-6}$ m/day.

The soil properties in the experimental plot, such as soil moisture tension, soil classification, porosity, linear shrinkage coefficient, dry bulk density and clay type were examined in the laboratory of the Royal Irrigation Department.

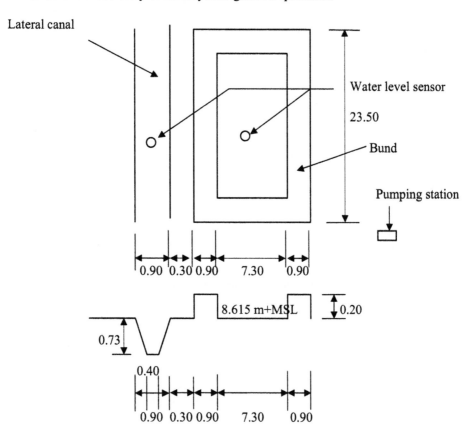

Note: All dimensions are in metre

Figure 7.33 Schematic layout and cross-section of the area for the rainfed rice experiment

Discussion on the observation results

The groundwater table in the experimental area during the dry season was 1.2 to 1.3 m-surface. The small amount of rainfall in the dry season had no effect on the groundwater table in the study area.

The clay in the soil has little Montmorillonite, which means that there is not much swelling or shrinking when the soil moisture increases or decreases. The irreversible process of shrinkage of clay soil, which is caused by Kaolinite and Illite dominated in this soil mass. This implied that the permeability of the soil changed in a specific range when it was wet or dry.

The permeability of the soil in the laboratory is low, but in fact the water can flow through the cracks. Therefore, the field permeability can be high.

The crop yield for rainfed rice was 2.2 ton/ha/crop season, which it is a quite common yield for rice under rainfed conditions.

7.4.4 Hydrological and water management data

For this case study the following hydrological and water management data have been used:
-      the data from 1 January 2003 to 31 December 2003 for the model calibration:
    •   class A pan daily evaporation in mm, Samchuk Irrigation Water Use Research Station;
    •   hourly rainfall in mm, Samchuk Irrigation Water Use Research Station;
    •   hourly pumping discharge to the drainage canal of the experimental plot;
    •   surface water level or groundwater table in the field;
    •   surface water level or groundwater table in the drainage canal;
-      the data for validation and computation:
    •   hourly rainfall in mm at Ladkrabang station, Bangkok, 1991 to 1999;
    •   class A pan evaporation, Bangkok Metropolitan, 1991 to 1999.

Characteristics of a typical rainfed rice polder

In this study it has been assumed that the typical polder has a total area of 10,000 ha. The soil storage was assumed to exist in the upper one metre, which consisted of 0.25 m of puddled layer (layer I) and 1.15 m of layer II. Only rice was grown in this area. The area of the plot was about 1.6 ha. The dimension of the plot was about 100 * 160 m and it has been assumed that there were six plots in one parcel. The lay out of the parcel is shown in Figure 7.34. The side slopes of the watercourses were:

Lateral drains   = 1:1.5;
Sub-main drains   = 1:3;
Main drains   = 1:3.

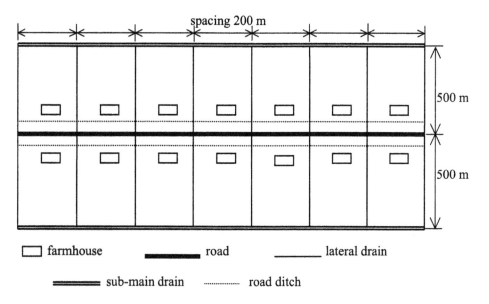

Figure 7.34 Schematic layout of parcels in the rainfed rice polder in Thailand

Soil moisture tension

The soil moisture tension computation in the model was based on the empirical relation as given by Equation 7.2. The constants a, b and c for the different soil profiles (Table 7.73) have been calculated and are shown in Tables 7.74 and 7.75 (see Appendix VI).

Table 7.73 Clay content in the soil profile under conditions of a rainfed rice polder at Samchuk Irrigation Water Use Research Station

Range from surface level in m	Typical soil type	Description	Clay content in %
0.00 to 0.15	puddled soil	Clay loam	38.0
0.15 to 1.30	subsoil	Clay loam	36.3
> 1.30	subsoil	Clay loam	32.4

Table 7.74 Soil moisture tension in pF and soil moisture content in percentage for a typical rainfed rice polder at Samchuk Irrigation Water Use Research Station

Layer \ pF	2.01	2.5	3.49	4.18
I	54.16	43.06	31.17	24.20
II	55.25	44.81	32.41	24.82
III	52.36	41.88	26.40	21.05

Note: Soil characteristics obtained by sampling from the experimental plot and analysed by the laboratory of the Royal Irrigation Department

Table 7.75 Constants and moisture contents for a typical rainfed rice polder at Samchuk Irrigation Water Use Research Station

Layer	Constants			Moisture content in %		
	a	b	c	Sat	FC	WP
I	5.200	0.798	-0.299	54.7	51.8	23.8
II	5.240	0.722	-0.255	55.8	53.6	24.3
III	5.180	0.561	-0.169	53.0	51.9	20.4

The saturated hydraulic conductivity from the results obtained by the Royal Irrigation Department laboratory was used. The saturated hydraulic conductivity of the cracked soil is assumed as function of the dynamic crack width given by equation 6.172. The unsaturated hydraulic conductivity above the phreatic line is dependent on the clay content, porosity, and instantaneous moisture status. The model computes the unsaturated hydraulic conductivity and capillary flow using Equations 7.2 to 7.4.

Potential evapotranspiration was based on the class A pan evaporation in mm/day at Sirigiti National Conference Centre station, Bangkok. The crop factors multiplied with the class A pan evaporation were taken from literature (Royal Irrigation Department, 1994).

Rice growing method for rainfed rice

The starting date of the sowing depends on the rainfall. Farmers wait for heavy rains to puddle before transplanting seedlings, no matter how long rains are delayed. It is estimated that around 30 to 40% of the total area under rainfed lowland rice in the northeastern part of Thailand is either not planted at all or planted, but this results in the crop failing completely after transplanting (Naklang, 1996).

Table 7.76 Different rice-growing methods for rainfed rice in areas in the northeastern part of Thailand (Naklang, 1996)

Components	Transplanting	Broadcasting		
		Dry-seed dry soil	Dry-seed	Germinated-seed (by 24 h soaking +24 h incubation)
Cultivars	RD6,KDML105,RD15	Same	Same	Same
Land preparation	2 ploughings +1 harrowing	2 ploughings +1 harrowing	2 ploughings +1 harrowing	2 ploughings +1 harrowing + smoothed surface
Time of seeding Soil surface at seeding or transplanting time	June-August Standing water preferred	May-mid August Dry	May-mid August Moist	May-mid August Saturated
Seeding rate	25 to 30 kg/ha	50 to 100 kg/ha	50 to 100 kg/ha	50 to 100 kg/ha
Spacing	0.20x0.20 m (3 seedling/hill)	Scattered	Scattered	Scattered
Labour	25 to 30 man-day/ha	1 man-day/ha	1 man-day/ha	1 man-day/ha
Weed management	Hand-weeding or post-emergence herbicide	Pre or post emergence herbicide	Same as dry soil	Same as dry soil
Fertilizer Basal rate: Time: Top-dressing at panicle initiation stage	150 to 185 kg/ha of fertilizer grade Transplanting 45 kg/ha of urea or 100 kg/ha ammonium sulphate	Same 30 DAE Same	Same 30 DAE Same	Same 30 DAE Same
Insect pest management	Apply insecticide if necessary	Same	Same	Same
Time of harvesting	Nov. to Dec.	Same	Same	Same
Rice stubble and crop residue management	Plough under after harvesting or at the first ploughing	Same	Same	Same
Rice straw management	Feed cattle, mulching field crops or incorporate into soil after harvesting	Scatter over the area after dry rice seed is broadcasted	Same as dry soil	Same as transplanting

Note: DAE   = Days after emergence

There are two main methods of growing rice in rainfed lowland conditions in Thailand, which are transplanting and direct seeding. For transplanting, seedlings are first grown in a seedbed for a month while the main fields are being ploughed along the bunds of each parcel. When each parcel has been prepared, the seedlings are pulled from the seedbed and then transplanted in the field by hand.

In the dry-seed broadcasting method, as soon as rainfall has moistened the soil enough then ploughing is possible, the land is ploughed (late of April, May or June). In the heavy clay soil of the Central Plain, the seed is broadcasted within a few days after ploughing, before the next shower. Much of the seed falls into cracks between, around or under the overhang of large clods. When the next shower comes, the heavy clay clods slake down, so that the clod fragments cover most of the seed. In light-texture soil, such as loamy sand, sandy loam and sandy clay loam, in the lowland rice areas of the northeastern part of Thailand; the method of land preparation is differing between farmers. Many farmers plough the field only once, but some plough twice before broadcasting the seed and then the field is harrowed so that the seed is covered by the loosen soil to retain moisture and protect them from birds.

Therefore the crop factor is estimated based on the average conditions of the seeding rice (is assumed in first of August). In practice farmers will prepare the land in a dry situation and then use dry seeding and wait for the rain. If it is raining the rice seed will germinate. After that if there is rain the rice will grow, but if there is no rain afterwards for a long time the young rice will die. The farmers will spread dry seed again and wait for the next rain. Some farmers use Germinated-seed (24 h soaking + 24 h incubation) to be safe from drought at the early stage of growth of the young rice. Many farmers nowadays use this method. However, in some years there is no yield at all if the length of the growth period is too short, because the rice doesn't come at full growth for photo-sensitive rice in rainfed conditions. If the situation is modelled for the second condition, it means that land preparation is assumed to be done in early June.

Crop factors for rainfed rice

The crop factors for rainfed rice are shown in Table 7.77.

Table 7.77 Crop factors (Kp) for evapotranspiration for KDML105 rice under rainfed conditions (Royal Irrigation Department, 1994)

Month	Decade		
	I	II	III
January	0.49	0.49	0.49
February	0.49	0.49	0.49
March	0.49	0.49	0.49
April	0.49	0.49	0.49
May	0.49	0.49	0.49
June	0.49	0.49	0.49
July	0.49	0.49	0.49
August	0.61	0.74	0.90
September	1.31	1.37	1.47
October	1.47	1.42	1.29
November	1.07	0.89	0.55
December	0.49	0.49	0.49

7.4.5 Economical data

Costs for the water management system

The costs for the water management system include the following:
- costs for the surface field drainage system;
- costs for the main drainage system;
- costs for pumping.

The costs for the surface drainage system have been computed using the formula as given in Chapter 6 by using the unit costs as given in Table 7.78.

The formulas used for the estimation of the costs for construction of the main drains, sub-main drains and lateral drains have been explained in Chapter 6. The estimation of costs for the drainage system is based on the overall area of the polder, parcel length and the distance between the lateral drains. The interest rate is estimated at 5% Basic values for the estimation of the water management system are given in Table 7.79.

Table 7.78 Unit costs for the surface drainage system in a typical rice polder (Royal Irrigation Department, 2002)

Item	Construction cost in €	Annual maintenance cost in €	Lifetime in years
Cost of earth movement per m³	1.20	0.05	50
Cost of construction bund per m³	2.50	0.25	50
Cost of PVC pipe per m	1.50	0.15	25

Table 7.79 Unit costs for the determination of the costs for the drainage system in a typical rice polder (Royal Irrigation Department, 2002)

Item	Construction cost in €	Annual maintenance cost in €	Lifetime in years
Earth movement /m³	1.20	0.05	50

The pumping head is based on an average of the main drain water level of 3.0 m-MSL and an average water level in the collecting system of 0.50 m-MSL. The annual rainfall surplus and seepage are estimated to be 367 mm and 73 mm for the estimation of the operation cost for pumping. The Drainage modulus is estimated at 46 mm/day. The installation, and operation and maintenance cost are given in Table 7.80. Costs for field pumping for drainage have been estimated from the Royal Irrigation Department in 2001, which is 0.017 €/m³.

Table 7.80 Pumping costs for a rainfed rice polder

Installation in million €	Operation in million €/year	Maintenance in million €/year	Lifetime in years
11.841	0.2033	0.0861	50

Costs for buildings, infrastructure and crops

The average farm size has been estimated at 5 ha (it is a fact that there are many plots in a farm, average size of a plot is between 0.8 and 3.2 ha). The value of the buildings is estimated at 12,500 €/farm (Royal Irrigation Department, 2002). The area of buildings and yards is estimated at 1,000 m²/farm. For facilities such as shops, factories, schools, etc. 20% of the value of the farms has been taken.

The infrastructure in the rural area is mainly composed of roads. In order to compute the cost of infrastructure, the density of paved roads is determined at 25 m/ha (Figure 7.34) at 5 m width. The estimated cost of roads is 3.89 €/m².

The investment costs of rice are split into the activities of agricultural practices and other activities related to productivity of crops such as land acquirement, transportation, etc. Normally the investment costs are dependent on the time of the growing seasons. The investment costs and time schedule for rice are shown in Figure 7.35 and Table 7.81.

Table 7.81 Agricultural activities, investment costs and scheduling for rainfed rice

Activity	Investment cost in €/ha	Starting date
1 lease/rent	17.85	1 January
2 ploughing	23.74	10 May
3a fertilizers	35.72	10 June
b spreading	9.61	15 June
4a harrowing/tillage	10.18	8 July
b seed rice	13.38	12 July
c planting	7.40	15 August
5a fertilizers/pesticides	11.61	20 Aug. - 10 Nov.
b spraying	13.65	20 Aug. - 10 Nov.
6a harvesting	56.68	21 November
b transportation	24.29	27 November
Total	224.11	
7 gross profit	274.71	

Source: Asian Engineering Consultants Co. Ltd and Macro Consultants Co. Ltd, 1996 and
http://www.aciar.gov.au/ publications /proceedings/77/04.pdf

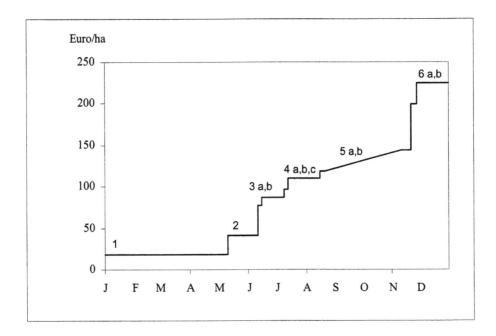

where:

1   = lease/rent
2   = ploughing
3 a,b   = fertilizers and spreading respectively
4 a,b,c = harrowing/tillage, seed rice and planting respectively
5 a,b   = fertilizers/pesticides and spraying respectively
6 a,b   = harvesting and transportation respectively

Figure 7.35 Investment costs and scheduling for rainfed rice

Cost of damage to crops, buildings and infrastructure for a rice polder

Crop damage occurs when the water depth is too high in the field or when there are too dry conditions. The method of calculation and the parameters used for estimating the cost of crop damage have been described in Chapter 6.

Damage to buildings and infrastructure is estimated as a percentage of the value of buildings and infrastructure in relation to the groundwater table as shown in Table 7.11.

7.4.6 Calibration of the parameters

Calibration of the parameters of model in the rural area for rainfed rice condition involved the determination of the hydrological parameters, which have been explained in Chapter 6 as follows:
- maximum interception in mm E_{max};
- potential evapotranspiration from layer I in mm/m V_{ep};
- parameters f_r and d_a in the discharge formula;
- maximum moistening rate in layer I in mm/m/hour V_{max1};
- maximum moistening rate in the columns of layer II in mm/m/hour V_{max2}.

Table 7.82 Calibrated parameters of the hydrological model for a rainfed rice polder

Year	Maximum interception in mm	Upper limit for potential evapotranspiration in mm	Parameter discharge model		Maximum moistening rate in mm/m/hour		Goal function	Annual discharge computed in mm
			f_r	d_a	Layer I	Layer II		
2003	1.5	77.4	0.00054	0.57	1.45	2.98	8.786	0

In Table 7.82 annual computed discharge out of the plot is 0 m^3 and also the annual measured discharge is zero for 2003. The model efficiency, based on the observed and computed groundwater table, is 0.704.

Sensitivity of the parameters for a rainfed rice polder

The canopy coverage of the rice is assumed to be constant all over the year. Any excess beyond E_{max} and the absorption capacity of the underneath layer will be recharged to the groundwater or stored on the soil surface. Therefore the higher the interception the lower the values of recharge, the lower the groundwater table and water depth above the ground surface will be. The optimal value of the maximum interception capacity has been found at 1.50 mm. In reality the maximum interception will change throughout the year due to the growth stages and the cropping pattern.

The upper limit for the potential evapotranspiration has been influenced by the water losses from the soil through evapotranspiration during the whole period because the soil moisture has been almost all the time above saturation. The calibrated parameter value for this case was the average value from Table 7.82, which is 77.4 mm.

The empirical reduction factor accounts for tortuousity and necking of the cracks as described by Equation 6.172. This factor has strongly influenced to the permeability of soil and the discharge from the groundwater zone. A small increase in this factor will cause a significant change in groundwater table. An increase of this parameter results in a higher permeability and will result in a higher discharge from the soil to the lateral canal. Therefore this parameter will affect to the groundwater table during wet periods. The calibrated parameter value for this case is the value shown in Table 7.82, which is 0.00054.

The parameter described the diameter of aggregates (d_a) characteristic the property of the soil for the slit model as described by Equation 6.172. It can be observed from the soil in the field when the soil is dry. In this experiment the diameter of the cods or aggregates was observed from the field after the rice was harvested and the soil was dry. The diameter of the cods was found to be 0.50 m. Increase of this parameter will result in a lower permeability and a higher groundwater table due to a lower discharge to the lateral canal during wet periods. This parameter was found through the calibration and had a value of 0.57 m, which was almost the same as the observed value.

The value of maximum moistening rate gives an effect to the discharge to the lateral canal and the soil storage, but has little effect on the annual discharge (the less moistening rate the more discharge). The higher the maximum moistening rate the less time is need to recharge the subsoil. Hence the maximum moistening rate also affects the soil moisture, potential evapotranspiration and the groundwater table in the soil (the lower the moistening rate the higher the groundwater table)

However in this study, the maximum moistening rate in layer I has less influence than in layer II because the depth of layer I is smaller than of layer II. The calibrated value for the maximum moistening rate in layer I and layer II are the average values, which are 1.45 and 2.98 respectively.

7.4.7 Model validation with fixed calibrated parameters

The model has been validated with the calibrated parameters in Table 7.82 for 2003, as shown in Figures 7.36 to 7.38.

In Figure 7.37 the computed groundwater tables fitted well with the observations during the rainy season but during the dry season they did not fit so well because the fluctuations of the groundwater table were influenced by the water level in the river near the experimental area (Figure 7.31).

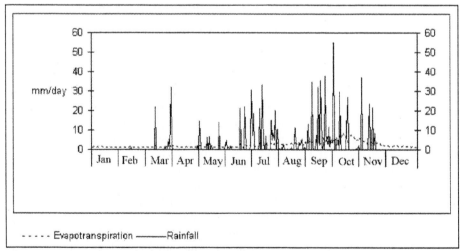

Figure 7.36 Rainfall and evapotranspiration of rainfed rice in the experimental plot at Samchuk Irrigation Water Use Research Station, 2003

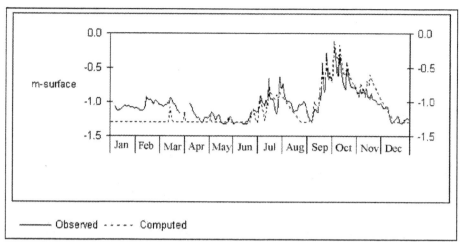

Figure 7.37 Computed and observed of groundwater table for rainfed rice in the experimental plot at Samchuk Irrigation Water Use Research Station, 2003

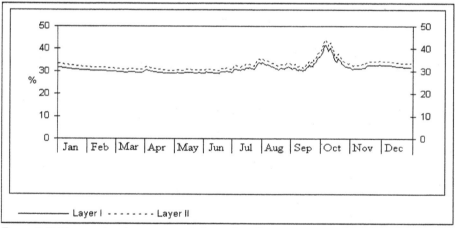

Figure 7.38 Computed moisture content in layers I and II for rainfed rice in the experimental plot at Samchuk Irrigation Water Use Research Station, 2003

In Figure 7.38 computed moisture content in layer II is higher than in layer I throughout the year, because the moisture in layer I will decrease more rapidly than in layer II during the dry period may be due to a higher evapotranspiration than the capillary rise from layer II to layer I. However, during the rainy season, moisture content in layer II is still higher than in layer I because of the cracks in layer I (soil moisture content in layer I is between 24% to 51%). During the dry season the soil may be dry enough for the development of cracks. The soil moisture in layer II will increase more rapidly than in layer I after heavy rain, because the rainfall falls also directly into the cracks. The soil moisture in layer I increases more rapidly than in layer II because the moistening rate in layer I is higher. The soil moisture content is most of the time lower than field capacity in both layers even during the rainy season. During the dry season the soil moisture content is about 30% while the wilting point of the soil is about

24%. The moisture was higher than the wilting point, because the soil has a high clay content, which has a high moisture retention capacity.

7.4.8 Analysis with the optimization model

Hydrological analysis

The hydrological analysis with the validated model for the experimental plot has resulted in the data as given in Table 7.83.

Table 7.83 Review of annual data in mm for rainfed rice in the experimental plot

Year	Rainfall in mm	Evapotranspiration in mm	Discharge in mm		Capillary rise in mm	
			Into the plot	Out of the plot	II → I	III → II
2003	870	867	0	0	174	854

Economic computation

The economic computation was done based on the total area of the polder of 10,000 ha. The results and review computations for optimal conditions are shown in Tables 7.84 to 7.87.

Table 7.84 shows that the value of buildings is about 81% of the total value in a rice polder under rainfed conditions. Therefore the damage to buildings may be the most important in a rice polder under rainfed conditions.

Table 7.84 Review of the total values of crops, buildings and infrastructure in a rice polder under rainfed conditions at optimal water management

Item	Area in ha	Highest value		Lowest value	
		in € per ha	in € * 10^6	in € per ha	in € * 10^6
Buildings	233	90,000	20.98	90,000	20.98
Infrastructure	41	38,900	1.58	38,900	1.58
Rice	9,483	357	3.39	18	0.17
Total	9,757		25.95		22.73

Table 7.85 Optimization analysis of the main components of the water management system for a rice polder under rainfed conditions

Polder water level in m-s.	Open water in %	Pumping capacity in mm/day	Field drain capacity in mm/day	Costs in € * 10^6	Damage in € * 10^6	Value of the goal function
-0.09	1.1	7.50	6.7	0.978	2.413	3,391,000

In Table 7.85 the simulated optimal values for the main components of the water management system are shown. The polder water level at optimal conditions is 0.09 m+surface. At this level the rice in the plot can benefit from the water that flows to the plot during the dry season and excess water can be drained by gravity during the rainy season. The deeper the polder water level in rainfed rice conditions the more the water loss through the cracks to the lateral canals. Therefore, there will be higher damage to the crops due to drought. In rainfed rice conditions the loss of this water sometimes is of benefit to the farmers because of the low water depth in the field in case of heavy rainfall. While a shallow polder water level will increase the probability of inundation of the cropped area and overtopping the bund, which may cause a large damage to the area.

Table 7.86 Total costs for the water management system under conditions of a rainfed rice polder at optimal water management

Item	Construction costs in € * 10^6	Annual maintenance costs in € * 10^6	Annual equivalent costs in € * 10^6
Field drainage	1.19	0.39	0.48
Open water	4.73	0.20	0.46
Pumping	0.23	0.03	0.04
Total	6.15	0.62	0.98

Table 7.86 shows that the overall annual maintenance costs in a rainfed rice polder are about 10.1% of total of the construction costs, while the overall annual equivalent costs are about 15.9%. The open water takes the highest amount, which is about 76.9% of the total construction costs. While the pumping costs are the lowest amount, which are about 3.0%. This may be caused by a small pumping capacity that required for pumping water outof the polder under rainfed conditions and low head of pumping.

Table 7.87 Specification of average annual damage and yield reduction under conditions of a rainfed rice polder at optimal water management

Item	High polder water level		High water level in the field		Drought	
	in €	in % of the value	in €	in % of the value	in €	in % of the value
Buildings	0	0.0	0	0.0	0	0.0
Infrastructure	86,900	5.5	0	0.0	0	0.0
Rice	211,700	6.3	1,262,200	37.3	851,900	25.2
Total	298,600	1.2	1,262,200	37.3	851,900	25.2

In Table 7.87 the simulated average annual damage costs and average annual damage in percentage of the total value are shown. From Table 7.87 it can be seen that in rainfed rice polder at optimal conditions a too high water level in the rice field has more influence to yield reduction than too dry conditions. The water in the rice field can be lost to the lateral canal because of the high permeability of the cracks. When there is heavy rainfall there will be a high water level in the field because of a high water level in the polder at the same time when the water level in the rice field is high. In this simulation it is assumed that there is no pumping from the field. Therefore the water has to be only drained by gravity. Moreover, the damage will be caused by inundation and overtopping of the bund at a high water level in the polder. The model takes into account that when the water level increases to 0.05 m below bund the crest of the bund it is assumed that there will be damage by a certain percentage of the crops due to overtopping. In reality the damage due to overtopping is dependent on the topography of the area.

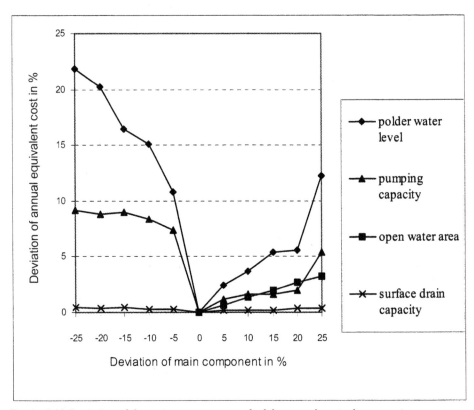

*Figure 7.39 Deviation of the main components and of the annual equivalent costs in percentage
for a rice polder under rainfed conditions*

In Figure 7.39 the influence of deviations from the simulated optimal values for a
rice polder under rainfed conditions are shown. It shows that the polder water level
(lateral water level) has most influence on the annual equivalent costs. If the polder
water level is 25% smaller (shallower) in comparison with the optimal value the annual
equivalent costs will increase approximately 22%, because of the increased of damage
to crops due to a too high water level in the field and a higher probability of inundation
and overtopping of the bund. However, when the polder water level is deeper, there will
be more construction costs and more damage due to drought because water will be lost
through the cracks to the lateral canal. A decrease in pumping capacity by 25% from the
optimal value will increase the annual equivalent costs by approximately 9%, because
damage due to a high polder water level in the rice field increases. An increase in the
surface drain capacity has no significant influence on the annual equivalent costs,
because of construction cost of surface drains is not a large portion of the water
management costs in this polder system.

Discussion

Based on the results as shown above the following points can be discussed.
The polder water level has most influence on the annual equivalent costs because
this component has the highest influence on damage of rainfed rice in both too dry and

too wet conditions. Moreover, it can also influence the damage caused by inundation of the bund and the rice field.

The surface drain capacity is the least sensitive parameter for the rainfed rice conditions (Figure 7.39) because the water in the plot can flow through the cracks to the lateral canals. The influence of this parameter can be higher when the soil is less permeable and the size of the plot or the distance between the lateral canals is increased.

A shallow polder water level will prevent flow of water from the rice field in case of heavy rainfall. When the polder water level is higher than the culvert level in the rice field, the damage to the crops due to both a too high water level in the field and in the lateral canal increases. However, when the polder water level is lower the damage will be dominated by drought.

The lateral canal water level in a rainfed rice field may be kept high to prevent loss of water and to get benefit for irrigation water that will flow through the culvert to the rice field when the water level in the rice field is low. However, this will create a higher risk of inundation of the rice field due to heavy rainfall. In practice the water in the polder will be drained before a storm comes.

A deep polder water level (Figure 7.39) will increase the damage due to drought, which can be explained as follows. When the polder water level is deeper it will cause loss water from the rainfed rice field to the main drainage system through the cracks. However, this condition is dependent on the properties and moisture condition of the soil. In Montmorillonite clay soil the cracks will close when the soil moisture content increases and flow through the cracks will be low, whereas for the other clay types the cracks will remain open.

The values in Table 7.87 are only indicative for water management in rainfed rice. In practice for individual cases, land use, gradient of land, structure, and soil condition will have to be considered. However, a rice polder under rainfed conditions may not really exist in practice in Thailand, because farmers always pump water from the canals to irrigate their area.

7.5 A rice polder under irrigated conditions

7.5.1 General

The experiment for the irrigated rice polder was done from 1 January to May 2004 at Samchuk Irrigation Water Use Research Station. The start growing crops was at 22 January 2004 and harvested at 20 May 2004. This is the period of growing dry season crops in Thailand. Most of the time in this period water management involves irrigation. General information on the experimental area has been given in Section 7.4.1. The layout of the experimental area is shown in Figure 7.31.

The aim of the experiment was to find the responds of ground/open water and discharge related to the crop schedule, soil moisture conditions, hydrological conditions, drainage system and water management conditions in an irrigated rice polder.

7.5.2 Physical conditions

In the irrigated rice in this experiment area was in the low part of the area (Figure 7.40). In the area also rice was previously grown. The soil is wet throughout the year because of the high groundwater table. The groundwater table was around 0.40 m-surface during the dry season. There is a lined canal for conveying the water to the irrigated rice and besides the plot there is the lateral canal to drain excess water from the rice field or other crops. The land in the experimental plot was prepared by puddling in order to reduce the macro pores in the upper layer, consequently to reduce the permeability in the upper layer of the soil. The soil in this area is clay loam to loam.

Soil properties

The soil properties were as follows:
-   the soil in the study area is loam, which is approximately composed of sand 45%, silt 32% and clay 23%;
-   the clay mineral analysis has shown that the clay in the study area is composed of Kaolinite 32.5%, Illite 54.7%, and Montmorillonite 12.8%;
-   average saturated permeability at 0.30 m-surface is $0.0135*10^{-3}$ m/day; at 0.60 m-surface it is $0.312*10^{-3}$ m/day.

Figure 7.40 Irrigated rice experimental area at Samchuk Irrigation Water Use Research Station, 2004

7.5.3 Set-up of the experimental plot

The schematic layout of the experimental area is shown in Figure 7.41. The net crops area for irrigated rice is 315 m². Both an irrigation and a drainage system are available. The water management criteria in this area are to control the water level in the field at the preferred water level in the dry periods but in the wet periods the water level in the field was controlled in such a way that the water level in the plot will not exceed the maximum acceptable water level for the crops (Figure 5.3 and Tables 5.2 and 5.3). Non-photo sensitive rice is normally grown in the irrigated rice area. Patumtani-1 rice was used in this study.

Water management and water control

In the experimental plot water management for irrigated rice was intended to keep water level in the field at the preferred water layer (Figure 5.3 and Table 5.2). The water was pumped to the plot during dry periods when the water level in the field was 5 mm lower than the preferred water level and stopped when the water level in the field was 5 mm above the preferred water level. In case of rain, when the water level in the field was higher than 20 mm above the preferred water level the pumping is starting to pump the excess water out of the plot.

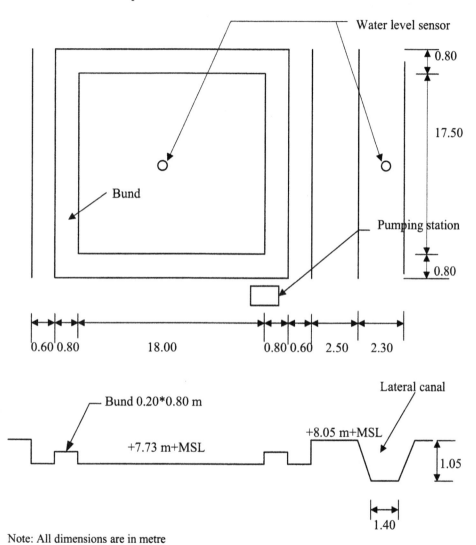

Note: All dimensions are in metre

Figure 7.41 Schematic layout and cross-section of the area for the irrigated rice experiment

Monitoring and data recording for the irrigated rice experimental plot

The monitoring and data recording for irrigated rice was done from 1 January 2004 to 31 May 2004. The following data were recorded:
-   hourly rainfall in mm;
-   daily pan evaporation in mm;
-   discharge pumping out of the plot;

- hourly irrigation input to the plot;
- surface water level or groundwater table in the field;
- surface water level or groundwater table in the lateral canal;
- crops and agricultural practice schedule;
- crop growth and growth stage;
- investment costs and time schedule for agricultural practice;
- production.

The data mentioned above were used for calibration and verification of the model.

Discussion on the observation results

The groundwater table in the rice field during early stage of growth period fluctuated between 0.15 to 0.50 m-surface. At this period the groundwater table may be influenced of water level in the drainage canal and the water level in the river.

In April the ground water table dropped from 0 to 0.70 m-surface in three days after water layer on the ground surface was drained from the field at the full head development stage of rice. The ground water table decreased very rapidly due to high evapotranspiration because this period was the hottest and driest period in Thailand.

The crop yield for irrigated rice is 5 ton/ha/crop season, which is about 2.3 times of the rice under rainfed conditions.

7.5.4 Hydrological and water management data

For this case study the following hydrological and water management data have been used in the analysis:
- the data from 1 January 2004 to 31 May 2004 for the model calibration:
  - class A pan daily evaporation in mm, Samchuk Irrigation Water Use Research Station, Suphanburi;
  - hourly rainfall in mm, Samchuk Irrigation Water Use Research Station, Suphanburi;
  - hourly discharge pumping out of the experimental area;
  - hourly irrigation input to the plot in mm;
  - surface water level or groundwater table in the field;
  - surface water level or groundwater table in the lateral canal;
- the data for validation and computation:
  - hourly rainfall in mm at Ladkrabang station, Bangkok, 1991 to 1999;
  - class A pan evaporation, Bangkok Metropolitan, 1991 to 1999.

Characteristics of a typical irrigated rice polder

In this study it has been assumed that the irrigated rice polder has a total area of 10,000 ha. The soil storage exists in the upper 0.40 m, which consists of 0.15 m of puddle layer (layer I) and 0.25 m of layer II. Only rice was grown. Based on average conditions the cultivated area of the plots was about 1 ha. The dimensions of plots were 100 * 100 m and there were six plots in one parcel. The lay out of the parcel was shown in Figure 7.42.

The ground surface level in this area was set at 0.5 m+MSL, which is a common level of rice polders. The side slopes of the watercourses are:

Lateral canals   = 1:1.5;
Sub-main canals   = 1:3;
Main canals   = 1:3.

The soil moisture tension computation in the model is based on an empirical relation as given in Equation 7.2. The constants a, b and c for the different soil profiles are shown Table 7.88 (see Appendix VI).

Table 7.88 Clay content in the soil profile for irrigated rice conditions in Samchuk Irrigation Water Use Research Station

Depth in m-surface	Typical soil type	Description	Clay content in %
0.00 - 0.30	Ploughed layer	Loam	21.7
0.30 - 0.60	Subsoil	Loam	22.7
> 0.60	Subsoil	Loam	22.5

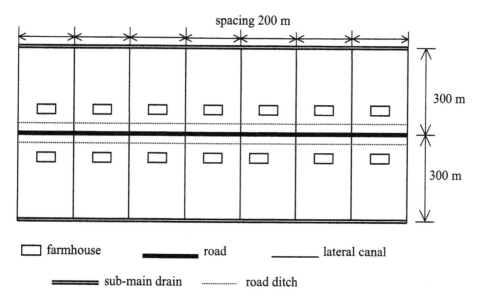

farmhouse   road   lateral canal

sub-main drain   road ditch

Figure 7.42 Schematic layout of a parcel in a typical irrigated rice polder in Thailand

Table 7.89 Soil moisture tension in pF and soil moisture content in percentage for irrigated rice conditions at Samchuk Irrigation Water Use Research Station

Layer \ pF	2.01	2.53	3.01	3.49	4.18
I	44.12	29.96	23.41	18.76	10.35
II	45.52	31.87	23.75	19.91	12.90
III	44.50	31.62	25.18	20.50	12.87

Note: Soil characteristics obtained by sampling from the experimental plot and analysed by the laboratory of Royal Irrigation Department

The saturated hydraulic conductivity from the result obtained by the Royal Irrigation Department Laboratory was used. The saturated hydraulic conductivity of the cracked soil has also been calculated by Equation 6.172. The model computes the unsaturated hydraulic conductivity and capillary flow using Equations 7.2 to 7.4.

Table 7.90 Constants and moisture contents for irrigated rice conditions at Samchuk Irrigation
          Water Use Research Station

Layer	Constants			Moisture content in %		
	a	b	c	Sat	FC	WP
I	4.760	0.488	-0.181	45.0	44.1	9.8
II	5.000	0.850	-0.304	36.9	33.9	12.6
III	5.140	0.979	-0.300	36.6	32.6	12.6

The crop factors that have been applied to compute the potential evapotranspiration
from the class A pan were based on two crops of rice per year and are given in Table
7.91. The rainy season crops begin in 1 - 15th August and are harvested in 15th
November - 1st December. The dry season crops begin in 15 - 31st January and are
harvested in 1 - 15th May.

Table 7.91 Crop factors (Kp) for evapotranspiration for HYV rice in the dry and rainy season for
          irrigated conditions (Royal Irrigation Department, 1994)

Month	Decade		
	I	II	III
January	0.70	0.79	0.88
February	1.09	1.18	1.34
March	1.49	1.54	1.50
April	1.37	1.20	0.90
May	0.55	0.21	0.21
June	0.21	0.21	0.21
July	0.70	0.70	0.70
August	0.79	0.88	1.09
September	1.18	1.34	1.49
October	1.54	1.50	1.37
November	1.20	0.90	0.55
December	0.21	0.21	0.70

Note: The above values are based on a coefficient by assuming two crops of rice in the rainy
      season and the dry season.

7.5.5 Economical data

The costs for the water management system include the following:
- costs for the surface field drainage system;
- costs for the canal system;
- costs for pumping.

The costs for the surface drainage system were computed using the formulas as
given in Chapter 6 by using the unit costs as given in Table 7.92.

The formulas used for the estimation of the costs for construction of the main canals,
sub-main canals and lateral canals have been explained in Chapter 6. The estimation of
costs for the canal system are based on the overall area of the polder, parcel length and
the distance between the lateral canals. Basic values for the estimation of the costs for
the water management system are given in Table 7.93. The interest rate is estimated at
5%.

Table 7.92 Unit costs for the surface drainage system under conditions of an irrigated rice polder (Royal Irrigation Department, 2002)

Item	Construction cost in €	Annual maintenance cost in €	Lifetime in years
Cost of earth movement per m³	1.20	0.05	50
Cost of construction of the bund per m³	2.50	0.25	50
Cost of PVC pipe per m	1.50	0.15	25

Table 7.93 Unit costs for the determination of the costs for the canal system under conditions of an irrigated rice polder (Royal Irrigation Department, 2002)

Item	Construction cost in €/m³	Annual maintenance cost in €/m³	Lifetime in years
Earth movement	1.20	0.05	50

The pumping head is based on an average of the main canal water level of 3.0 m-MSL and an average water level in the lateral canals of 0.50 m-MSL. The annual rainfall surplus and seepage are estimated to be 370 mm and 183 mm respectively. The drainage modulus is assumed at 46 mm/day. The installation and operation and maintenance costs are given in Table 7.94. The costs for field pumping for irrigation and drainage have been estimated from unit cost of the Royal Irrigation Department in 2001, which is 0.017 €/m³.

Table 7.94 Pumping costs under conditions of an irrigated rice polder

Installation in million €	Operation in million €/year	Maintenance in million €/year	Lifetime in years
11.841	0.2033	0.0861	50

Investment costs for crops

In the typical rice polder there is only rice. The investment costs of this crop are based on the activities of agricultural practices and other activities related to the productivity, such as land acquirement, transportation, etc. Normally the investment costs are dependent on the time of the growing seasons. The investment costs and time schedule for irrigated rice are shown in Table 7.95 and Figure 7.43.

Costs for buildings and infrastructure

The average farm size has been estimated at 5.0 ha (There are many plots within a farm, average size of plot is 0.8 - 3.2 ha). The value of the buildings is estimated at 15,000 €/farm (Royal Irrigation Department, 2002). The area of buildings and yards is estimated 1000 m²/farm. For facilities such as shops, factories, schools, etc, 20% of the value of the farms has been taken.

The infrastructure in rural area is mainly composed of roads. In order to compute the cost of infrastructure, the density of paved roads is determined at 100 m²/ha (Figure 7.42) and 6 m width. The estimated cost of roads is 6 €/m².

Cost of damage to crops, buildings and infrastructure for a rice polder

The method of calculation and the parameters used for estimating cost of crop damage have been described in Chapter 6. Damage to buildings and infrastructure is estimated as percentage of value of buildings and infrastructure in relation with the groundwater table as shown in Table 7.11.

Table 7.95 Agricultural activities, investment costs and scheduling for irrigated rice

Activity	Investment cost in €/ha	Starting date
1 lease/rent	52.08	1 May
2 ploughing	23.74	10 May
3a fertilizers	35.72	10 June
b spreading	9.61	15 June
4a harrowing/tillage	10.18	8 July
b seed rice	13.38	12 July
c planting	7.40	15 August
5a fertilizers/pesticides	11.61	20 Aug. - 10 Nov.
b spraying	13.65	20 Aug. - 10 Nov.
6a harvesting	56.68	21 November
b transportation	24.29	27 November
7 ploughing	22.22	30 November
8a fertilizers	51.95	2 December
b spreading	13.51	5 December
9a harrowing/tillage	9.52	8 December
b seed rice	14.73	15 December
c planting	9.93	15 January
10a fertilizers/pesticides	12.85	21 Feb. - 5 April
b spraying	16.31	21 Feb. - 5 April
11a harvesting	61.18	15 April
b transportation	26.22	22 April
Total	496.76	
12 gross profit	793.29	

Source: Asian Engineering Consultants Co. Ltd and Macro Consultants Co. Ltd, 1996
http://www.moc.go.th/thai/dbe/research/rice_st.html#tb14
http://www.oae.go.th/Price/MonthPrice/hommali.htm

7.5.6 Calibration of the parameters

Calibration of the parameters of model in the rural area for irrigated rice condition involved the determination of the hydrological parameters:
- maximum interception in mm E_{max};
- potential evapotranspiration from layer I in mm/m V_{ep};
- parameters f_r and d_a in the discharge formula;
- maximum moistening rate in layer I in mm/m/hour V_{max1};
- maximum moistening rate in the columns of layer II in mm/m/hour V_{max2}.

Sensitivity of the parameters for an irrigated rice polder

The canopy coverage of the rice is assumed to be constant all over the year. Any excess beyond E_{max} and the absorption capacity of the underneath layer will be recharged to the groundwater or stored on the soil surface. Therefore the higher the interception the lower the values of recharge, the lower the groundwater table and water depth above the ground surface will occur. The optimal value of the maximum interception capacity has been found at 1.1 mm. In reality the maximum interception in this will change throughout the year in relation to growth period.

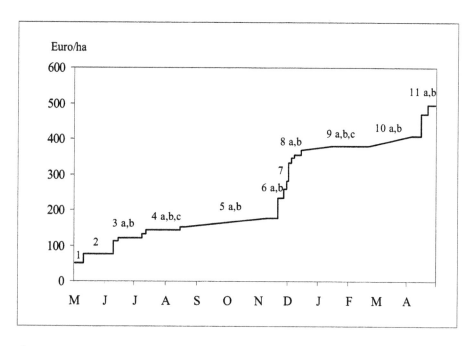

where:

1   = lease/rent
2   = ploughing for rainy season rice
3 a,b   = fertilizers and spreading for rainy season rice respectively
4 a,b,c = harrowing/tillage, seed rice and planting for rainy season rice respectively
5 a,b   = fertilizers/pesticides and spraying for rainy season rice respectively
6 a,b   = harvesting and transportation for rainy season rice respectively
7   = ploughing for dry season rice
8 a,b   = fertilizers and spreading for dry season rice respectively
9 a,b,c = harrowing/tillage, seed rice and planting for dry season rice respectively
10 a,b = fertilizers/pesticides and spraying for dry season rice respectively
11 a,b = harvesting and transportation for dry season rice respectively

Figure 7.43 Investment costs and scheduling for irrigated rice

The calibrated parameter for the upper limit for potential evapotranspiration value in this case was the average value from Table 7.96, which is 70.3 mm. This parameter may be influenced when the water level is below the ground surface in the rice field.

The calibrated parameter value for the empirical reduction factor is shown in Table 7.96, which is 0.00053. The diameter of cod or aggregates has been found through the calibration as being 0.53 m, which is almost the same as the observed value.

The calibrated values for the maximum moistening rate in layer I and layer II are the average values, which is 1.79 and 2.98 respectively.

Table 7.96 Calibrated parameters of the hydrological model for an irrigated rice polder

Year	Maximum interception in mm	Upper limit for potential evapotranspiration in mm	Parameter discharge model		Maximum moistening rate in mm/m/hour		Goal function	Annual discharge computed in mm
			f_r	d_a	Layer I	Layer II		
2004	1.1	70.3	0.00025	0.53	1.79	2.98	0.6038	0

7.5.7 Model validation with fixed calibrated parameters

The model has been validated with calibrated parameters in Table 7.96 for 2004 from January to May 2004 is shown in Figures 7.44 to 7.46.

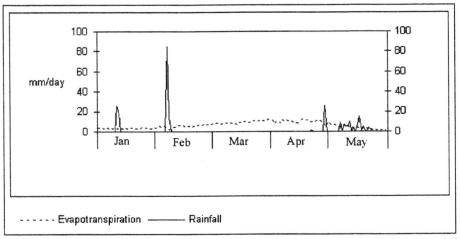

Figure 7.44 Rainfall and evapotranspiration of irrigated rice in the experimental plot at Samchuk Irrigation Water Use Research Station, 2004

During the period of the experiment with irrigated rice, there was little rainfall and high evapotranspiration. In irrigated conditions the soil in the rice field is most of the time saturated and the evapotranspiration is at the potential rate. Pan evaporation can be as high as 9 mm/day between March and April.

In Figure 7.45 it can be seen that the irrigation water supply to the experimental plot is high compared to crop water use in Figure 7.44. The main reason is loss through the experimental plot and deep percolation because the area adjacent the experimental plot was dry and the soil in the experimental area is loam with a small portion of clay (Table 7.88). Moreover, due to the clay soil properties in this area there are permanent cracks, especially in the second layer. The water is lost through these cracks very rapidly. It has been tested and estimated at the field that when there is a water layer in the rice field plot the loss through the bund and percolation is about 19 mm/day. The irrigation input was 30 to 35 mm/day and the water use by the crop was 8 to 10 mm/day. But in this simulation for optimal conditions deep percolation was estimated at 3 mm/day (Charnvej, 1999).

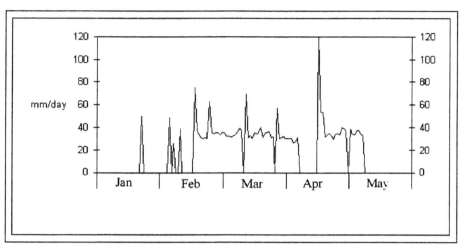

Figure 7.45 Irrigation input for irrigated rice in the experimental plot at Samchuk Irrigation Water Use Research Station, 2004

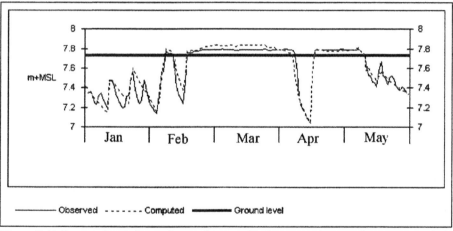

Figure 7.46 Computed and observed groundwater table for irrigated rice in the experimental plot at Samchuk Irrigation Water Use Research Station, 2004

In Figure 7.46 the computed groundwater table fitted well with the observations during the experimental period. However, the level of the groundwater table was influenced by external factors such as water level fluctuations in the river (Figure 7.31), even though the distance is rather far from the river, while the soil is very permeable. There is no computed discharge outof the plot, as well as measured discharge out for 2004. Model efficiency, which is calculated, based on the observed and computed groundwater table, is 0.923.

7.5.8 Analysis with the optimization model

Hydrological analysis

The hydrological analysis with the validated model for the experimental plot has resulted in the data as given in Table 7.97.

Table 7.97 Review of data in mm for irrigated rice in the experimental plot for the period from 1 January to 31 May 2004

Year	Rainfall in mm	Evapotranspiration in mm	Discharge in mm	
			Into the plot	Out of the plot
2004	246	957	2,736	0

In the Table 7.97 it can be seen that there is a high irrigation water supply to the plot. The loss of water through the bund, outflow of groundwater and surface water from the field to the adjacent area and deep percolation is in total 2,025 mm.

Economic computation

The economic computation was done based on a total area of the polder of 10,000 ha. The results and review computations for optimal conditions are shown in Tables 7.98 to 7.101.

Table 7.98 Review of the total value of crops, buildings and infrastructure in a rice polder under irrigated conditions at optimal water management

Item	Area in ha	Highest value		Lowest value	
		in € per ha	in € * 10^6	in € per ha	in € * 10^6
Buildings	218	150,000	32.71	150,000	32.71
Infrastructure	92	60,000	5.51	60,000	5.51
Rice	8,868	793	7.04	10	0.09
Total	9,178		45.26		38.31

Table 7.98 shows that the value of buildings is about 72% of the total value in a rice polder under irrigated conditions. Therefore the damage to buildings may be the most important.

Table 7.99 Optimization analysis of the main components of the water management system for a rice polder under irrigated conditions

Polder water level in m-s.	Open water in %	Pumping capacity in mm/day	Field drain capacity in mm/day	Costs in € * 10^6	Damage in € * 10^6	Value of the goal function
0.84	2.5	32.9	44.8	5.659	0.538	6,197,000

In Table 7.99 the simulated optimal values for the main components of the water management system are shown. The polder water level at optimal conditions is 0.84 m-surface. At this level the water has to be pumped for irrigated rice. The model takes that throughout the year that the polder water level was kept at the preferred water level. However, the irrigation water, which was supplied to the polder, will cause a higher probability of inundation due to overflow of the bund to the crops area in case of heavy rain. In reality irrigation water was supplied based on the weather forecasting, which is not included in the model. In practice the polder water level may be kept higher at the end of the rainy season to store water in the polder area for irrigation during the dry season.

Table 7.100 Total costs for the water management system under conditions of an irrigated rice polder at optimal water management

Item	Construction costs in € * 10⁶	Annual maintenance costs in € * 10⁶	Annual equivalent costs in € * 10⁶
Field drainage	1.38	0.46	0.56
Open water	18.65	0.78	1.80
Pumping	0.87	0.52	0.57
Total	20.90	1.76	2.93

Table 7.100 shows that the overall annual maintenance costs in a rice polder under irrigated conditions are about 8.4% of the construction costs, while the overall annual equivalent costs are about 14.0%. The open water constitutes the highest part of the total construction costs, which is about 89.2%. While the pumping costs are the lowest amount, which are about 4.2%. This may be caused by low head of pumping.

Table 7.101 Specification of average annual damage and yield reduction under conditions of an irrigated rice polder at optimal water management

Item	High groundwater tables		High water level in the field		Drought	
	in €	in % of the value	in €	in % of the value	in €	in % of the value
Buildings	73,200	0.2	0	0.0	0	0.0
Infrastructure	50,200	0.9	0	0.0	0	0.0
Rice	0	0.0	385,200	5.5	12,500	0.2
Total	123,400	0.3	385,200	5.5	12,500	0.2

In Table 7.101 the simulated average annual damage costs and average annual damage in percentage of the total value are shown. The main damage may be caused by too high water levels in the field and very small damage due to drought because of the very small chance that evapotranspiration is less than potential. The damage due to drought may occur in the first stage and some later stage when water is drained from the field when the soil is dry.

Economical analysis for a water management system at the present situation

In practice the control criteria for the water level in polder vary throughout the year. The cropping areas and crop types are also mixed. The simulation was done based on an area of 10,000 ha. It is assumed that only rice is grown in this area.

The general values of the main components of the water management system of a typical polder under irrigated rice conditions for water management in practice may be approximated as follows:
- polder water level in the rainy season   = 0.80 m-surface;
- open water area   = 2.4%;
- field drain capacity   = 46.0 mm/day;
- pumping capacity   = 26.0 mm/day.

The results and review computations for practical water management conditions are shown in Tables 7.102 to 7.105.

Table 7.102 Review of the total values of crops, buildings and infrastructure in a rice polder under irrigated conditions for water management in practice

Item	Area in ha	Highest value in € per ha	Highest value in € * 10^6	Lowest value in € per ha	Lowest value in € * 10^6
Buildings	218	150,000	32.76	150,000	32.76
Infrastructure	92	60,000	5.52	60,000	5.52
Rice	8,883	793	7.05	10	0.09
Total	9,194		45.33		38.37

Table 7.103 Optimization analysis of the main components of the water management system for a rice polder under irrigated conditions for water management in practice

Polder water level in m-s.	Open water in %	Pumping capacity in mm/day	Field drain capacity in mm/day	Costs in € * 10^6	Damage in € * 10^6	Value of the goal function
0.80	2.4	26.0	46.0	5.424	2.296	7,721,000

Table 7.103 shows that damage is larger than in Table 7.99 and also the costs are higher than in Table 7.99. The damage is higher due to the higher probability of inundation of the rice area when the polder water level is kept shallower and there is lower pumping capacity.

Table 7.104 Total costs for the water management system under conditions of an irrigated rice polder for water management in practice

Item	Construction costs in € * 10^6	Annual maintenance costs in € * 10^6	Annual equivalent costs in € * 10^6
Field drainage	1.39	0.46	0.56
Open water	18.01	0.75	1.74
Pumping	0.71	0.42	0.46
Total	20.11	1.63	2.76

Table 7.104 shows that the overall annual maintenance costs for water management in practice in a rice polder under irrigated conditions are about 8.1% of total of the construction costs, while the overall annual equivalent costs are about 13.7%. The open water is the largest component of the total construction costs, which is about 89.6%. The pumping costs are the lowest amount, which are about 3.5%. From Table 7.104 it can be derived that the costs for open water are lower than at optimal conditions (Table 7.100) because of the lower percentage of open water. Also the pumping costs are lower than under optimal conditions because less water from the field flows to the lateral canal and lower pumping head. However, the probability of inundation of the bund increases when water level is higher. The costs for drainage are higher than at optimal conditions because of the larger area to be drained and the higher field drainage capacity.

In Table 7.105 the damage due to a too high groundwater table is higher compared to optimal conditions (Table 7.101) because of there is inundation and overtopping of the bund. In this case the model calculates damage for the cultivated area as a certain portion of the cropped area that is damaged when the water level in the polder reaches

0.05 m below the crest of the bund. Actually this value will be dependent on the topography of the polder area.

Table 7.105 Specification of average annual damage and yield reduction under conditions of an irrigated rice polder for water management in practice

Item	High groundwater tables		High water level in the field		Drought	
	in €	in % of the value	in €	in % of the value	in €	in % of the value
Buildings	73,300	0.2	0	0.0	0	0.0
Infrastructure	137,500	2.5	0	0.0	0	0.0
Rice	440,400	6.3	385,600	5.5	12,200	0.2
Total	651,200	1.4	385,600	5.5	12,200	0.2

Table 7.106 Comparison between the optimal and present values of the main components of the water management system in a rice polder under irrigated conditions

Item	Polder water level in m-s.	Open water in %	Pumping capacity in mm/day	Field drain capacity in mm/day	Costs in € * 10^6	Damage in € * 10^6	Value of the goal function
Optimal	0.84	2.5	32.9	44.8	5.659	0.538	6,197,000
Practice	0.80	2.4	26.0	46.0	5.424	2.296	7,721,000
% differ.	+5	+3	+26	-3	+4	-76	-20

In Table 7.106 the pumping capacity at optimal conditions is higher than in practice while the other components are more or less the same. Damage under water management conditions in practice is higher than at optimal conditions due to the shallow polder water level and insufficient pumping capacity. Increase of the pumping capacity can be considered as a first priority and a deeper polder water level may have to be further studied.

In Figure 7.47 the influence of deviations from the simulated optimal values for a rice polder under irrigated conditions are shown. The polder water level and the pumping capacity have most influence on the annual equivalent costs in the system. If the polder water level is 25% shallower compared to the optimal value the annual equivalent costs will increase approximately 23%, because of the increased probability of inundation and overtopping of the bund. It will be also more difficult to drain by gravity when the polder water level is high. A lower polder water level will increase the loss of irrigation water from the rice field. This will require more pumping for irrigation and also higher construction costs. A pumping capacity which is 25% lower than the optimal value increases the annual equivalent costs by approximately 22%, because of the higher damage due to inundation of the area, overtopping of the bund and the increased difficulty to drain by gravity due to the high polder water level. Changes in the surface drain capacity can affect the amount of water that can be drained by gravity. Due to this more pumping will be needed. This is the reason for the increase in annual cost, which increases around 15% when this parameter decreases 25%. On the other hand when the surface drain capacity is higher the costs increase and there may also be over drainage because the model doesn't consider operation of the water level in case of the expectation of rainfall or a drought.

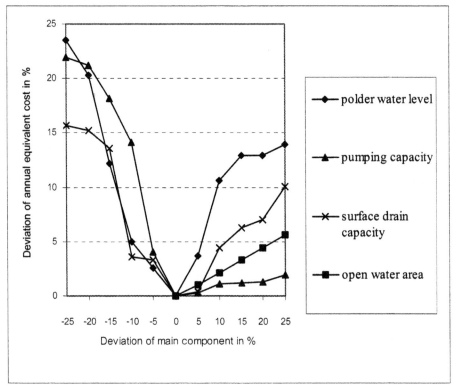

Figure 7.47 Deviation of the main components and of the annual equivalent costs in percentage for a rice polder under irrigated conditions

Discussion

Based on the results as shown above the following points can be discussed.

In Figure 7.47 it is shown that the component, which has most influence on the annual equivalent costs, is the polder water level, hence a lower polder water level during the rainy season is recommended, but during the dry season it can be kept higher to store water for irrigation.

An increase in pumping capacity and in the height of the bund at the side adjacent to the open water area are recommended.

A shallow polder water level will be beneficial for irrigation water supply by gravity to the crops, but the probability of overtopping of the surrounding dike is higher.

The values in Table 7.106 are only indicative for improvement of the water management in irrigated rice conditions. In practice land use, gradient of land, structure, soil conditions, etc. have to be taken into account for individual cases.

7.6 A dry food crops polder

7.6.1 General

The experiment for the dry food crops polder was done from 1 January 2003 to 31 December 2003 at Samchuk Irrigation Water Use Research Station. General in formations of the experimental area have been given in section 7.4.1. Layout of the experimental area is shown in Figure 7.31.

7.6.2 Physical conditions

The experiment area was in the lower part of the area, where rice was usually grown. By excavation ditches and beds were made by filling soil between the ditches. This resulted in a raised bed system for the growth of dry food crops.

7.6.3 Set-up of the experimental plot

The distance between the ditches and bed system was made according to the general practice in Thailand (Figures 7.48 and 7.49).

Figure 7.48 Dry food crops experimental field at Samchuk Irrigation Water Use Research Station, 2003

Water management and water level control

The ditch water level in the plot was manually controlled by pumping for both irrigation and drainage.

The water management in this area was done for both irrigation and drainage by pumping. The water level in the ditches was kept at 0.60 m-surface (bed level). During dry periods the water was pumped to the ditches when the ditch water level decreased to 0.65 m-surface and pumping was stopped when the ditch water level had increased to 0.55 m-surface. In case of rainfall the pump started when the ditch water level increased to 0.55 m-surface and stopped when the ditch water level decreased to 0.60 m-surface.

Irrigation water was given when the soil was dry and the moisture content of the soil in the upper layer was lower than field capacity, by spraying water over the crops area. The amount of irrigation water was controlled by the spraying time and the capacity of the sprinkler.

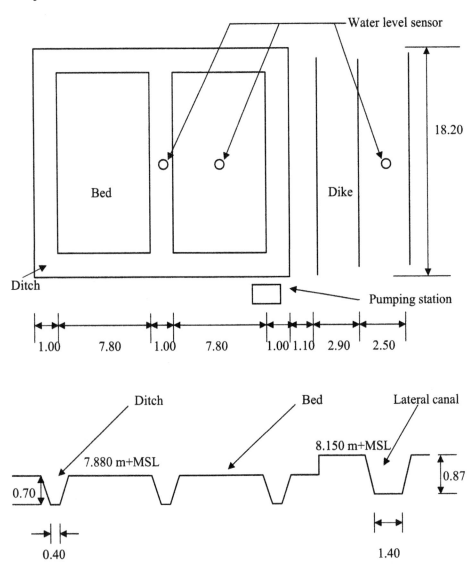

Note: All dimensions are in metre

Figure 7.49 Schematic layout and cross-section of the area for the dry food crops experiment

Monitoring and data recording for the dry food crops experimental plot

The following data were recorded from 1 January 2003 to 31 December 2003:
- hourly rainfall in mm;
- daily pan evaporation in mm;

-   discharge pumping outof the plot to the drainage canal;
-   pumping into the plot;
-   pumping to irrigate the crops area;
-   groundwater table in the beds;
-   water level in the ditches;
-   surface water level or groundwater table in the drainage canal;
-   crops and agricultural practice schedule;
-   crops growth and growth stage;
-   investment costs and time of agricultural practice;
-   production.

The data for hydrological and surface water level or groundwater table was generally recorded by using automatic recording sensors and the data are stored in the data logger. The data can be downloaded to the microcomputer by connecting directly to the data locker or connecting with modem, which interface to the telephone line. In this experiment, the method for data collection was done by loading data from data logger through telephone modem. For other data such as discharge, crops growth, irrigation and agricultural practice were manually recorded.

The crops which grown in the area was Chinese kale (Brassica oleracea var. alboglabra) because this crop is commonly grown in polder area in Thailand. The crops season was divided into 4 crops for 2003. The start of the first crops was on 1st of February and harvested on 21st of March. The start of the second crops was on 1st of May and harvested on 24th of June. The start of the third crops was on 1st of August and harvested on 15th of September. The start of the forth crops was on 1st of November and harvested on 21st of December.

In order to get the required information for calibration and verification of the model the groundwater table in bed, the ditch water level, the lateral canal water level in dry food crops were observed during the whole year.

The soil properties are the same as indicated in the irrigated rice polder because the experimental plot for the dry food crops is adjacent to the irrigated rice plot (Figure 7.31 and Tables 7.89 and 7.90).

Discussion on the observation results

The groundwater table in the bed during the dry period is less than water level in the ditch in summer due to evapotranspiration from the soil more than water flow from the ditch to the soil. In the wet period the groundwater table in the soil is higher than the water level in the ditch due to rainfall much more than evapotranspiration from soil. The dry period the irrigation by spraying water was needed especially if the moisture in the upper layer is less than field capacity. But in the wet period discharge from the soil to the ditches and pumping out of the plot was needed. In these 2 situations, the soil permeability, soil hydraulic conductivity, the distance between the ditches and ditch water level play an important role for both irrigation and drainage.

The clay mineral of the soil in the study area compose of little portion of Montmorillonite, that means that the swell is not high for this soil mass. The irreversible process of shrinkage of clay soil caused by Kaolinite and Illite will be dominated in this soil mass. This implies that the permeability of soil will little change when it becomes wet or dry especially in the sub soil.

The groundwater table responded fast to the ditch water level implied that the permeability of soil is high. The permeability of soil from the laboratory test results was low, but in fact the water can flow through cracks.

The average crops yield was 22 ton/ha/crop season, which is common yield for this crops.

7.6.4 Hydrological and water management data

Data for the hydrological analysis and water management computation

For this case study the following hydrological and water management data have been used in the analysis:
-   the data from 1 January 2003 to 31 December 2003 for the model calibration:
    -   daily class A pan evaporation in mm, Samchuk Irrigation Water Use Research Station, Suphanburi;
    -   hourly rainfall in mm, Samchuk Irrigation Water Use Research Station, Suphanburi;
    -   hourly discharge pumping to the lateral canal from the experimental plot;
    -   hourly irrigation input by pumping into the plot in mm;
    -   hourly spraying water to the crops;
    -   hourly groundwater table in the bed;
    -   hourly water level in the ditches;
    -   hourly surface water level or groundwater table in the lateral canal;
-   the data for validation and computation:
    -   soil characteristics from Laboratory, Royal Irrigation Department, 2003;
    -   hourly rainfall in mm at Ladkrabang station, Bangkok, 1991 to 1999;
    -   daily class A pan evaporation in mm, Bangkok Metropolitan,1991 to 1999.

Characteristics of a typical dry food crops polder

In this study it has been assumed that the typical polder has a total area of 10,000 ha for rural area. The soil storage is assumed to exist in the upper at 0.60 m, which consists of 0.15 m of ploughed layer (layer I) and 0.45 m of layer II. Only Chinese kale was grown. Based on average conditions the area of the plots was about 1.0 ha. The dimensions of plot were 100 * 100 m and there were six plots in one parcel as shown in Figure 7.50. The ground surface level at centre of bed in this area is set at 7.850 m+MSL. Bed level of the ditch is 7.144 m+MSL. The side slopes of the watercourses are:

Ditch   = 1: 0.5;
Lateral canals   = 1:1.5;
Sub-main canals   = 1:3;
Main canals   = 1:3.

The soil moisture tension computation in the model is based on an empirical relation as given in Equation 7.2. The constants a, b and c for the different soil profiles are shown Tables 7.107, 7.89 and 7.90 (see Appendix VI).

Table 7.107 Clay content in the soil profiles of a dry food crops plot at Samchuk Irrigation Research Station

Range from surface level in m	Typical soil type	Description	Clay content in %
0.00 - 0.30	Ploughed layer	Loam	21.7
0.30 - 0.60	subsoil	Loam	22.7
> 0.60	subsoil	Loam	22.5

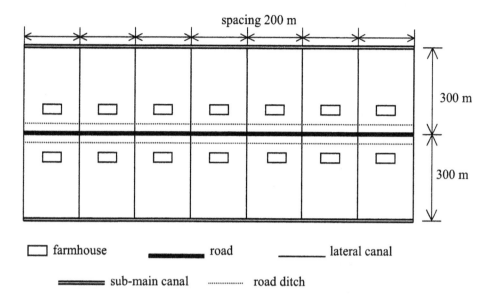

farmhouse   road   lateral canal

sub-main canal   .... road ditch

Figure 7.50 Schematic layout of a parcel in a typical dry food crops polder in Thailand

Potential evapotranspiration is based on the class A pan evaporation in mm/day at Sirigiti National Conference Center station. The crop factors multiplied with the class A pan evaporation are taken from literature (Royal Irrigation Department, 1994). The crop factors that have been applied to compute the potential evapotranspiration from the class A pan were based on four crops of rice per year. The start of the first crops is on 1st of February and harvested on 21st of March. The start of the second crops is on 1st of May and harvested on 21st of June. The start of the third crops is on 1st of August and harvested on 21st of September. The start of the forth crops is on 1st of November and harvested on 21st of December. The crop factors are given in Table 7.108.

Table 7.108 Crop factors (Kp) for evapotranspiration for Chinese kale (Royal Irrigation Department, 1994)

Month	Decade		
	I	II	III
January	0.21	0.21	0.21
February	0.53	0.58	0.62
March	0.64	0.61	0.25
April	0.21	0.21	0.21
May	0.53	0.58	0.62
June	0.64	0.61	0.25
July	0.21	0.21	0.21
August	0.53	0.58	0.62
September	0.64	0.61	0.25
October	0.21	0.21	0.21
November	0.53	0.58	0.62
December	0.64	0.61	0.25

Note: The above values are based on a coefficient by assuming four crops of Chinese kale per year.

The saturated hydraulic conductivity from the result obtained by the Royal Irrigation Department Laboratory was used. The saturated hydraulic conductivity of the cracked soil has also been calculated by Equation 6.172. The model computes the unsaturated hydraulic conductivity and capillary flow using Equations 7.2 to 7.4.

7.6.5 Economical data

Costs for the water management system

The costs for the water management system include the following:
- costs for ditch system;
- costs for the canal system;
- costs for pumping.

The costs for the ditch and canal system are computed using the formulas as given in chapter 6 by using the unit costs as given in Table 7.109. The interest rate is estimated at 5%.

Table 7.109 Unit costs for the ditches and the canal system under conditions of a dry food crops polder (Royal Irrigation Department, 2002)

Item	Construction cost in €	Annual maintenance cost in €	Lifetime in years
Cost of earth movement per m^3	1.20	0.05	50
Cost of construction dike per m^3	2.50	0.25	50
Cost of PVC pipe per m	1.50	0.15	25

The pumping head is based on an average of main canal water level of 3.0 m-MSL and an average water level in the lateral canals of 0.50 m-MSL. The annual rainfall surplus and seepage are estimated to be around 370 mm and 183 mm for the estimation of the operation cost for pumping. The Drainage modulus is estimated at 46 mm/day. The installation and operation and maintenance cost are given in Table 7.110.

Table 7.110 Pumping costs under conditions of a dry food crops polder

Installation in million €	Operation in million €/year	Maintenance in million €/year	Lifetime in years
11.841	0.2033	0.0861	50

Costs for field pumping for irrigation and drainage to the plots have been estimated from unit cost of the Royal Irrigation Department in 2001 at 0.017 €/m^3. The costs for spraying water to crops on the bed by movable pumping on a small boat have been estimated at 0.021 €/m^3.

Costs for buildings, infrastructure and crops

The average farm size has been estimated at 5 ha. The value of the buildings is estimated at 12,500 €/farm. The area of buildings and yards is estimated 1,000 m^2/farm. For facilities such as shops, factories, schools, etc. 20% of the value of the farms has been taken.

The infrastructure in rural area is mainly composed of roads. In order to compute the cost of infrastructure, the density of paved roads is determined at 100 m^2/ha (Figure 7.50) and 6 m width. The estimated cost of roads is 6.0 €/m^2.

In the typical dry food crops polder there is only Chinese kale. The investment costs of this crop are separated based on the activities of agricultural practices and other activities related to the productivity of crops such as land acquirement, transportation, etc. Normally the investment costs are dependent on the time of the growing seasons. The investment costs and time schedule for the crops are shown in Table 7.111 and Figure 7.51.

Table 7.111 Agricultural activities, investment costs and scheduling for Chinese kale

Activity	Investment cost in €/ha	Starting date
1 lease/rent	148.81	1 January
Crop1		
2 ploughing	79.52	15 January
3 a fertilizers	90.70	15 January
b spreading	67.17	15 January
4 a seeding/spreading seed	126.10	18 January
b planting selection	100.45	1 February
5a fertilizers/pesticides	192.56	5 Feb. - 10 March
b spraying	128.34	5 Feb. - 10 March
6 harvesting & transportation	268.73	17 March
Crop2		
7 ploughing	79.52	15 April
8a fertilizers	90.70	15 April
b spreading	67.17	15 April
9 a seeding/spreading seed	126.10	18 April
b planting selection	100.45	1 May
10a fertilizers/pesticides	192.56	5 May - 8 June
b spraying	128.34	5 May - 8 June
11 harvesting & transportation	268.73	15 June
Crop3		
12 ploughing	79.52	15 July
13a fertilizers	90.70	15 July
b spreading	67.17	15 July
14 a seeding/spreading seed	126.10	18 July
b planting selection	100.45	1 August
15a fertilizers/pesticides	192.56	5 Aug. - 7 Sep.
b spraying	128.34	5 Aug. - 7 Sep.
6 harvesting & transportation	268.73	15 September
Crop4		
16 ploughing	79.52	15 October
17a fertilizers	90.70	15 October
b spreading	67.17	15 October
18 a seeding/spreading seed	126.10	18 October
b planting selection	100.45	1 November
19a fertilizers/pesticides	192.56	5 Nov. - 8 Dec.
b spraying	128.34	5 Nov. - 8 Dec.
20 harvesting & transportation	268.73	15 December
Total	4,363.09	
8 gross profit	7,631.26	

Source: Asian Engineering Consultants Co. Ltd and Macro Consultants Co. Ltd, 1996
   http://www.doae.go.th/library/html/detail/paddy/A1.htm

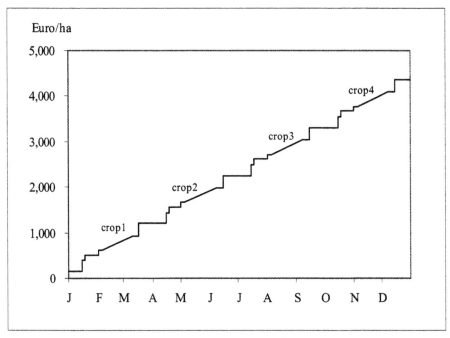

Figure 7.51 Investment costs and scheduling for Chinese kale

Cost of damage to crops, buildings and infrastructure for a dry food crops polder

The method of calculation and the parameters used for estimating cost of crop damage are described in Chapter 6. Damage to buildings and infrastructure is estimated as a percentage of value of buildings and infrastructure in relation with the groundwater table as shown in Table 7.11.

7.6.6 Calibration of the parameters

Calibration of the parameters of model in a dry food crops polder area for Thailand conditions involved the determination of the hydrological parameters, which was explained in Chapter 6 as follows:
- maximum interception in mm E_{max};
- potential evapotranspiration from layer I in mm/m V_{ep};
- parameters f_r and d_a in the discharge formula;
- runoff coefficient C,
- maximum moistening rate in layer I in mm/m/hour V_{max1};
- maximum moistening rates in the columns of layer II in mm/m/hour V_{max2}.

Sensitivity of the parameters for a dry food crops polder

Table 7.112 Calibrated parameters of the hydrological model for a dry food crops plot

Year	Maximum intercept in mm	Upper limit for potential evapotranspiration in mm	Parameter discharge model		Runoff coefficient	Maximum moistening rate in mm/m/hour		Goal function	Annual discharge computed in m³
			f_r	d_a		Layer I	Layer II		
2003	2.7	21.5	0.00029	0.53	0.51	2.52	2.93	0.4296	192

The interception is a function of canopy coverage, type of vegetation, the crop growth stage and the agricultural practice during the year. The canopy coverage of the Chinese kale is assumed to be constant all over the year. The optimal value of the maximum interception capacity has been found between 2.7 mm. In really the interception in this has changed through the whole year due to cropping pattern.

The upper limit for potential evapotranspiration has been influenced by the water losses from the soil through evapotranspiration especially during the wet period when the soil moisture is almost above field capacity. However, this parameter has little effect in the computation of yearly discharge. The calibrated parameter value for this case was the result of the calibration in the year 2003 as shown in Table 7.112, which is 21.5 mm.

The empirical reduction factor accounts for tortuousity and necking of the cracks as described by Equation 6.172. This factor has strongly influenced to the permeability of soil and the discharge from the groundwater zone in the bed. A small increase in this factor will cause a significant change in groundwater table in the bed. An increase of this parameter the more of permeability and will result in a higher discharge from the soil to the ditches or a higher recharge into the soil dependent on condition of the ditch water level and the groundwater table in the bed. The calibrated parameter value was found through the calibration of 0.00029.

The diameter of aggregates is property of soil for slit model was described by Equation 6.172. Increase of this parameter will give less permeability and effect to less fluctuation of groundwater in the bed due to less discharge out or recharge into the soil in the bed. This parameter can be observed at the adjacent plot when soil is dry but it was found through the calibration of 0.53 m.

Runoff coefficient is the fraction of rainfall, which transforms to surface and recharge to the groundwater in the soil in the bed system during the wet period. This parameter depends on moisture status in the soil surface, cracks, permanent macrospores in the soil, coefficient of linear extension of soil and moistening rate of the soil. Hence this factor will give also much influence to groundwater table. The higher value of this factor, the higher the runoff into the ditch and the less the recharge the groundwater zone in the bed. However, this parameter was taken as constant value through whole period during year. The parameter was found through calibration of 0.51. Actually this factor is varying during the period dependent on soil moisture condition.

In this study the maximum moistening rate in layer I has less sensitivity than in layer II because the depth of layer I is less than layer II. The calibrated value for maximum moistening rate in layer I and layer II was the result of the calibration in the year 2003, which is 2.52 and 2.95 respectively.

7.6.7 Model validation with fixed calibrated parameters

The model has been validated with calibrated parameters in Table 7.112 for 2003 as shown in Figures 7.52 to 7.55.

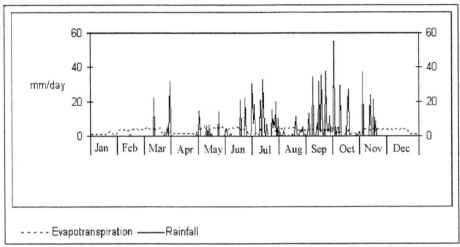

Figure 7.52 Rainfall and evapotranspiration of dry food crops in the experimental plot at Samchuk Irrigation Water Use Research Station, 2003

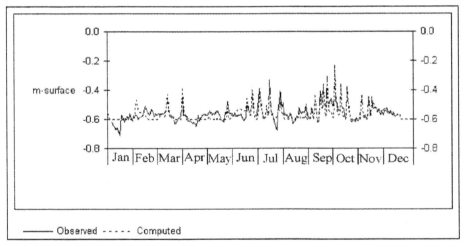

Figure 7.53 Computed and observed of groundwater table at the centre of the bed of dry food crops in the experimental plot at Samchuk Irrigation Water Use Research Station, 2003

From Figure 7.53 the groundwater from computed and observed was well fit with each other during the wet period but during the dry period was not so well fit because of the influence of fluctuations of the ditch water level due to manual control.

In Figure 7.54 computed and observed of discharge out of the plot is shown. Total computed discharge out of plot is 192 m³ or 573 mm while total measure discharge out of plot was 288 m³ or 848 mm for 2003. Total computed irrigation water of 460 m³ or

1,359 mm was pumped into the plot while total observed amount of pumping of water into plot was 545 m^3 or 1,610 mm for 2003. Model efficiency is 0.578. The observed irrigation water which was pumped into the plot is higher than the computed because during the dry period there was loss of water from the plot to the adjacent area. While pumping out of the water from the plot is higher than the computed because there was some runoff from the road near the plot flow into the plot during heavy rainfall and the water level in the adjacent area was high in this period so there was also seepage flow into the plot.

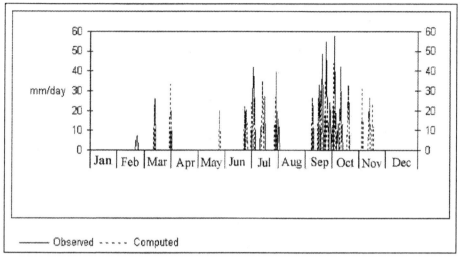

Figure 7.54 Computed and observed of discharge out of the dry food crops experimental plot at Samchuk Irrigation Water Use Research Station, 2003

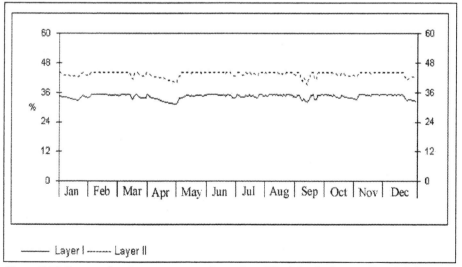

Figure 7.55 Computed moisture content in layers I and II of dry food crops, Samchuk Irrigation Water Use Research Station, 2003

In Figure 7.55 moisture in layer I is higher than in layer II because there is irrigation water given to the soil surface by sprinkler. But during the period when there was no crop and no irrigation water was sprayed, the moisture in layer I decreased more rapidly than in layer II. The soil may dry enough during the dry period for the cracks to develop. Due to cracks the moisture soil in layer II increased more rapidly than in layer I after suddenly heavy rain because soil in layer II receive both rain water that fall into cracks and runoff that fall into the cracks. After the soil layer I has more moisture the moisture in the soil layer II increase slowly due to the cracks area on the surface of layer I is decrease or close.

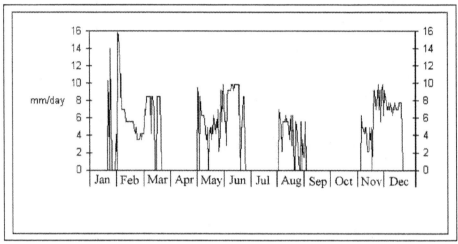

Figure 7.56 Irrigation by spraying water to the bed of dry food crops in the experimental plot at Samchuk Irrigation Water Use Research Station, 2003

From Figure 7.56 In the first crops season the irrigation by spraying into the field is high because the operator did not know the right amount of water that need by the crops. Therefore there was over irrigated in the early of the growing period of the first crop. The second crops season was grown during summer with dry and high temperature period, some amount water was given to the crops to reduce the temperature otherwise the crops will not properly grow. The third crops season the amount of spraying was small because there is a lot of rainfall especially from mid to late of crops season. While the forth crops season is not so high temperature but rather dry and low humidity so there was relatively more water is needed to spray to the crops compared to the third crops season.

7.6.8 Analysis with the optimization model

Hydrological analysis

The hydrological analysis with the validated model for the experimental plot has resulted in the data as given in Table 7.113.

Table 7.113 Review of annual data in mm for dry food crops in the experimental plot

Year	Rainfall in mm	Evapotranspiration in mm	Discharge in mm		Capillary rise in mm	
			Into plot	Out of plot	II → I	III → II
2003	870	1,097	1,358	567	221	86

Economic computation

The computation was done based on the total area of polder is 10,000 ha. The results and review computations for optimal conditions is shown in Tables 7.114 to 7.117.

Table 7.114 Review of the total values of crops, buildings and infrastructure in a polder under conditions of dry food crops at optimal water management

Item	Area in ha	Highest value		Lowest value	
		in € per ha	in € * 10^6	in € per ha	in € * 10^6
Buildings	232	150,000	34.74	150,000	34.74
Infrastructure	98	60,000	5.85	60,000	5.85
Chinese kale	9,420	1,908	17.97	149	1.40
Total	9,751		58.56		41.99

Table 7.114 shows that the value of buildings is about 59% of the total value in a polder under conditions of dry food crops. Therefore the damage to buildings may be the most important in a dry food crops polder.

Table 7.115 Optimization analysis of the main components of the water management system for a polder under conditions of dry food crops

Distance between ditches in m	Ditch water level in m-s.	Polder water level in m-s.	Open water in %	Pumping capacity in mm/day	Field drainage capacity in mm/day	Costs in € * 10^6	Damage in € * 10^6	Value of the goal function * 10^6
8.45	0.74	-0.06	2.4	36.5	34.8	4.883	8.535	13.419

Table 7.115 shows that at optimal conditions the polder water level is above the ground surface, which means that irrigation water, can flow to the plot by gravity.

Table 7.116 Total costs for the water management system under conditions of a dry food crops polder at optimal water management

Item	Construction costs in € * 10^6	Annual maintenance costs in € * 10^6	Annual equivalent costs in € * 10^6
Field drainage	22.53	1.28	2.88
Open water	17.89	0.75	1.73
Pumping	0.68	0.18	0.22
Total	41.10	2.21	4.83

Table 7.116 shows that the overall annual maintenance costs in a polder under conditions of dry food crops are about 5.4% of the construction costs, while the overall annual equivalent costs are about 11.7%. The drainage system constitutes the highest part of the total construction costs, which is about 54.8%. The drainage costs are high because construction the ditches required a large amount of excavation. While pumping

costs are the lowest amount, which are about 1.7%. This may be caused by low operation head of pump.

Table 7.117 Specification of average annual damage and yield reduction under conditions of a dry food crops polder at optimal water management

Item	High groundwater tables		Drought	
	in €	in % of the value	in €	in % of the value
Buildings	2,976,000	8.6	0	0.0
Infrastructure	178,000	1.4	0	0.0
Chinese kale	2,929,000	4.1	2,385,000	3.3
Total	6,083,000	5.4	2,385,000	3.3

In Table 7.117 it can be seen that even through there is irrigation was given in this simulation, but due to high moisture tension in soil layer the drought was occurring. The ditch water level is one of the important factors that cause yield reduction and this phenomenon was reported by Chaudhary, et al., 1975. It can be described that the moisture tension in soil is increased before irrigation water was given when there is a deeper groundwater table.

Economical analysis for a water management system at the present situation

In practice the control criteria for polder water level is varied throughout the year. The cropping area and crop types in the polder area are also mixed.

The general values of the main components of the water management system of a dry food crops polder may be approximated as follows:

- distance between the ditches = 6 to 8 m;
- ditch water level = 0.60 m;
- polder water level = 0.50 m-surface;
- open water area = 1.10%;
- field drain capacity = 46 mm/day;
- pumping capacity = 26 mm/day.

The results and review computations for the practical conditions are shown in Tables 7.118 to 7.121.

Table 7.118 Review of the total values of crops, buildings and infrastructure in a polder under conditions of dry food crops for water management in practice

Item	Area in ha	Highest value		Lowest value	
		in € per ha	in € * 10^6	in € per ha	in € * 10^6
Buildings	227	150,000	34.11	150,000	34.11
Infrastructure	96	60,000	5.74	60,000	5.74
Chinese kale	9,249	1,908	17.65	149	1.38
Total	9,572		57.50		41.23

Table 7.118 shows that the value of buildings is about 59.3% of the total value in dry food crops polder. Therefore the damage to buildings may be the most important in a dry food crops polder.

Table 7.119 Optimization analysis of the main components of the water management system for a polder under conditions of dry food crops for water management in practice

Distance between ditches in m	Ditch water level in m-s.	Polder water level in m-s.	Open water in %	Pumping capacity in mm/day	Field drainage capacity in mm/day	Costs in € * 10⁶	Damage in € * 10⁶	Value of the goal function * 10⁶
7.0	0.60	0.50	2.4	26.0	46.0	5.044	26.616	31.661

Table 7.119 shows that damage is higher than optimal conditions in Table 7.115 and also the costs are higher than optimal conditions in Table 7.115. Damage is higher due to higher groundwater tables in the bed than optimal conditions due to shallower ditch water level. The costs are higher than optimal conditions due to higher construction and maintenance cost of the ditches due to less distance between the ditches, more construction and maintenance cost of canal system due to lower polder water level and more field drain capacity than optimal conditions.

Table 7.120 Total costs for the water management system under conditions of a dry food crops polder for water management in practice

Item	Construction costs in € * 10⁶	Annual maintenance costs in € * 10⁶	Annual equivalent costs in € * 10⁶
Field drainage	22.25	1.26	2.84
Open water	20.24	0.84	1.95
Pumping	0.60	0.16	0.19
Total	43.09	2.26	4.98

Table 7.120 shows that the overall annual maintenance costs for water management in practice in a polder under conditions of dry food crops are about 5.3% of the construction costs, while the overall annual equivalent costs are about 11.6%. The drainage system constitutes the highest part of the total construction costs, which is about 51.6%. While pumping costs are the lowest, which are about 1.4%. From Table 7.120 the open water costs are higher than at optimal conditions (Table 7.116) because depth of excavation of main canal increases due to polder water level at practical is lower than at optimal conditions. The pumping costs are lower than optimal conditions because of lower pumping capacity. But the cost of the drainage system at optimal conditions is higher than at practical due to lower the ditch water level increases excavation cost.

Table 7.121 Specification of average annual damage and yield reduction under conditions of a dry food crops polder for water management in practice

Item	High groundwater table		Drought	
	in €	in % of the value	in €	in % of the value
Buildings	3,522,000	10.3	0	0.0
Infrastructure	1,734,000	13.7	0	0.0
Chinese kale	15,935,000	22.6	2,254,000	3.2
Total	21,191,000	19.2	2,254,000	3.2

In Table 7.121 the damage due to high groundwater table in crease compare to at optimal conditions (Table 7.117) because of the ditch water level at practical conditions is shallower than optimal conditions. But the drought at practical conditions is less than optimal conditions because at practical conditions the crops can be more benefit by capillary rise due to shallower ditch water level. While at optimal conditions no need to pump for irrigation. However, the polder water level may be cannot kept so high due to high probability of overtopping the surrounding dike around the dry food crops plot and the damage may be very high. The model calculated the damage of the surrounding dike in the same way as damage of infrastructure due to high groundwater tables under influence of the polder water level.

Table 7.122 Comparison between the optimal and practical values for the main components of the water management system in a polder under conditions of dry food crops

Item	Distance between ditches in m	Ditch water level in m-s.	Polder water level in m-s.	Open water in %	Pumping capacity in mm/day	Field drainage capacity in mm/day	Costs in € * 10⁶	Damage in € * 10⁶	Value of the goal function * 10⁶
Optimal	8.45	0.74	-0.06	2.4	36.5	34.8	4.883	8.535	13.419
Practice	7.00	0.60	0.50	2.4	26.0	46.0	5.044	26.616	31.661
% differ.	26	21	112	0	40	-24	-4	-68	-57

In Table 7.122 it could be considered to lower the ditch water level from 0.60 to 0.74 m-surface at first priority for existing areas. For the new developments the distance between the ditches may be increased from 6 to 8.45 m. Increasing the pumping capacity could also be considered. The polder water level, especially during the dry period, can be kept higher than the bed level of the ditch to enable taking of water to the plot by gravity.

In Figure 7.57 the influence of deviations from the simulated optimal values is shown. It shows that the ditch water level has most influence on the annual equivalent costs in the system. If the ditch water level is 25% shallower compared to the optimal value the annual equivalent costs will increase approximately 225%, because of the increased damage to the crops due to too high groundwater tables. A ditch water level, which is 25% deeper than the optimal value increases the annual equivalent costs by approximately 148%, because of the extra construction cost, damage due to drought and more pumping water to spray the crops due to less capillary rise. When the distance between the ditches increases 25% from the optimal value, the annual equivalent costs increase approximately 60% because the damage due to too high groundwater tables increases. A decrease of the distance between the ditches has less effect, because only the costs for construction of the ditches increase. Changes in the field drain capacity, polder water level, and pumping capacity have only a limited effect.

Discussion

Based on the results as shown above the following points can be discussed.

The component, which has most influence on the annual equivalent costs, as shown in Figure 7.57, is the ditch water level (below the level of the raised beds), hence a lower water level is recommended for the improvement of the water management conditions for dry food crops.

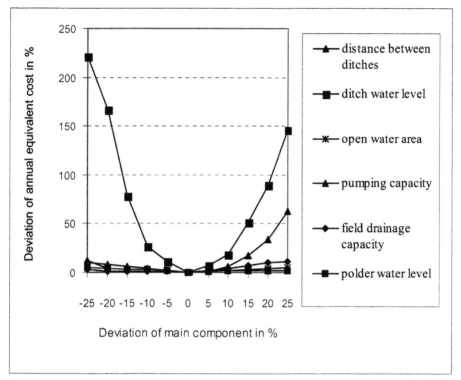

Figure 7.57 Deviation of the main components and of the annual equivalent costs in percentage for a polder under conditions of dry food crops

A deeper ditch water level and a smaller distance between the ditches can reduce the yield of the crops because the moisture tension before irrigation may be higher and there is less benefit of capillary rise.

A shallow polder water level will be beneficial for irrigation water supply by gravity to the crops, but the probability to overtopping of the surrounding dike of the plot will be higher.

At present the pumping capacity is low (Table 7.122). There has been an attempt to increase the pumping capacity in some polders in Thailand. The simulation results show that this is useful.

The distance between the ditches and the ditch water level is the most sensitive factor to crops damage, the more the distance between the ditches and the less ditch water level the higher the groundwater table during the wet period. Hence the more damage will occur.

The water level in the lateral canal in a dry food crops polder may be kept high to prevent loss of water from the plots and to get benefit for irrigation water that will flow through the culvert to the plots when the water level in the plot is low. However, this will create more risk to inundation of the field due to heavy rainfall. In practice the water level in the polder will be lowered before a storm comes.

In reality the water level in the canal system may be affected by topography and the gradient in the water when pumping, but the model determines the water level by the volume of water that is stored in the canals. This may be a reason as well why the polder water level in practice and in the simulations is different.

The higher pumping capacity, the higher the polder water level can be kept and the farmers can have more benefit from water that flows to the plot by gravity.

7.7 The rural polder

7.7.1 General

The study is located at intersection between latitude 13° 30' North and longitude 100° 50' East. The project area occupied Prakanong region in Bangkok, Bangplee district, Muang district in Samut Prakan province (Figure 7.58). It is bordered by Klongpravetburirom canal to the North, The Chao Phraya river dike to the West, King Initiated dike to the East and Klongchythale canal to the South.

Figure 7.58 Location of the rural polder in Thailand

7.7.2 Physical conditions

Topography

The study area is flat lowland situated between the Chao Phraya river mouth and the Bangpakong river mouth. The area is sloped to the Gulf of Thailand (Figure 7.59). The ground level is between 0.50 to 1.50 m+MSL and declined from the North to the South. However, the ground level is always changed due to settlement caused by pumping of groundwater in the area and landfill in the urban area.

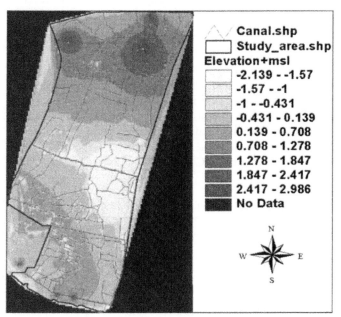

Figure 7.59 Topography of the rural polder in Thailand (Source: Royal Irrigation Department, 2001)

The report of land subsidence from 1983 to 2001 has been estimated at 0.10 - 0.20 m (Public Work Department, 1988) and landfill in the area ranges from 1.35 m+MSL to 0.50 m+MSL the average fill level is estimated at 0.97 m+MSL for the urban area. Based on average conditions the different of the ground surface in urban and rural area consider of land subsidence is 0.689 m for present situation.

Agricultural practice

Rice was grown in this area by both spreading and transplanting method. In rainy season the rice in rather low area, the farmer grows floating rice by spreading rice directly to the field. Normally floating rice is grown in the adjacent the main canals. While in the higher land, rice is grown by transplanting. The main season crops is grown in mid August harvest in late November, while the second crops is grown in late of January harvest in early May. Due to the second crops rice is grown in the dry season the spreading seedling rice method is applied to avoid water shortage in the early stage of growth. The seedling rice is prepared by soaking rice seed about 12 hours and after that keep it in a humid place 24 to 36 hours. This seedling rice should have a length of the root about 1 to 2 mm when it is spread.

Fishery is occupied a large portion in the study area. The fishpond normally was constructed by modification of the rice field by excavating and making the pond deeper than rice field and using the excavated material to enforce the bund. The water depth in the pond is between 1.0 to 1.50 m. The parcel area of the fishpond is about 0.60 - 0.80 ha. The most fish in this area is called 'Plasalit'. The time for growing fish farm is usually between 8 to 10 months starting from July and harvesting between March and April. After harvesting the fish farmer will let the pond dry for 1 month to clean the pond by sunlight and dredging the pond bed, then the water will be pumped to the pond again. The time schedule for fish farm can be changed throughout the year. However, it can be seen that fishpond requires a large amount of water through out a year and it can behave as reservoir to store rainwater in the polder.

Land use

The most area is used for fishpond or rice field (Figures 7.60 and 7.61). The land use area composition is presented Table 7.123. There is small part of the land where it is used for fruit trees by making of raised bed system. In this area there are a lot of canals connecting each other as a network, which are used for both water supply and drainage. In the southern part is affected by sea. Most of the land in this area was irrigated. The soil in the area is clay in both topsoil and subsoil. Consequently the drainage capacity of the soil is poor.

Table 7.123 Land use in the rural polder in Thailand, 1994

Type of land use	Area in ha	Area in %
Shrimp pond	103	0.6
Urban and commercial area	3,335	17.8
Developing land	1,230	6.6
Grassland	87	0.5
Rice field	4,726	25.1
Fish pond	6,771	36.0
Forest	117	0.6
Swamp	282	1.5
Industrial area	1,614	8.6
Recreation area	187	1.0
Fruit trees	217	1.2
Miscellaneous area	91	0.5
Total	18,760	100.0

Source: Geographic Information Center, Royal Irrigation Department

This area is located near the sea. The weather is warm humid, the rain is normally reliable in rice crops season. Although the land near sea shore is rather saline, but due to drainage of saline water from the area and receiving fresh water from the northern part of the area, the yield of the rice in this area is still good.

Urban and commercial area is located in the northern part of the area. Most of the urban areas are surrounded by rice fields and fishponds.

The area is simplified into 2 categories are urban area of 6,367 ha and agriculture area of 12, 394 ha as percentage as follows: rice field is 39.8% and fishpond is 60.7% of total agricultural area. Because the fishpond gave more benefit to the farmer so most of the rice field in the study was changed into fishpond. Some agricultural area was also changed into urban due to increase in population. From the field observation and image from satellites image in the year 2001 (Figure 7.61) as follows urban was estimated to 7,390 ha and agricultural remained 11,370 ha and rice area occupied 12.8% and fishpond occupied 87.2%. The type of house for the urban area is shown in Table 7.124.

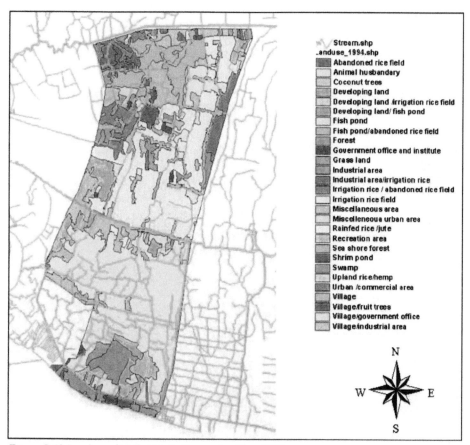

Figure 7.60 Land use distribution in the rural polder, 1994

Table 7.124 Composition of the residence area in the rural polder in Thailand

Type of residence	Percentage (%)
Single house	37.3
Twin house and town house	50.4
High buildings	12.1
Other	0.2
Total	100.0

Source: National Statistical Office, 1997

Soil characteristics

Soil in the area is alluvial soil or Typic Tropaquepts (USDA, 1970). Most of the soil in the area is Bangkok series (Figure 7.62). There are very little areas the soil belongs to Bang Nam Prieo series. Soil texture is clay, dark grey or dark greyish with mottles over grey with brown mottles over greenish grey. Organic matter content at depth from 0 to 0.30 m is between 1.0 to 1.5%. Effective depth is more than 1.50 m. Poorly drainage as described in USDA soil survey manual. Permeability is slow and the hydraulics conductivity is of less than 0.12 m/day.

Soil surface runoff characteristics is slow. Soil is subjected to little or no erosion hazards due to slow runoff.

Soil moisture at surface is above field capacity about 5 months due to rain and seepage (excluding irrigation) and 8 - 10 months for subsurface soils. Groundwater is 1.50 m-surface for 1 - 3 months.

Figure 7.61 Satellite image of the rural polder, 2001

Canal characteristics in the area

All canals in the area are earth canal with grass at bank. Some canals have floating weeds in the canals. The main canals width at the ground surface is between 20 to 40 m and the bottom level is between 2.50 to 3.00 m-MSL. The lateral canals width at the ground surface between 10 to 15 m and the bottom level is about 1.80 m-MSL. The side slope of the canal is about 1:1.5 and it is almost no slope at bed slope. The main and lateral canals are mostly natural canals and the distance between the lateral canals is about 1,000 m.

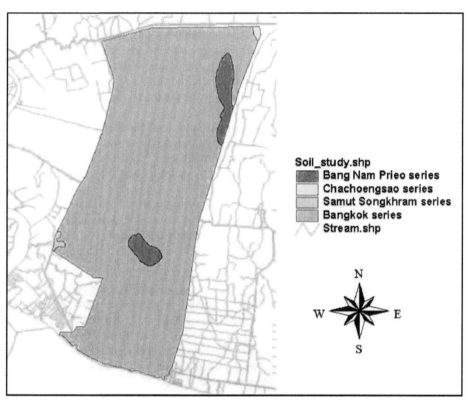

Figure 7.62 Soil map in a rural polder in Thailand. Source: Geographic Information Center, Royal Irrigation Department, 2001

7.7.3 Present situation of operation and maintenance

Water management

Water management in this area is called 'water conservation'. The water will be stored in the main and lateral canals before end of rainy season for the irrigation and other purposed. If the water stored in the canals is not enough for crops, the irrigation water has to be supplied through the dikes and the canal system (Figure 7.63). The water supply system normally composes of main canals, sub-main canals, later canals, and dikes. The irrigated water is distributed to the area through the regulating gates along the dikes for crops and other requirements. On the other hand in the rainy season the excess water have to be drained from the area to prevent flood damage.

The irrigated water is distributed to the area through the dikes by regulators (box culvert with gate). The irrigation water is influenced to flooding the area outside dike because in case of heavy rain in the area the gates along the dike have to be closed. The water level outside the area will rise up which cause flood damage outside this area. The water management in this area is depending on rainfall distribution during the year in the area.

Normally from May to October, there will be a lot of rain in the area. Hence the irrigated which is supplied through the dike must be stop in order to reduce amount of the water in the area. However, due to changing in land use from agriculture to urban and industrial area there will be less water demand for irrigation. Hence since 1985 there was no irrigation water was supplied through the dike. In case of heavy rain in the area the excess of water in the area is pumped out through one pumping station in the south of the area. Sometimes in summer the water inside the area is polluted, therefore the good water from outside area is sent to dilute the pollute water inside and then the mixed water is pumped out to the sea or river. Therefore the water, which is supplied to the area and rainfall, is important to the costs for pumping in and the amount of water to be drained from this area.

The rainfall in the area is influenced by tropical monsoon and storm; the proper way to manage the drain is to follow the weather forecasting. If the water manager found that there would be heavy rain caused by storm and the velocity of the storm is known, the water will be drained from the area, in order to increase the storage volume inside the area before heavy rain.

However, the water should be stored in the main and the lateral canals before the dry season for the irrigation and other purposed. In general between October and November is the period, which is starting to store water in the canals in this area.

Water management at field level

The farmers normally pump the water from main canals to lateral canals and then pump to irrigate the crop in the area. In rainy season usually the water level in the main canal is high enough. The farmer can get the water through the gate by open the gate and let the flow to lateral canals and then pump to the rice field. There are very few of rice area which can get the irrigation by gravity because the water level in rainy season must be kept low from the ground surface to increase storage volume in canal system in case heavy rainfall. In the dry season normally the water level in the main canals is low. The farmers have to pump water from the main canals to the lateral canals and then pump to the rice field or fishpond. If the water level in the fishpond is high due to heavy rain the farmers drain the water to the lateral canals by overflowing across the bund. But the rice area there is very little volume of water drained by this way because the fully growth rice can tolerate the water depth up to 0.40 to 0.50 m above ground level. The water management at the lateral canals and field level was done and organized by farmers.

Polder water level

In practice the maximum water level in the main canals is usually kept as follows:
-   rainy season between May and June is at 0.50 m-surface;
-   rainy season between July to October is at 0.80 m-surface;
-   dry season between November to December is at 0.30 m-surface (storing period).

In the model taking the polder water level control between January to April at 0.30 m-surface and other periods is the same as practical above. However in practice, this polder water level can be manually changed due to operation and weather forecasting.

Operation of the pump

At Tamru pumping station (pump E4 in Figure 7.64), which is located at the southern part of the area, compose of 6 pump units. Each unit has full capacity about 3.0 m³/s and efficiency is about 70%. These pumps are manually operated by Royal Irrigation Department's officer and the operation and maintenance cost come from the government budget. The operation of the pump is depended on the water level and the decision of the operator. In general the operation rule during wet period is shown in Table 7.125.

Table 7.125 Operation rule of the pump in the rural polder in Thailand

Water level in m+surface	Water depth in m+polder water level	No. of pumps in a unit	Total capacity in m³/s	Total capacity in mm/day
0 to 0.10	0.45 to 0.55	5 to 6	11.55	5.3
-0.20	0.25	4	8.40	3.9
-0.30	0.15	2 to 3	5.25	2.4
-0.40	0.05	1 to 2	3.10	1.4
-0.50	-0.05	0	0.00	0.0

In Table 7.125 water level above polder water level is computed by setting reference polder water level at 0.45 m-surface then it can be used for operation for pump in the model (Figure 7.63).

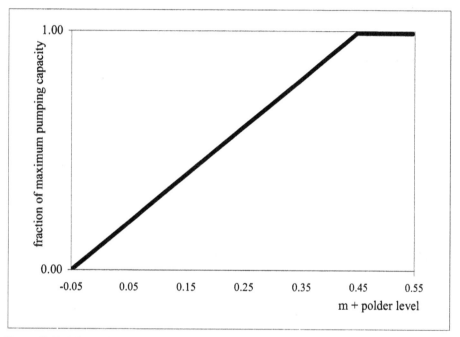

Figure 7.63 Relation between pumping operation and preferred polder water level in the rural polder in Thailand

The model application to this pumping capacity situation by following relation of the water depth above prefer water level and number of pump operation.

From Figure 7.63 the pumping capacity and water depth related to preferred polder water level is as follows:

If $I_w < -0.05$ then $G_c = 0$ $\hspace{4cm}$ (7.9)

If $-0.05 < I_w \leq 0.45$ then $G_c = (2 * I_w + 0.10) * G_{c\max}$ $\hspace{2cm}$ (7.10)

If $I_w > 0.45$ then $G_c = G_{c\max}$ $\hspace{5cm}$ (7.11)

where:
I_w $\hspace{1cm}$ = water depth related to preferred polder water level (m)
G_c $\hspace{1cm}$ = operation of pumping capacity (mm/day)
$G_{c\max}$ $\hspace{0.5cm}$ = maximum pumping capacity (mm/day)

7.7.4 Hydrological and water management data

The data were collected at the different locations as shown in Figure 7.64.

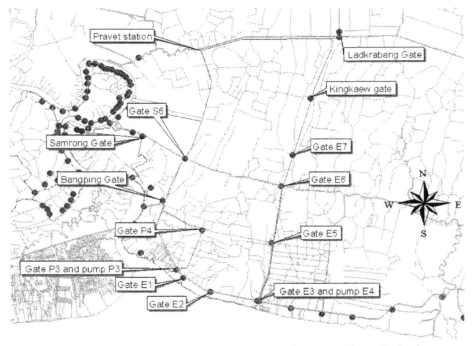

Figure 7.64 The location, name and type of the structure in the rural polder in Thailand

For this case study the following hydrological and water management data have been used in the analysis:
-   the data for 1998 and 2000 for the model calibration:
    -   daily class A pan evaporation in mm at Sirigiti National Conference Center station, Bangkok;
    -   daily average rainfall in mm at the station adjacent to area by Thiessen method;
    -   daily average rainfall in mm for 5 stations: Samrong Regulator station, Bang Tamru Regulator station (E4), Ladkrabang Regulator station, Pravetburirom canal station, and Bangping station for 1998;

- daily average rainfall in mm for 3 stations: Bang Tamru Regulator station (E4), Ladkrabang Regulator station, and Pravetburirom Canal station for 2000;
- hourly pumping discharge out of area at Tamru pumping station (E4);
- daily average polder water level at Tamru pumping station (E4);
- the data for validation and computation:
  - daily class A pan evaporation in mm at Sirigiti National Conference Center station, Bangkok, 1989 to 1998;
  - daily average rainfall average 3 stations: Bangping station, Ladkrabang Regulator station, Bang Tamru Regulator station by Thiessen method, 1989 to 1998.

Note: for 1999 rainfall data are not enough for calibration and for 2000 only 3 stations are available.

The soil profiles in the plot is a typical type of soil in Thailand selected a catalogue of major soil series of Thailand, Department of Soil Science, Kasetsart University, Kamphaengsean (Yingjajaval, 1993).

The soil moisture tension computation in the model is based on an empirical relation as given by Equation 7.2. The constants a, b and c for the different soil profiles are shown Tables 7.126 to 7.128 (see Appendix VI).

Table 7.126 Clay content in the soil profiles in the rural polder in Thailand (Yingjajaval, 1993)

Depth m-surface	Typical soil type	Description	Clay content in %
0.00 - 0.25	topsoil	clay	75.9
0.25 - 0.60	Sm (subsoil)	clay	73.4
> 0.60	Sm (subsoil)	clay	73.4

Table 7.127 Soil moisture tension in pF and soil moisture content in percentage in the rural polder in Thailand (Yingjajaval, 1993)

Layer \ pF	1.0	1.7	2.0	2.3	2.5	3.0	3.7	4.2
I	68.5	63.0	59.8	57.8	55.2	50.2	39.0	33.5
II	71.1	63.5	62.8	62.4	61.0	55.0	26.2	23.5
III	71.1	63.5	62.8	62.4	61.0	55.0	26.2	23.5

Table 7.128 Constants and moisture contents for the rural polder in Thailand

Layer	Constants			Moisture content in %		
	a	b	c	Sat	FC	WP
I	3.750	0.875	-1.375	64.4	59.3	31.8
II	3.600	0.442	-1.254	65.0	63.6	18.3
III	3.600	0.442	-1.254	65.0	63.6	18.3

Evapotranspiration and crop factor

Potential evapotranspiration is based on the class A pan evaporation in mm/day at Sirigiti National Conference Center station. The crop factors multiplied with the class A pan evaporation are taken from literature (Royal Irrigation Department, 1994). The crop factor in the area is based on two crops of rice per year given in Table 7.129. The rainy season crop begins between mid and end of July and harvest between mid of November and early of December. The second crop begins between early and mid of January and harvest early and late of May.

Table 7.129 Crop factors (Kp) for evapotranspiration for HYV rice in the dry and rainy season for the rural polder in Thailand (Royal Irrigation Department, 1994)

Month	Decade		
	I	II	III
January	0.83	1.03	1.14
February	1.22	1.34	1.44
March	1.53	1.55	1.55
April	1.52	1.40	1.21
May	0.77	0.49	0.49
June	0.49	0.49	0.54
July	0.70	0.76	0.97
August	1.11	1.19	1.30
September	1.42	1.52	1.55
October	1.55	1.53	1.42
November	1.23	0.81	0.49
December	0.49	0.61	0.70

Note: The above values are based on a coefficient by assuming two crops of rice in the dry season and rainy season.

7.7.5 Economical data

Costs for the water management system

The costs for the water management system include the following:
-    costs for surface field drainage system;
-    costs for the canal system;
-    costs for pumping.

The costs for the surface drainage system are computed using the formula as given in chapter 6 by using the unit costs as given in Table 7.92.

The formulas used for the estimation of the costs for construction of the main canals, sub-main canals and lateral canals have been explained in Chapter 6. The estimation of costs for the canal system is based on the overall area of the polder, parcel length and the distance between the lateral canals. Basic values for the estimation of the water management system are given in Table 7.93. The interest rate is estimated at 5%.

The pumping head is based on the average of main canal water level of 3.0 m-MSL and the average water level in the collecting system of 0.50 m-MSL. The annual rainfall surplus and seepage are estimated at 367 mm and 183 mm for the estimation of the operation costs for pumping. The drainage modulus is estimated at 46 mm/day. The installation and operation, maintenance costs are given in Table 7.130.

Table 7.130 Pumping costs for the rural polder

Installation in million €	Operation in million €/year	Maintenance in million €/year	Lifetime in years
11.841	0.2033	0.0861	50

Costs for field pumping for irrigation and drainage have been estimated from unit cost of the Royal Irrigation Department in 2001 at 0.017 €/m^3.

Costs for buildings, infrastructure and crops

The average farm size has been estimated at 12.8 ha. The value of the buildings is estimated at 12,500 €/farm (Royal Irrigation Department, 2002). The area of buildings and yards is estimated 1,000 m²/farm. For a facility such as shops, factories, schools etc 20% of the value of the farms has been taken.

The infrastructure in rural area is mainly composed of roads. In order to compute the cost of infrastructure, the density of paved roads is determined at 150 m²/ha. The cost of roads is estimated at 6.0 €/m² (Royal Irrigation Department, 2002).

The crops in the polder are assumed to be only rice. The investment costs of this crop are separated due to the activities of agricultural practices and other activities related to productivity of crops such as land acquirement, transportation etc. Normally the investment costs are dependent on the time of the growing seasons. The investment costs and time schedule for the crops are shown in Figure 7.43 and Table 7.95.

Cost of damage to buildings, infrastructure and crops for a rice area in the rural polder

The method of calculation and the parameters used for estimating cost of crop damage are described in Chapter 6. Damage to buildings and infrastructure is estimated as percentage of value of buildings and infrastructure in relation with the groundwater table as shown in Table 7.11.

7.7.6 Calibration of the parameters

The area is composed of urban and rural areas. Due to the agricultural practice most of the times the soil moisture in the area is above field capacity or saturated. The main land uses in the rural area are fishponds and rice. Therefore the model takes into account the water level in the fishponds and in the rice fields in combination with the time schedule.

The following four parameters are found during calibration of the model:
- maximum interception for the rice area in mm;
- maximum storage depression for the urban area in mm;
- runoff coefficient for the urban area;
- soil storage in the urban green area in mm.

The results of the calibration of the parameters are shown in Table 7.131. It has been simulated by approximating that there is no lagtime between rainfall in urban area and no interception for the fishpond. In the urban green area is assumed evapotranspiration from soil is about 1.1 of the evaporation of the pan due to some tree in this area, infiltration rate of soil is taken as constant rate for this area of 1.5 mm/hour. If rainfall is more than infiltration rate it will flow to the canal system. Also if soil has been filled to full storage capacity the excess rain will become surface runoff to the canal system. The field pumping capacity of rice field is based on average conditions is 1.00 mm/day. Seepage to the area is taken as 0.5 mm/day. Combination of flow from the urban and the rural area, physical dimensions of the polder canal system and pumping data are available for calibration the parameters as mentioned above.

The area of polder is rather large. Therefore the effect of peak rainfall has been reduced by average rainfall using the Thiessen method in order to get a real condition for calibration. Due to data availability for 1998 the average for five stations which are Samrong Regulator station, Bang Tamru Regulator station (E4), Ladkrabang Regulator station, Pravetburirom Canal station, and Bangping station (Figure 7.63). For 1999 data of rainfall station is not sufficient for calibration. For 2000 the average rainfall from only three stations is available, which are Bang Tamru Regulator station (E4), Ladkrabang Regulator station, and Pravetburirom Canal station.

Table 7.131 Calibration of the parameters for the rural polder

Year	Interception for rice in mm	Storage in depressions for urban in mm	Runoff coefficient	Soil storage in urban green area in mm	Goal function	Annual pumping discharge in mm
1998	1.54	1.05	0.85	54.2	164	162
2000	1.50	1.20	0.82	96.6	368	170
Average	1.52	1.12	0.83	75.4	-	166

7.7.7 Model validation with fixed calibrated parameters

The model has been validated with average calibrated parameters in Table 7.131 for 1998 and 2000 as shown in Figures 7.65 to 7.70.

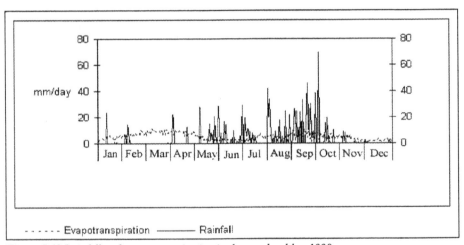

Figure 7.65 Rainfall and evapotranspiration in the rural polder, 1998

From the calibration it is found that the fishpond and rice field perform as reservoir; most of the rainfall was retained in the fishpond, which will influence flow to the canal system runoff. The most influence to inflow to the system is runoff, which come from urban area.

In Figures 7.66 and 7.70 from January to April the polder water level becomes low due to pumping water from the canal system. The pumping from the canal system use for fishpond and irrigation rice during the dry period was the main factor that influence to polder water level in this period. On the other hand water level will be increase rapidly if there is discharge out from fishpond due to high water layer in the fishpond. However, from this simulation it is found that this situation was not occurred in the 1998 and 2000. It means that fishpond behaviour as retention pond. In reality the fish farmer will drained the pond for different time during the whole year.

Assuming of initial water level in the fishpond is important for water level in the polder area due to pumping activity.

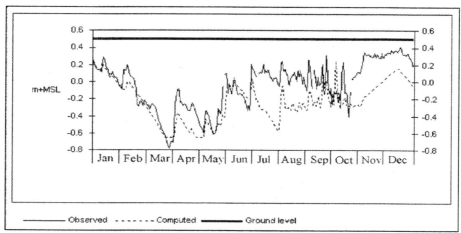

Figure 7.66 Polder water level validated with the calibrated parameters, 1998

Between end of October to December is the period to stored water in the polder canal system, the water level increases in this period due to the water flow from the north of polder to the South due to slope of the area and stagnant at the Tamru pumping station. But the model calculated the average water level due to volume of water in the canal system did not consider this flow, that is the reason for the calculation polder water level is lower from observed at Tamru pumping station (E4) (Figures 7.66 and 7.70)

From Figure 7.66 the pumping in the simulation is more than the observed value while the water level in the simulation is lower than the observed value. This is because of the simulation computed pumping capacity based on water level in polder water level and prefer polder water level. But in reality the water level in this polder was controlled manually. The most important factor is expecting of the rainfall or storm that make difference from general operation rule, which was used for this simulation.

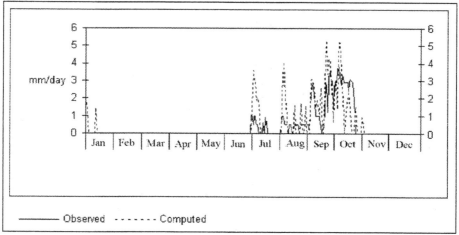

Figure 7.67 Computed and observed pumping discharge validated with the calibrated parameters, 1998

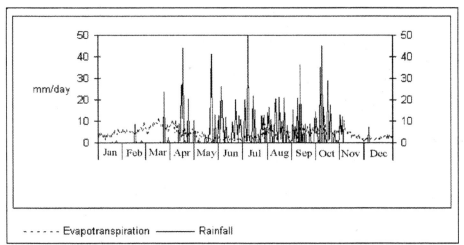

Figure 7.68 Rainfall and evapotranspiration in the rural polder, 2000

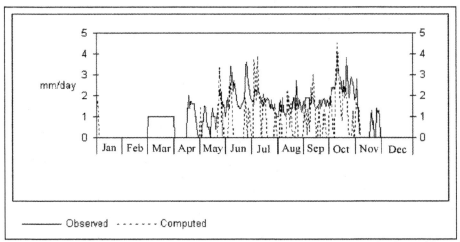

Figure 7.69 Computed and observed pumping discharge validated with the calibrated parameters, 2000

However, there are many factors to get different pumping discharge between observed and computed such as spatial rainfall, gradient of flow during the peak discharge, obstruction of flow, operation rule may be adjusted during the period according to weather forecasting (set level of polder, computation of storage volume) and also estimation pump capacity is not so accurate.

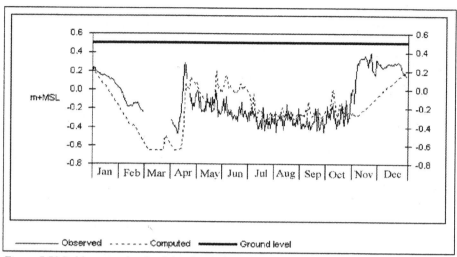

Figure 7.70 Polder water level validated with the calibrated parameters, 2000

7.7.8 Analysis with the optimization model

Hydrological analysis

The hydrological analysis with the validated model for the rural polder has resulted in the data as given in Table 7.132.

Table 7.132 Review of annual data in mm for the rural polder

Year	Rainfall in mm	Evapotranspiration in mm	Discharge out of area in mm
1998	1,283	1,599	162
2000	1,220	1,611	170

Table 7.132 shows the rainfall and evapotranspiration in the rice area. It can be seen that evapotranspiration is higher than the rainfall. Therefore in the rice area irrigation is needed, especially during the dry season. On the other hand during the wet season the drainage will be required.

Economic computation

The results and review computations for optimal conditions are shown in Tables 7.133 to 7.136. This simulation used hydrological data of available rainfall from average 3 stations: Bangping station, Ladkrabang Regulator station and Bang Tamru Regulator station by Thiessen method for 1989 to 1998. Data for buildings and infrastructure costs is the average price of buildings and infrastructure in polder area based on price in the year 2002.

Table 7.133 Review of the total values of crops, buildings and infrastructure in the rural polder at
optimal water management

Item	Area in ha	Highest value		Lowest value	
		in € per ha	in € * 10⁶	in € per ha	in € * 10⁶
Urban one story houses	477	800,000	381	800,000	381
Urban two stories houses	4,375	1,444,000	6,190	1,444,000	6,191
Urban high buildings	658	5,800,000	3,814	5,800,000	3,814
Urban infrastructure	1,821	240,000	437	240,000	437
Rural buildings	96	150,000	14	150,000	14
Rural infrastructure	155	60,000	9	60,000	9
Fish ponds	8,813	-	-	-	-
Rice	1,290	793	1	10	0.01
Total	17,685		10,848		10,847

Table 7.133 shows that the value of houses and buildings is about 96% of the total
value in this polder area. Therefore the damage to urban area may be the most important
in this polder area. It is assumed that there is no damage to fishponds so the value of
fishponds is not included in this calculation. Rice is very small portion value in this
polder area only 0.009% of the total value.

Table 7.134 Optimization analysis of the main components of the water management system in
the rural polder

Polder water level in m-s.	Open water in %	Pumping capacity in mm/day	Surface drainage capacity in mm/day	Costs in € * 10⁶	Damage in € * 10⁶	Value of the goal function
0.24	2.0	6.3	62.9	2.200	0.312	2,512,000

Table 7.134 shows that at optimal conditions the irrigation for rice has to be pumped
to the rice field and fishpond due to the polder water level is lower than the ground
surface. But it is no need to pump for the drainage. This polder water is the water level
in the area during rainy season.

Table 7.135 Total costs for the water management system for the rural polder at optimal water
management

Item	Construction costs in € * 10⁶	Annual maintenance costs in € * 10⁶	Annual equivalent costs in € * 10⁶
Field drainage	0.17	0.06	0.07
Open water	10.78	0.45	1.04
Pumping	0.31	0.21	0.23
Total	11.26	0.72	1.34.

Table 7.135 shows that the overall annual maintenance costs in a rural polder are
about 6.4% of the construction costs, while the overall annual equivalent costs are about
11.9%. The open water is the highest amount of the total construction costs, which is
about 95.7%. While the drainage costs are the lowest cost, which is about 1.5%,
because only surface drainage system in the rice area is considered.

Table 7.136 Specification of average annual damage and yield reduction for the rural polder at optimal water management

Item	High groundwater tables		High water layer		Drought	
	in €	in % of the value	in €	in % of the value	in €	in % of the value
Buildings	84,900	0.001	0	0.0	0	0.0
Infrastructure	263,600	0.06	0	0.0	0	0.0
Rice	0	0.00	68,100	6.7	20,500	2.0
Total	348,500	0.003	68,100	6.7	20,500	2.0

Table 7.136 shows that rice may be have more damage due to inundation more than drought while damage to buildings and infrastructure is very little.

Economical analysis for water management system at the present situation

In practice the control criteria for the polder water level varies throughout the year. Take only the critical period between July to October then the polder water level is kept at 0.80 m-surface.

The values of the main components of the water management system of a rural polder may be approximated as follows:

- polder water level   = 0.80 m-surface;
- open water area   = 3.3%;
- field drain capacity   = 46.0 mm/day;
- pumping capacity   = 5.3 mm/day.

The results and review computations for present conditions is shown in Tables 7.137 to 7.141.

Table 7.137 shows that the value of houses and buildings is about 95.7% of the total value in this polder area. Therefore the damage to urban area may be the most important in this polder area. It is assumed that there is no damage to fishpond so the value of fishponds is not included in this calculation. Rice is very small portion value in this polder only 0.009% of the total value.

Table 7.137 Review of the total values of crops, buildings and infrastructure in the rural polder for present water management

Item	Area in ha	Highest value		Lowest value	
		in € per ha	in € * 10^6	in € per ha	in € * 10^6
Urban one story houses	470	800,000	376	800,000	376
Urban two stories houses	4,232	1,444,000	6,111	1,444,000	6,111
Urban high buildings	649	5,800,000	3,765	5,800,000	3,765
Urban infrastructure	1,797	240,000	431	240,000	431
Rural buildings	91	150,000	14	150,000	14
Rural infrastructure	147	60,000	9	60,000	9
Fish ponds	8,363	-	-	-	-
Rice	1,224	793	1	10	0.01
Total	16,973		10,707		10,706

Table 7.138 Optimization analysis of the main components of the water management system for the rural polder at present water management

Polder water level in m-s.	Open water in %	Pumping capacity in mm/day	Surface drainage capacity in mm/day	Costs in € * 10^6	Damage in € * 10^6	Value of the goal function
0.80	3.3	5.3	46.0	3.252	0.386	3,638,000

From Table 7.138 it can be seen that costs increase from optimal conditions (Table 7.134) due to the percentage of open water is higher and costs for pumping irrigation to the field have increase due to more water loss from the field. Also the damage has increase due to drought in the rice field because more water loss from the rice field due to lower polder water level and less amount of water that is stored in canal system.

Table 7.139 Total costs for the water management system for the rural polder at present water management

Item	Construction costs in € * 10^6	Annual maintenance costs in € * 10^6	Annual equivalent costs in € * 10^6
Field drainage	0.16	0.05	0.06
Open water	22.06	0.92	2.13
Pumping	0.31	0.22	0.24
Total	22.53	1.19	2.43

Table 7.139 shows that at present conditions the overall annual maintenance costs in the rural polder are about 5.3% of total of the construction costs, while the overall annual equivalent costs are about 10.7%. The open water is the highest amount, which is about 97.9% of the total costs for water management. While the surface drainage in the rice area is the least amount, which is about 0.7%.

From Table 7.140 it can be seen that at present conditions rice have more damage due to inundation and drought than optimal conditions (Table 7.136). Because the model was model in such a way that when there is water less than prefer water level at certain level the pumping of irrigation water is pumped to the rice field and stop when water level in rice field increase to certain level. While water loss from the field is more than optimal conditions, then more tendency to pump water into the field. If there is rain to the area then more probability to get inundation in the rice field in present conditions than optimal conditions. In practice the farmer will pump by follows weather forecasting. While drought is increased because more water loss from the field during the dry periods and less water store in the canal system also.

Table 7.140 Specification of average annual damage and yield reduction for the rural polder at present water management

Item	High groundwater tables		High water layer		Drought	
	in €	in % of the value	in €	in % of the value	in €	in % of the value
Buildings	91,900	0.0009	0	0.0	0	0.0
Infrastructure	133,400	0.03	0	0.0	0	0.0
Rice	0	0.00	77,500	8.0	26,200	2.7
Total	225,300	0.002	77,500	8.0	26,200	2.7

Table 7.141 Comparison between the optimal and present values for the main components of the water management system in the rural polder

Item	Polder water level in m-s.	Open water in %	Pumping capacity in mm/day	Field drain capacity in mm/day	Costs in $€ * 10^6$	Damage in $€ * 10^6$	Value of the goal function
Optimal	0.24	2.0	6.3	62.9	2.200	0.311	2,512,000
Present	0.80	3.3	5.3	46.0	3.252	0.386	3,638,000
% differ.	+70	-39	+20	+36	-32	-19	-31

From Table 7.141 it can be seen that at present the pumping capacity and surface field drain capacity should be increased. The water level during the rainy season can be kept higher than present conditions. However, the model did not consider gradient of flow, so it may be get inundation because of this gradient, so it is suggest that in the dry period or storing water period water level can be kept higher or above the ground surface level in the rural area but in the urban area water level can be the same at optimal conditions. During the wet periods the polder water level may be kept the same as or little higher than the present practice. While the open water area may be enough for this polder area.

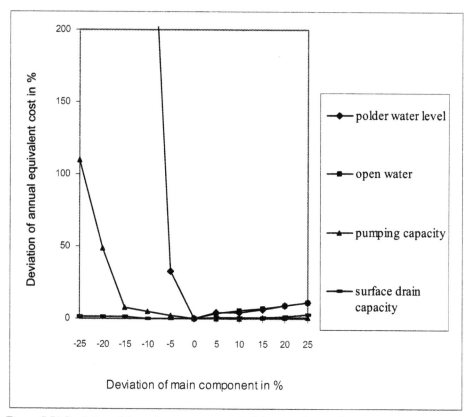

Figure 7.71 Deviation of the main components and of the annual equivalent costs in percentage in the rural polder

In Figure 7.71 the influence of deviations from the simulated optimal values is shown. It shows that the polder water level and the pumping capacity have most influence on the annual equivalent costs in the system. If the polder water level is 25% shallower compared to the optimal value the annual equivalent costs will increase approximately 2,330%, because of increase in the damage to the urban area due to too high groundwater tables and inundation, which is very high value in this polder. The pumping capacity, which is 25% lower than the optimal value increases the annual equivalent costs by approximately 110%, because of increase in the damage due to too high groundwater tables and inundation to the urban area. While the changes in the surface field drain capacity and the open water area have only a limited effect.

Discussion

Based on the results as shown above the following points can be discussed.

A shallow polder water level will cause a higher chance of inundation. The damage to the urban area is much larger compared to the damage in the rural area because there are costly buildings and infrastructure in this polder area. The polder water level in the urban and the rural area may be separated by a weir or inner dike to control the water level between the rural and the urban area.

At present conditions the open water area is larger than at optimal conditions, but the pumping capacity is lower than at optimal conditions. Therefore an increase of the pumping capacity in this area should be considered.

The polder water level and the pumping capacity have a high influence on damage of the urban area (Figure 7.71). The shallow polder water level will give a sharp increase in damage because the damage due to inundation and too high groundwater tables increases. A decrease in the pumping capacity will give a higher chance that too high water levels in the polder occur.

The damage due to drought and floods for fishponds is very small, therefore it has been neglected in this study.

Fishponds can be used as storage in this polder area. In the rainy season it can be used for storing excess rain and supplying water to the crops area and for other purposes during the dry season.

In the simulation of this optimization only one polder water level in the rainy season was used. It can be seen that the open water area at present is larger than at optimal conditions and the polder water level is deeper than at optimal conditions (Table 7.141). In the real situation the safety, return period and gradient of flow during heavy rain will have to be considered as well. The surface drain capacity is lower than at optimal conditions, so it would have to be considered to increase this.

7.8 The urban polder Sukhumvit

7.8.1 General

The study area of the case study in Thailand is located in the urban area of Bangkok Metropolitan at the eastern side of Bangkok. This part of Bangkok is divided to 10 polders. The polder Sukhumvit has been selected for study. It is one of inner urban polders of Bangkok Metropolitan, which shown in Figure 7.72.

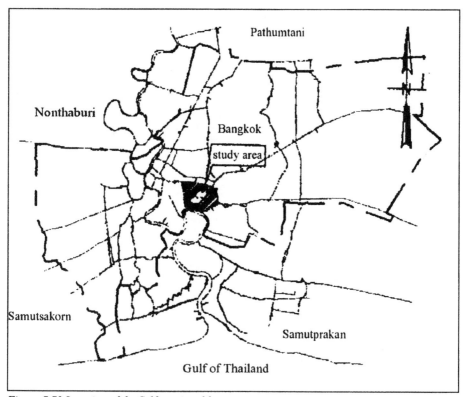

Figure 7.72 Location of the Sukhumvit polder

Sukhumvit polder has the total area of 2,500 ha and the boundaries are as follows. The North is up to Seanseab canal. The East is up to Ton canal and Prakanong canal. The West is up to Chongnonsri railway. The South is up to The Chao Phraya river in the Klongtoei harbour. The polder is located in the Klongtoei regional office and Watana regional office.

7.8.2 Physical conditions

Topography

The area has a declination from the North to the South, the ground surface in the North is around 1.5 m+MSL and the ground surface in the middle and the South is 0.50 and 1.0 m+MSL. In some areas the level is at 0.00 m+MSL, which means that the gravity drainage may be not possible especially the area location far from river. The riverbank level is 1.5 m+MSL and dike embankment level is 1.5 to 2.0 m+MSL.

Land use

The majority of land is used for residence (Figure 7.73); there are 18,280 residence houses (Social Development Office, 1996). Moreover, study area is located adjacent to the east of Silom road, Sathorn road and Peornchit road business centre. This area is easy to access therefore there is a lot of industry. In the year 1994 there are 661 factories (Factory Department, 1997) in this area and also a lot of high rising buildings.

There is no agricultural area. Land use composition in Sukhumvit polder is shown in Table 7.142.

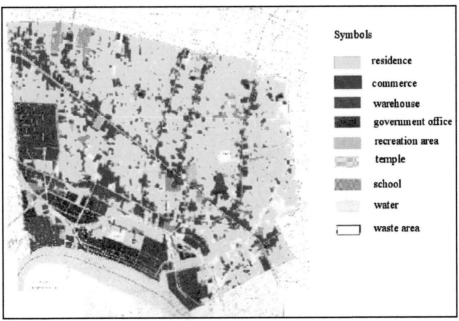

Figure 7.73 Land use in the Sukhumvit polder(Landscape Office, Bangkok Metropolitan Administration, 2002)

Table 7.142 Land use composition in the Sukhumvit polder

Type of land use	Area in ha
Residential [2]	1,839.72
Industrial [2]	100.80
Ware house [2]	62.40
Commercial area [2]	96.70
Officer buildings [2]	283.84
Park/garden [1]	16.54
Water area (include retention pond)	-
Road	-
Total	2,400

Source: [1] Social Welfare Office, Bangkok Metropolitan Administration, 2003
        [2] Landscape Office, Bangkok Metropolitan Administration, 1992, 2003

The composition of residences in Bangkok and adjacent area is as shown in Table 7.143.

Table 7.143 Composition of residence in Bangkok and vicinity

Type of residence	Percentage (%)
Single house	34.5
Twin house	0.9
Town house	25.0
Flat/Apartment	8.8
Dormitory	23.9
Improvised quarters	4.8
Other	2.1
Total	100.0

Source: National Statistical Office, 1992

The total number of houses is 87,683 units and population density is 103.74 persons/ha (Landscape Office, BMA, 2002).

7.8.3 Present situation of operation and maintenance

The drainage system in the area is composed of secondary sewer pipes mostly with a diameter 0.30 m with a distance between pipes of 40 to 50 m receive water from the adjacent field areas. The water discharges to the main sewers which is mostly a pipe with a diameter 0.60 m with an average distance between sewers around 100 m which located beneath the footpath along a secondary road and the water flow to transport pipes which are located beneath the main roads, usually box culvert of 1.50 * 1.50 m to 2.0 * 2.0 m, then the excess water discharges into the main canals which surround the polder area by pumping from sump at the end of road adjacent to 3 surrounding canals, Seanseab canal, Ton canal, Prakanong canal. Finally the water was pumping by main pumping at Phraram IV pumping station, which has pumping capacity 22.80 m^3/s or 82.08 mm/day to the Chao Phraya river. This pumping station shared function of drainage with other polder area.

The area inside this polder can be separated into 7 sub-polders dependent on flow direction. Sub-polder DF, which is more or less clearly separated from the surrounding areas, has been used for calibration and validation (Figure 7.74). This sub polder has a total area of 368 ha.

In practice the polder water level in the main canals level is between 0.50 to 1.0 m-MSL (1.05 to 1.65 m-surface) and the highest water level is at 0.50 to 1.0 m+MSL. (average ground surface 0.55 m+MSL) the crest heights of the dike/or retaining wall at Ton canal or Prakanong canal is between 0.6 to 1.2 m+MSL.

Water level control in the sub-polder DF

Water level control in the sub-polder DF is as follows:
- the ground surface elevation is between 0.40 to 0.70 m+MSL. Average ground surface level is at 0.55 m+MSL.
- alarm water level 34.70 BMA level or 0.33 m-MSL (0.88 m-surface);
- water level at critical 34.90 BMA level or 0.13 m-MSL (0.68 m-surface)
- water level at expect to rain 33.50 to 34.00 BMA level or 1.53 to 1.03 m-MSL (2.08 to 1.58 m-surface);
- water level at normal condition 34.00 to 34.50 BMA level or 1.03 to 0.53 m-MSL (1.58 to 1.08 m-surface);

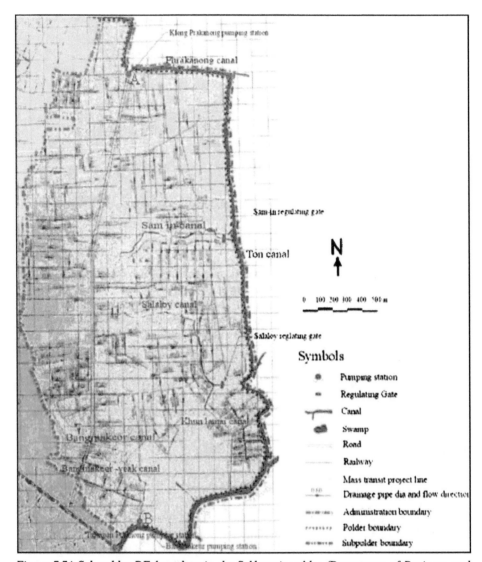

Figure 7.74 Sub-polder DF boundary in the Sukhumvit polder (Department of Drainage and Sewerage, Bangkok Metropolitan Administration, 1998)

-    water level for draining polluted water not exceed 34.60 BMA level or 0.43 m-MSL (0.98 m-surface).

The coefficient suggested by Water Development Consultants Company, 1998 is as shown in Table 7.144:

Table 7.144 Runoff coefficients according to different land use in the sub-polder DF

Index	Area description	Runoff coefficient
1	Commercial	0.63
2	Dense residence	0.58
3	Medium residence	0.45
4	Industrial	0.60
5	Ware house	0.64
6	Officer buildings	0.31
7	Park	0.17

Source: Department of Drainage and Sewerage, Bangkok Metropolitan Administration, 1998

Canal in the sub-polder DF

There are five canals in this area as follows: (Figure 7.74)
- Bangmakeor canal, this canal has retaining wall throughout the length, the canal bottom is lining with concrete. The dimension is 1,400 m long, 5.0 m width and 2.5 m depth.
- Bangmakeor-yeak canal, this canal has retaining wall throughout the length, the canal bottom is lining with concrete. The dimension is 240 m long, 3.0 m width and 2.0 m average depth.
- Salaloy canal, this canal is natural canal, non-lining. The dimension is 1,140 m long, 5.0 m width and 2.0 m average depth.
- Sam-in canal, this canal is natural canal, non-lining. The dimension is 930 m long, 5.0 m width and 2.0 m average depth.
- Khun Lamai canal, this canal is natural canal, non-lining. The dimension is 750 m long, 5.0 m width and 1.5 m average depth.

Most of the canal in this area is shallow by garbage and some houses invasion to the canal.

7.8.4 Hydrological and water management data

In this case study the following data analysis:
- data for model calibration:
  • hourly rainfall at Sirigiti National Conference Center station, Bangkok, for, 2003;
  • daily class A pan evaporation at Sirigiti National Conference Center station, Bangkok, for, 2003;
  • daily pumping discharge at Bangmakeor pumping station (point B in Figure 7.74), for 2003.
- data for validation and computation:
  • daily class A pan Evaporation data at Sirigiti National Conference Center station, Bangkok, in mm for 1971 to 2003;
  • hourly rainfall data at Sirigiti National Conference Center station, Bangkok, in mm, 1971 to 2003.

The data used in the model are based on the data of Bangmakeor pumping catchments area. The wastewater disposal system and seepage has been estimated by water balance during rainfall and pumping data during the dry period. This amount was estimated between 6.6 to 8.5 mm/day depend on the period of tide level of the Chao Phraya river.

Parameter values for the simulation of the unpaved area according to the formula of Horton use the same as given in Section 7.3.3: The storage in depressions is set at 3.0 mm, which is a normal value for unpaved areas in flat urban areas.

Catchment characteristics

The catchments area in the sub-polder DF is about 368 ha. The catchment in the urban area is divided into 248 ha paved and 120 ha unpaved area. The discharge from the paved area is directly conveyed by the sewer system while the unpaved area discharges into the subsurface drain pipes.

The levels simulation based on average conditions are as follows. The level of houses is set at 0.50 m+surface. The level of square and path is set at 0.40 m+surface. The quarter road is at 0.30 m+surface and main road is at 0.00 m+surface.

Water depth in the canal is 1.20 m.

Canal profile:

Under water slope	= 1:2.0;
Bank slope	= 1:2.0;
Timbering	= 0.30 m.

The paved area may comprise the following:

- housing area;
- parking lot;
- commercial centre;
- industrial area;
- roads.

The unpaved area of the urban area is provided with a surface drainage system as in the rural area. It is assumed that the discharge from this area is behave in the same manner as in the rural area.

The paved area is composed of flat or slope roof houses, asphalt roads, brick roads, etc. The composition of paved areas, which is used in the hydrological analysis, is given in Table 7.145.

Squares, paths and quarter roads are classified according to their capacity to infiltrate the rainfall. The composition of the different paved surfaces is described in Table 7.146.

Table 7.145 Composition of paved areas in the sub-polder DF

Paved area	Area in ha
Roofs:	
-sloping	127
-flat	46
Squares and paths	32
Quarter roads	25
Main roads	16
Green areas and gardens	120

Table 7.146 Percentage of paved surface areas for different types of roads in the sub-polder DF

Type of road	Tiles in %	Bricks in %	Asphalt in %
Squares and paths	0.0	90.0	10.0
Quarter roads	0.0	80.0	20.0

7.8.5 Economical data

Data for the economic computation

For the economic computation for urban areas the percentage of different paved areas have been taken into account as shown in Table 7.147.

Table 7.147 Percentage of different paved areas for the cost estimation for the sub-polder DF area

Type of road	Tiles in %	Bricks in %	Asphalt in %
Squares and paths	0.0	100.0	0.0
Quarter roads	0.0	100.0	0.0
Main roads	0.0	10.0	90.0

The costs for the water management system may be divided into the following:
- surface drainage;
- sewer system;
- open canal system.

Data of construction and maintenance costs of the water management system are given in Tables 7.148 to 7.153. Lifetime for the subsurface drains is taken as 30 years for the economic analysis.

Table 7.148 Costs for surface drainage in the sub-polder DF (Royal Irrigation Department, 2002)

Item	Cost in €/m
Construction	3.20
Maintenance	0.32

Construction cost of gutters is taken as 95.81 €/piece (RID, 2002). The cleaning of sewers consists of flushing, which is estimated at one time in two years. Annual cost is taken as 0.23 €/year/m. The lifetime for the sewers is taken as 30 years for the economic analysis.

Table 7.149 Construction costs for sewers in the sub-polder DF (Royal Irrigation Department, 2002)

Diameter in m	Cost in €/m
0.30	8.78
0.40	12.71
0.50	16.98
0.60	21.25
0.70	25.94
0.80	30.63
1.00	44.48

Table 7.150 Construction costs for the urban canals (Royal Irrigation Department, 2002)

Item	Unit cost
Digging	1.20 €/m³
Profiling	5.00 €/m
Fence	9.80 €/m

A standard canal profile in the sub-polder DF area is as follows: under water slope of 1:2.0 and bank slope of 1:2.0. Annual maintenance cost for the transport pipe in the sub-polder DF is 0.375 €/m. Lifetimes for the economic analysis for the sub-polder DF area are as follows: canals of 100 years and pump of 50 years. The interest rate has been set at 5.0% for the determination of the present or capital value in the sub-polder DF area.

Data for the damage computation

Data for damage computation are given in Tables 7.151 to 7.152. Damage function and indirect damage used are shown in Tables 7.60 and 7.64. The cost of buildings and infrastructure are estimated based on average price from RID, 2002.

Table 7.151 Data on values of buildings for the sub-polder DF

Item	Number/ha	Average value in €/number	Furniture value in %
Houses with one floor	5	11,500	50
Houses with two floors	20	18,400	50
Shops and offices	15	28,800	80
Industrial buildings	2	97,900	70

Table 7.152 Data on the value of infrastructure in the sub-polder DF

Type of infrastructure	Value of infrastructure
Squares and paths	11.39 €/m^2
Quarter roads	18.06 €/m^2
Main roads	36.57 €/m^2
Public facilities	10,520 €/ha
Utilities	10,520 €/ha

7.8.6 Calibration of the parameters

Due to data at Klongprakanong pumping station (point A in Figure 7.74) is not available. The area of polder DF is divided by topography and direction of flow in the pipe for calibration. The area influenced by Bangmakeor pumping station (point B in Figure 7.74) is 252 ha was used for calibration.

The calibration for the urban area involves the following parameters:
- storage in depressions;
- runoff coefficient;
- parameters k_1 and n_1 for transformation of rainfall into sewer inflow;
- parameters k_2 and n_2 for transformation of sewer inflow into sewer discharge.

The calibration of these parameters was done by the Rosenbrock method as described before. The calibration in this area was done only in the year 2003 because only pumping discharge is available in this year. The calibrated parameters, which are obtained from the year 2003, are shown in Table 7.153.

Discharge from the urban area to the canal in the urban area comes from both storm sewers and surface drains. It has been approximated that there is no lagtime between rainfall and storm sewer discharge, while the surface drains behave in the same manner as in the rural area.

Table 7.153 Calibrated parameters of the hydrological model for the sub-polder DF area

Storage in mm	Runoff coefficient	Rainfall> sewer inflow		Sewer inflow > sewer discharge		Pumping discharge in mm		Goal function
		k_1	n_1	k_2	n_2	observed	computed	
0.82	0.54	2.50	0.211	2.545	0.210	3,570	3,609	13,570

Sensitivity in sub-polder DF

The value of 0.82 mm from the calibration process has been used for storage in depression.

The value of 0.54 from the calibration process has been used as a representative runoff coefficient to simulate rainfall into runoff. This value is quite well get along with Table 7.144.

Transformation of runoff into sewer inflow and from inflow to sewer discharge in the model is being assumed that it behaves as a non-linear function as described in Section 7.3. The values of transformation parameters from the calibration process are shown in Table 7.153.

The model was run for various time steps 1 hour due to only hourly data is available. Owing to the fact that during the computation of sewer inflow and sewer discharge no lagtime is taken into account and the input of hydrological data is 1 hour, there may be difference when the observed values and computed values are compared. The model efficiency at this time step is 0.593, which is low may because by as mentioned above. Moreover at this time step simulation peak canal water level may be not so accurate, which results in the inaccurate calculation of the pumping discharge.

7.8.7 Model validation with fixed calibrated parameters

The model has been validated with data from the calibrated parameters in Table 7.153 for 2003 as shown in Figures 7.75 to 7.77.

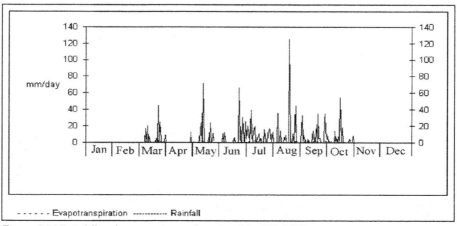

Figure 7.75 Rainfall and evaporation at the sub-polder DF, 2003

In Figures 7.75 and 7.76 there is pumping when there is no rainfall because there is some leakage of water to the polder area, it is estimated that the leakage water and seepage into the sub-polder DF between January and March is 8.6 mm/day, May and October is 9.6 mm/day and November to December 6.7 mm/day. While there is no

seepage and leakage in April because of a low water level in the surrounding canals and
in the Chao Phraya river. Also it is the driest period in Thailand. In Figure 7.76 the
computed pumping discharge is higher than observed pumping discharge during high
rainfall may be due to effect of spatial rainfall in the area.

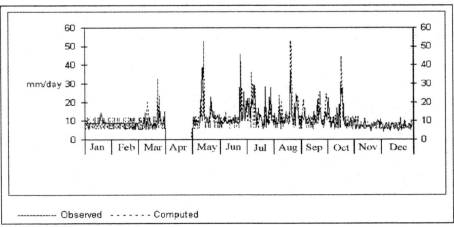

Figure 7.76 Computed and observed pumping discharge at pumping station Bangmakeor, 2003

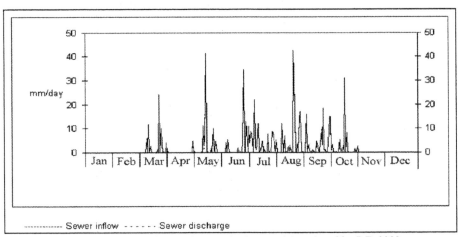

Figure 7.77 Computed sewer inflow and sewer discharge of the sub-polder DF, 2003

In Figure 7.77 there is small lag time between sewer inflow and sewer discharge
because there is a small volume in the sewer pipe to store water before discharging it
into the canal.

7.8.8 Analysis with the optimization model

Hydrological analysis

The hydrological analysis with the validated model for the sub-polder DF has resulted
in the data as given in Table 7.154.

Table 7.154 Review of annual data in mm for the sub-polder DF

Year	Rainfall in mm	Evaporation in mm	Infiltration in mm	Sewer discharge in mm	Pumping discharge in mm
2003	1,358	129	566	669	3,570

In Table 7.154 it can be seen that there is a lot of pumping in this area compare to discharge from sewer due to the fact that there is leakage to the polder area.

Economic computation

The economic computations are based on the values and function described above. The main components of urban area are shown in Table 7.155 and results are shown in Tables 7.156 to 7.158.

Table 7.155 Optimal values for the main components of the water management system in the sub-polder DF

Diameter sewer in m	Distance between transport pipes in m	Canal Water level in m-s.	Open Water area in %	Pumping capacity in mm/day	Costs in € * 10^6	Damage in € * 10^6	Goal function
1.00	1,480	2.56	1.44	117	0.589	0.065	654,000

Table 7.156 Costs for the drainage system in the sub-polder DF at optimal water management

Item	Living areas € * 10^6	Shopping- and office centres € * 10^6	Industrial areas € * 10^6	Total € * 10^6
Drains				
- construction costs	0.523	0.081	0.034	0.652
- maintenance/year	0.013	0.002	0.001	0.016
- annual equiv. costs	0.047	0.007	0.003	0.058
Sewers				
- construction costs	1.316	0.231	0.116	1.662
- maintenance/year	0.012	0.002	0.001	0.014
- annual equiv. costs	0.097	0.017	0.008	0.123
Canals				
- construction costs				4.447
- maintenance/year				0.012
- annual equiv. costs				0.302
Pump				
- construction costs				0.237
- maintenance/year				0.093
- annual equiv. costs				0.106

Table 7.156 shows that the costs for canals are the highest value in the urban area, which are about 63.5% of the total construction costs at optimal conditions. Because the depth of canal bed is deep due to water level in canal is kept at 2.56 m-surface and most of canals in this area have concrete retaining walls. While the costs of the sewers and drainage system in urban areas are about 33.1% of the total construction costs. Operation and maintenance costs for pumping station is 69% of total operation and maintenance costs, due to leakage of water outside to the sub-polder DF area.

Table 7.157 Review of the total value of buildings and infrastructure in the sub-polder DF at optimal water management

Item	Area in ha	Value including furniture	
		in €/ha * 10^6	in € * 10^6
Houses with one floor	49	0.086	4.23
Houses with two floors	196	0.552	108.19
Shops and offices	38	0.777	29.55
Industrial buildings	16	0.332	5.33
Infrastructure	63	0.189	11.92
Public facilities	299	0.015	3.14
Public utilities	299	0.015	3.14
Total	362		165.50

Table 7.157 shows that the value of infrastructure and facilities is relatively small, which is about 11.0% of the total costs in urban area. Therefore main damage may be expected especially to the houses, shops, offices and the industrial area.

Table 7.158 Specification of average annual damage in € in the sub-polder DF at optimal water management

Item	High groundwater table	Water at street	Exceedence canal water level
Living areas			
- Houses with one floor	2,330	25,200	
- Houses with two floors	2,330	25,200	
Shops and offices	1,090	8,360	
Industrial areas	280	28,590	
Green areas	0		
Urban area			0
Total			93,3800

Table 7.158 shows that water at the street is the more harmful to the urban area than a too high groundwater table and an exceedence of canal water level because rainfall intensity is high in the area and also fluctuation of the canal water level due to flash floods was not included in this study while only hourly data were used. The damage to house with one floor and with two floors is equal because harmful occurred only at the street, not at the houses due to the level of houses is higher than the level of street. The damage due to water at the street was high in industrial area due to less green area.

Economical analysis for water management system at the present situation

The values of the main components of the water management system of the sub-polder DF at present may be approximated as follows:
- diameter of sewer   = 0.60 m;
- distance between the transport pipes   = 600 m;
- canal water level during the wet period   = 2.08 m-surface;
- open water area   = 0.60%;
- pumping capacity   = 137.1 mm/day.

The results and review computations for present conditions is shown in Tables 7.159 to 7.162.

Table 7.159 Analysis with optimization of the values of the main components of the water management system at the present in the sub-polder DF

Diameter sewer in m	Distance between transport pipes in m	Canal water level in m-s.	Open water area in %	Pump capacity in mm/day	Costs in € * 10^6	Damage in € * 10^6	Goal function
0.60	600	2.08	0.60	137	0.395	0.400	795,000

Table 7.160 shows that the costs for canals is the highest value in urban area, which is about 52.7% of the total construction costs at present conditions. Operation and maintenance costs for pumping station is 72.3% of total operation and maintenance cost, due to leakage of water outside to sub-polder DF area. Therefore prevent this leakage of water should be done to decrease the cost of pumping.

Table 7.160 Costs for the drainage system in the sub-polder DF at present water management

Item	Living areas € * 10^6	Shopping- and office centres € * 10^6	Industrial areas € * 10^6	Total € * 10^6
Drains				
- construction costs	0.523	0.081	0.034	0.653
- maintenance/year	0.013	0.002	0.001	0.016
- annual equiv. costs	0.047	0.007	0.003	0.059
Sewers				
- construction costs	0.849	0.318	0.063	1.050
- maintenance/year	0.012	0.002	0.001	0.015
- annual equiv. costs	0.067	0.011	0.005	0.083
Canals				
- construction costs				2.179
- maintenance/year				0.007
- annual equiv. costs				0.148
Pump				
- construction costs				0.248
- maintenance/year				0.099
- annual equiv. costs				0.112

Table 7.161 shows that the value of infrastructure and facilities is relatively small, which is about 11.0% of the total costs in urban area. Therefore main damage may be expected especially to the houses, shops, offices and the industrial area.

Table 7.161 Review of the total value of buildings and infrastructure in the sub-polder DF at present water management

Item	Area in ha	Value including furniture in € * 10^6/ha	in € * 10^6
Houses with one floor	49	0.086	4.23
Houses with two floors	196	0.552	108.19
Shops and offices	38	0.777	29.55
Industrial buildings	16	0.332	5.33
Infrastructure	63	0.189	11.92
Public facilities	299	0.015	3.14
Public utilities	299	0.015	3.14
Total	362		165.50

Table 7.162 Specification of average annual damage in € in the sub-polder DF at present water
management

Item	High groundwater table	Water at street	Exceedence canal water level
Living areas			
- houses with one floor	13,220	61,010	
- houses with two floors	13,220	256,090	
Shops and offices	5,130	23,380	
Industrial areas	2,380	48,070	
Green areas	0		
Urban area			0
Total			422,500

Table 7.162 shows that at present conditions water at the street is the more harmful
to the urban area than damage due to the high groundwater table and exceedence of the
water level.

Table 7.163 Comparison between the optimal and present values for the main components of the
water management system in the sub-polder DF

Item	Diameter sewer in m	Distance between transport pipes in m	Canal water level in m-s.	Open water area in %	Pumping capacity in mm/day	Costs in € * 10^6	Damage in € * 10^6	Goal function
Optimal	1.00	1480.0	2.56	1.44	116.6	0.589	0.065	654,000
Present	0.60	600.0	2.08	0.60	137.1	0.395	0.400	795,000
% differ.	-40	-59	-19	-58	+18	-33	9467	-22

In Table 7.163 a comparison is shown between the simulated optimal water
management conditions and the present situation. From this table it can be derived that
open water area and sewer diameter in the present situation can be increased to have
less damage. The canal water level should be kept lower compared to the present
situation. The pumping capacity at present is enough to evacuate rainfall excess and
leakage water out of the area. It can be seen that the area of open water may have to be
increased, but in reality this area may be affected by many factors such as obstruction of
road, house, which have influence on the open water in the urban polder area.

In Figure 7.78 the influence of deviations from the simulated optimal values is
shown. It shows the canal water level has most influence on the annual equivalent costs
in the system. If the canal water level is 25% shallower compared to the optimal value
the annual equivalent costs will increase approximately 350%, because of the increase
of damage due to high groundwater tables and water at the street. If the sewer diameter
is 25% smaller than the optimal value the annual equivalent costs will increase
approximately 50%, because there will be less storage capacity of the sewer to store
water before discharging it into the canal. If the sewer diameter is increased by 25%, the
annual equivalent costs will increase approximately 75% due to the increase in
construction costs and increase in peak discharge to the canal in an urban area. When
the open water area is decreased by 25% from the optimal value, the annual equivalent
costs increase approximately 75%, because of the increase in damage due to high
groundwater tables and water at the street increase due to less volume of water can be
stored in the polder area. The change in the distance between transport pipes by ± 25%
has almost no effect on the annual equivalent costs, because only the costs for
construction change and there is a little change in storage volume of the drainage

system. Decrease in pumping capacity by 25% results in 50% increase in annual equivalent costs. While the 25% increase in the pumping capacity has almost no effect on the annual equivalent costs as only the construction costs increase.

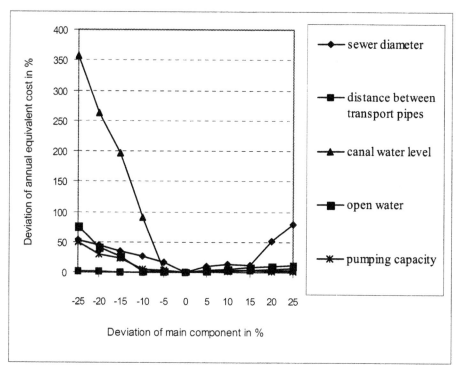

Figure 7.78 Deviation of the main components and of the annual equivalent costs in percentage in the sub-polder DF of polder Sukhumvit

Discussion

Based on the results as shown above the following points can be discussed.

A shallow canal water level will result in more chance to inundation in the urban area due to more stagnant water that cannot flow to the canal system due to the high water level. Therefore more water at the street may occur

At present the area of open water is less than at optimal conditions but the pumping capacity is larger than at optimal conditions in this sub urban polder DF. Therefore increase in open water area may have to be considered.

In the optimization only one canal water level in the rainy season is taken into account, but in practice the canal water level can be different in the different periods due to weather forecasts and operation purposes.

It can be seen that the open water area at present is smaller than at optimal conditions (Table 7.163) and the canal water level is shallower than at optimal conditions, this may be caused by the constraint of the residential buildings and the layout of the city. Also in former times a lot of canals in urban areas in Thailand have been filled to construct the roads. Instead of canals the transport pipes under the roads are used to transport the water.

8 Discussion

8.1 Discussion on the overall case study results

The clay and the peat polder in the Netherlands

In the case studies on the clay polder and the peat polder, the topsoil in both areas consists of peat with non-decomposed plants material. Normally a peat soil means a wet area. Therefore the land is generally used for meadow, while grass can grow in both areas. In the subsoil of the peat polder the soil is peat as well but in the clay polder the subsoil consists of heavy clay.

In peat soil has a high porosity, therefore discharge from the peat polder is faster than from the clay polder. The clay soil has a high moisture holding capacity, therefore the readily available moisture during dry periods is lower than in a peat soil. Hence there is higher tendency to yield reduction due to dry periods.

Drain depth and distance between the drains have a relationship to discharge from the area and the groundwater table. The peat soil requires a shallow drain depth to prevent oxidation. While in the clay soil a deeper drain depth may cause more subsidence due to the lower groundwater table. However, the shallower the drain depth the more damage may occur due to a too high groundwater table. The shallow drain depth will also cause damage due to too late sowing and too late harvesting, because there will be a higher probability to have a high moisture content in the topsoil. Due to this the soil may have not enough bearing capacity for fertilizing and harvesting of grass.

In the case studies, the peat polder (4.4%) there is a smaller area of open water area than in the clay polder (5.6%), because there may be more soil storage in a peat soil. Therefore the fluctuation of the water level is less than in a clay polder at the same open water area and hydrological conditions. Moreover, the groundwater table in a peat soil is shallower than in a clay soil due to a shallower drain depth and open water level. However, in the case study the open water area in the peat and the clay polder at optimal conditions are not so much different.

The pumping capacity at optimal conditions for the clay polder (16.2 mm/day) is higher than in the peat polder (4.9 mm/day). This is also caused by more soil storage in the peat polder.

A rice polder under rainfed and irrigated conditions in Thailand

In rainfed conditions the polder water level (0.09 m+surface) can be kept higher than in irrigated conditions (0.84 m-surface) to get benefit of the reverse water that flows into the rice field both through the culvert and through soil cracks and pores. In rainfed conditions the probability of inundation due to a too high water level is smaller than in irrigated conditions.

The optimal field drain discharge capacity (6.7 mm/day) can be smaller in rainfed conditions than in irrigated conditions (44.8 mm/day) due to reasons as follows. Firstly in rainfed conditions there is much more soil storage due to a lower groundwater table than in irrigated conditions. Secondly in irrigated conditions most of the time there is water ponding in the field, if there is rainfall the excess water will have to be pumped out of the field.

The optimal open water area under rainfed conditions is 1.1% while under irrigated conditions it is 2.5%. The open water area is dependent on the width and length of the parcels, size of plots, farm sizes, topography and social conditions. In rainfed conditions a larger size of the plots can be found, because a smaller amount of water has to be drained. In irrigated conditions more water needs to be drained from the fields because irrigation water is given to the area. The more open water the smaller the size of the plots. Moreover, under irrigated conditions, the storage in the soil is lower due to the high groundwater table. This is the other reason why more drainage capacity is needed in irrigated rice conditions.

The optimal discharge capacity of the pumping station for rainfed rice (7.5 mm/day) is much lower than under irrigated rice conditions (32.9 mm/day), because the soil storage in rainfed rice is more than under irrigated conditions. Moreover, in irrigated rice, the water for irrigation is pumped to be stored in the canals in the area, hence this also creates a higher chance that drainage will be required if there is heavy rainfall due to an unexpected storm. In practice the farmers will follow the weather forecasts to control the water level in their rice field. This is the reason why in irrigated conditions more surface drainage capacity is needed than in rainfed conditions.

A dry food crops polder in Thailand

The polder water level in a dry food crops polder is affected by the drainage capacity. The higher the polder water level, the higher the possibility of obstruction to discharge from the plots by gravity and the higher the requirement to pump out excess water during wet periods. Moreover, the higher the polder water level, the higher the probability for overtopping of the dikes around the plots.

The ditch water level and the distance between the ditches in a dry food crops polder have the most influence to damage of the crops in this polder area. The ditches in a polder function is to store water before pumping for irrigation and drainage. The smaller the distance between the ditches and the larger the depth of the ditches the more volume of water can be stored in the plots. The ditch water level affects the capillary rise in the beds of the dry food crops. The higher the ditch water level the higher the capillary rise in the bed and therefore the lower the requirement to pump water for spraying over the plots during dry periods. However, the higher the ditch water level, the higher the possibility of a high groundwater table in the plot will be, especially during wet periods and damage may increase.

In a dry food crops polder the optimal field drain capacity (34.8 mm/day) may be lower than for the irrigated rice (44.8 mm/day) conditions, because in the dry food crop plots there is some soil storage available in the beds.

The optimal open water area under the dry food crop conditions (2.4%), is smaller than for irrigated rice (2.5%) due to the same reason as mentioned above. Moreover, in dry food crop conditions the discharge from the field is smaller than for irrigated rice due to the soil storage in the plots.

The optimal discharge capacity of the pumping station for dry food crops (36.5 mm/day) is higher than for rice under irrigated conditions (32.9 mm/day), because the water area in the plot of the dry food crops is smaller than the water area in the plots of irrigated rice.

The rural polder in Thailand

The polder water level in the rural polder is influenced by the damage that may occur to the urban area in this polder. The polder water level is kept high in such a way that the fishponds and the rice area will have the benefit during the dry season, but if during the wet season if the polder water level is higher than the water inside the plot, the excess water will have to be pumped out. The polder water level at the end of the rainy season may be kept high (0.30 m-surface) to store water in the canal system in the polder area for irrigation and other uses.

The optimal discharge capacity of the pumping station for the rural polder is more or less equal to the capacity under rainfed conditions because in the simulation no irrigation water is applied to this area. Although there is some discharge of water from the urban area (33.9% of the total area) to the polder drainage system, most of the rainwater will be stored in the fishponds, which cover a large portion of the land use (52.9% of the total area) and can store water to a depth up to 1.0 - 1.5 m. Moreover, the average rainfall was used for the simulation, because the area is large, therefore a lower pumping capacity may be required than for a smaller area.

The urban polders in the Netherlands and in Thailand

In urban polders in Thailand the percentage of paved area is higher than in urban polders in the Netherlands especially in the area surrounding Bangkok. As in Thailand a lot of people move from the rural areas to the urban areas especially around Bangkok. Also in this study the urban polder in Delfland has more green area, such as parks, than the urban polder in Thailand. The type of road in the urban polder in the Netherlands has more bricks and tiles than in Thailand where most road surfaces are of concrete or asphalt. Therefore, a higher runoff discharge is expected, because asphalt is less pervious than bricks and tiles. The higher part of the urban polder in the Netherlands was drained over a weir. In Thailand most of the high part of the polder was separated from the lower part by a road and a culvert.

The canal water level has the highest influence on the damage in the urban area. The optimal canal water level in the urban polder in the Netherlands is 0.87 m-surface while in the urban polder in Thailand it is 2.56 m-surface. The canal water level in the urban polder in the Netherlands is shallower than in the urban polder in Thailand because the fluctuation of canal water level in the urban polder in Thailand is higher than in the urban polder in the Netherlands due to the higher sewer discharge.

In the urban polder in the Netherlands the optimal sewer diameter is 0.30 m while in the urban polder in Thailand it is 1.00 m. It can be concluded that in the urban polder in Thailand the damage due to surface water at the street is higher than in the urban polder in the Netherlands, therefore a bigger sewer diameter will be required to drain the water from the street.

The open water area in the urban polder in the Netherlands is larger than in the urban polder in Thailand. The urban polder in Thailand has a dense population and there is a rapid increase of the population. The canal in this area was filled and converted into a road with a culvert. A transport pipe was installed to carry the excess water to the pumping station. Therefore there is a smaller area of open water in this polder than in the urban polder in the Netherlands. At optimal conditions the required open water area in the urban polder in the Netherlands is only 0.2% and in the urban polder in Thailand it is 1.4%. Damage due to a too high canal water level will occur much more frequent in the urban polder in Thailand.

The optimal pumping capacity in the urban polder in the Netherlands is 5.0 mm/day, while in the urban polder in Thailand it is 117 mm/day. In Thailand's conditions a much higher pumping capacity is needed than in the Netherlands due to a higher intensity of the rainfall and a smaller percentage of green area in the urban polder.

Overall issues

The canal water level in the rural polders for all case studies can be kept higher during the dry period to enable the crops to benefit from capillary rise.

The case studies in the Netherlands and in Thailand show that for polders in the Netherlands the water control is done by pumping of the excess water to the collection and transport system. The function of the collection and transport system is to carry the water from the polders to the sea. In Thailand there is no clear a collection and transport system. Most polders in Thailand pump their water directly to the sea or the rivers. Therefore and due to the large subsidence over time the pumping head in the Netherlands may be higher compared to Thailand.

The water management in polders in Thailand involves both irrigation and drainage, while most of the time in the Netherlands it involves only drainage. In the Netherlands there is a low evapotranspiration, especially during the winter and rainfall is evenly distributed throughout the year. In Thailand, however, there is high evapotranspiration throughout the year and high rainfall in the rainy season, but in the dry season there is very few rainfall.

Water management in the rural areas of polders in the Netherlands is mainly concerned with groundwater table control, while in the rice polders in Thailand it concerns open water level control and for the dry food crop polders groundwater control.

8.2 Soil and water

Capillary flow and soil moisture in the unsaturated zone

The capillary flow in the unsaturated zone in or near the root zone is simulated in the model based on the soil moisture tension at the boundary conditions, which depend on the groundwater depth, soil properties, etc. In the temperate humid zone the soil moisture above the field drain level is above field capacity during winter, so there is no capillary rise. In this period the soil moisture content in layer II is higher than in layer I, because in layer I the model computes the soil moisture balance with evapotranspiration, while there is a small capillary rise from layer II. During the summer the soil moisture is below field capacity, therefore the capillary rise plays an important role. In the summer period the soil moisture content in layer I in the clay and the peat polder in the Netherlands decreases faster than the moisture content in layer II. This may be caused by the fact that the net result of the capillary rise from layer II to layer I, infiltration to layer I and the moisture taken by the crops takes more water than the net result of percolation from layer I to layer II and capillary rise from layer III to layer II supplies. Soils in the humid tropical zone can be drier than in the temperate humid zone due to the high evapotranspiration throughout the year. During dry conditions most of the clay in the humid tropical zone will crack and when sudden rainfall occurs water will be lost from the field to the canal system, especially in rainfed rice conditions. The capillary flow to the topsoil is also small under rainfed conditions during dry periods because of the deep groundwater table. Under irrigated rice conditions the soil moisture is almost saturated, therefore there will be no capillary rise. Deep percolation can play a role in the water management of irrigated system. With dry food crops the soil moisture was almost all the time at field capacity due to the irrigation water that was sprayed to the beds. The soil was only dry during a short period after harvesting. The capillary rise in dry food crop conditions can be of benefit to the crops and the farmers, while less pumping will be required for spraying of water during dry periods. During wet periods capillary rise may be small because the soil is almost above field capacity.

Soil storage

Soil storage may play an important role with respect to the water level fluctuations in the polder canal system, because the soil storage may be very high above the groundwater table, especially in peat soil. Soil storage is dependent on the soil moisture conditions and the soil properties. The soil storage in the humid tropical zone, especially with dry food crops plays an important role for the water level in the plot, because the lower the storage the higher the possibility of inundation of crops in the bed system. Soil storage also plays an important role in rainfed rice conditions because soils may be almost dry and it requires quite some rainfall to fill up this storage.

Groundwater fluctuation

Most polder areas are flat and the hydraulic gradient is negligible. Therefore to model the groundwater zone, a conceptualisation as reservoir storage and discharge to subsurface drains as a function of actual storage can be applied. For the temperate humid zone the discharge as non-linear relation with the storage in layer II can be used. The computed groundwater table in the clay and the peat polder is fluctuating above the field drain depth. In reality the groundwater, especially in peat soil, can drop in summer below the field drain depth. Therefore the lack of soil moisture for grassland in summer due to a low capillary rise may be more severe than in the simulation. In the humid tropical zone the water level involves both groundwater and open water. The observed groundwater table or open water level is needed to compare it with the simulated level. In the case studies in Thailand the results fit well with the observed groundwater table or open water level, especially in case of the dry food crops polder and the irrigated rice polder. The simulated groundwater tables in rainfed conditions were quite different from the observed data due to fluctuation of the water level in the river near the plot. In the experimental area in Thailand, the groundwater responds very fast to rainfall because the soil has a high permeability. Moreover, the groundwater in case of irrigated rice and dry food crops was affected by the water level in the lateral canal as well. In irrigated rice not only the fluctuation in the groundwater table but also in the open water level may have a significant influence on the adjacent area because of loss of water through the bunds and deep percolation.

8.3 Runoff relation with rainfall in the urban area

The two main functions of the sewer system in the urban area are disposal of storm water and wastewater from households and industries. The rainfall runoff relation is different for the temperate humid and the humid tropical zone, because of differences in temperature and rainfall pattern. In the case studies on the urban polders in the Netherlands as well as in Thailand, if the sewer diameter increases during optimization there will be a larger discharge to the canal system and a higher possibility to get damage due to too high water levels in the urban area. But in the Netherlands rainfall is not severe, so it is difficult to find a good rainfall runoff relation and damage due to water at the streets. Flash floods and water at the street can much more frequently occur in Thailand due to severe rainfall in a short time.

8.4 Water level in the main drainage system

The model assumption of a storage reservoir for the main drainage system during the computation of the water level is acceptable for a flat polder area. In the temperate humid zone the water level in the main drainage system can affect the flow from the subsurface field drainage system in case the water is higher than the subsurface drain depth. It may affect the soil moisture flow in the upper layers. However, a water level above the subsurface drain depth has little effect if the duration is short. In the humid tropical zone the water level in the main drainage system is affected by the discharge capacity from the plots for rice and dry food crops. During high water levels in the main drainage system in the wet season pumping will be needed, because excess water in the field cannot flow through the culvert and damage to the crops may occur. Moreover, the possibility to overtop the dike or bund increases if the water level in the main drainage system is high. During the dry season a high water level in the main drainage system is good for irrigation by letting water flow through the culvert by gravity or a smaller head for pumping. The water level in the main drainage system in polders in the humid tropical zone can be kept high at the end of the rainy season to store water for irrigation in the dry season.

8.5 Open water area

The open water area is directly concerned with storage of water in the polder. The smaller the open water area the less storage volume will be available. Normally reducing storage area will cause high water levels in the polder under the same drainage conditions. The possibility of damage to urban properties increases where the open water area is reduced. Also more pumping will be required for both rice and dry food crops due to higher water levels at the same drainage conditions. Moreover, the possibility to overtop the dike or bund will increase if the open water area in rice and dry food crop polder is reduced.

8.6 Discharge capacity

The discharge of polders is realised through pumping, or gravity flow through a culvert or over a weir. The higher the discharge capacity the less the polder water level rises up. In the Netherlands the discharge is mostly realised by pumping. In the humid tropical zone the discharge capacity can be divided into two categories: discharge from the field and discharge from the canal system. A lower discharge capacity from the field for rice will be more harmful due to more frequent high water levels in the field. A lower discharge capacity from the plots in case of dry food crops will result in a higher possibility of inundation of the beds in the plots and damage may be very high. Moreover, in a dry food crops polder a high water level in the ditches can create high groundwater tables that may damage the crops.

8.7 Conclusions and recommendations

Recommendations for optimal design, operation and maintenance in the Netherlands

Clay polder

In clay polders soil storage is generally not so high therefore the field drain depth may be deeper than in peat polders to avoid a too high groundwater table. Usually the field drain depth in a clay polder is between 1.00 to 1.50 m-surface and subsurface pipe drains are applied. At this depth an optimal distance between the drains of 23 m was found to obtain minimal damage due to too high groundwater tables. In this situation the discharge of excess water is not so high, therefore the fluctuation in the water level is not so high. According to the simulation the required pumping capacity is about 16.0 mm/day and open water area about 2%. For operation according to the simulation the polder water level would have to be kept low during the winter period, about 1.5 m-surface, the polder would have to be drained as much as possible before expected heavy rainfall to a low water level to avoid damage due to a too high water level. Therefore manual operation during expected rainfall may be useful. The subsurface drain depth has the most influence on the annual damage, therefore the subsurface drains would have to be installed at the optimal depth and well maintained to enable them to carry of the excess water to the collector drain in an efficient way.

Peat polder

In peat polders the soil usually has a high storage capacity and subsidence will occur if the open field drains are deep. Usually the open field drain depth is between 0.30 to 0.40 m-surface in order to avoid oxidation of peat soil. At this depth the simulated optimal distance between the drains would be 2.6 m to achieve low damage due to too high groundwater tables. In this situation the discharge is rather high, therefore the fluctuation of the water level in the polder is high. According to the simulation the required pumping capacity is about 18.0 mm/day and the open water area about 3.3%. According to the simulation the polder water level should be kept low during the winter period and the polder would have to be drained as much as possible before expected heavy rainfall to a lower water level to avoid damage due to a too high water level. Therefore manual operation during expected rainfall may be useful. The depth of the open field drains has the most influence on the annual damage, therefore this drain depth has to be well maintained.

Urban polder

The original purposed of land reclamation in the Netherlands was generally for agriculture, the urban area was supporting the local agricultural population. Only in the new polder Flevoland the new towns, Lelystad and Almere were developed directly after reclamation. In several of the existing polders, however, a large-scale urbanisation has taken place. Moderate rainfall together with a high percentage of green area have resulted in only incidentally high water layers at the street and a relatively low discharge to the canal system in the polder area, therefore a relatively slow rise of the water level in the canal system. The simulated optimal conditions are as follows: sewer diameter 0.30 m; distance between the urban canals 1,870 m; canal water level 0.87 m-surface; open water area 0.2% and pumping capacity 5.0 mm/day. These values may be not so accurate as a guideline for the design of water management systems, because they were the result of model simulations based on daily rainfall data. Thus short term peak discharge was eliminated in the simulation, therefore the simulated peak water level and highest water layer depth in the street may be lower than may be expected in practice. This may also have resulted in a too low pumping capacity. Moreover, in this simulation the high portion of the Hoge Abtswoudse polder performed as extra retention storage. Variations in the sewer diameter and pumping capacity have the most influence on the annual equivalent costs in this polder area. In the simulation it was found that an increase in the sewer diameter would result in a faster discharge and a more rapid water level rise. The damage mostly comes from water level rise not from water at the street, therefore it is not necessary to use a large sewer diameter. Application to other areas will have to take into account the ratio of the paved and unpaved area as well as the hydrological, physical and socio-economic conditions, which will be developed in the concerned urban polder.

Recommendations for optimal design, operation and maintenance in Thailand

Irrigated rice polder

In the irrigated rice polder the general principle for a good operation is that the polder water level can be kept lower than the ground surface during the rainy season to enable the farmers to drain the water from their field by gravity, but during the dry season the polder water level can be high to be enable them to irrigate their land by gravity. In practice water is stored in the canal system in the polder before the end of the rainy season, due to the cropping schedule there is normally no need to supply water in the rice field, but excess water needs to be drained before the harvesting of rice. In this period there are usually some conflicts between the authority that operates the polder system and the farmers who live in the relatively low part. If there is no water scarcity in this period the polder water level can be kept at least lower than the lowest ground surface level plus some safety freeboard to let farmers drain water out of their field by gravity. During the dry season water can be brought from storage reservoirs outside polder area. According to the simulation the optimal polder water level is 0.84 m-surface, open water area 2.5%, field drain capacity 44.8 mm/day and pumping capacity 32.9 mm/day. These values can change dependent on topography and socio-economic conditions. For example, the polder water level may be lower in a very flat area, because the plots for growing rice may be larger. Therefore more hydraulic gradient will be needed the when pump is operated. In Thailand, soils, physical and hydrological conditions will be different from place to place. Therefore, these values may be only guiding for the Central Plain area. The canals and outlets of the plots in these polders should be considered for both irrigation and drainage. The operation and maintenance should focus on the main components of the water management system that have the most influence on the annual equivalent costs, which are the polder water level and the pumping capacity.

Dry food crops polder

Dry food crop polders have usually modified plots of irrigated rice polders to be able to grow dry food crops. These polders have also flat land and are usually located near a river floodplain. In a dry food crops polder there are no constraints with respect to the cropping patterns. Farmers can grow crops any time if water is available. The basic principle for determining the polder water level is almost the same as for irrigated rice.

The polder water level should be kept lower than the ditch water level during the rainy season to enable the farmers to drain the water from their field by gravity, but during the dry season the polder water level should be higher than the ditch bed level to enable the farmers to supply water to their plot by gravity. In practice the water is stored in the canal system in the polder before the end of the rainy season. The water level in this period should not be too high to prevent overtopping of the surrounding dike. According to the simulation the optimal main components of the water management system are as follows: ditch water level 0.74 m-surface, distance between the ditches 8.45 m, polder water level 0.06 m+surface, open water area 2.4%, field drain capacity 34.8 mm/day and pumping capacity 36.5 mm/day. The optimal polder water level in this simulation is above the ground surface. This may be adopted for the dry season but in the determination the bed level of the ditches in the plot, the height of surrounding dike and the fluctuation of the water level have to be taken into account as well. Other values can be changed dependent on topography, physical and socio-economic conditions. For example, the open water level can be lower in the plots. In Thailand, the type of crops, soils, and hydrological conditions will be different from place to place. Therefore these values may be only guiding for the Central Plain. The canals and outlets of plots in these polders should be applicable to both irrigation and drainage.

Urban polder

Most urban polders in Thailand have a high population density, especially the polders in Bangkok and vicinity. Most of the area is paved. These are a small portion of gardens and green areas and also a small portion of open water area. The high intensity of rainfall results in a fast discharge to the canal system and also frequently to a high water layer at the street, and due to this also a fast increase of the water level in the canal system. According to the simulation the optimal main components of the water management system are as follows: sewer diameter 1.00 m, distance between the transport pipes 1,480 m, canal water level 2.56 m-surface, open water area 1.4% and pumping capacity 117 mm/day. These values may have to be modified based on the paved and unpaved area, as well as the hydrological, physical and socio-economic conditions in other places, which may have to be developed as an urban polder. However, these values may be not so accurate for the design water management system, because they were determined based on simulations based on daily rainfall data. The canal water level has most influence on the annual equivalent costs. Lowering of the water level in the rainy season can reduce damage due to water at the street and a high water level in the canal system. Also water can be discharged from the canals to provide more storage before expected heavy rainfall. This may reduce the damage in the urban area due to a too high groundwater table. The water level in the dry season can be kept high to prevent drying out of the soil and the development of cracks and shrinkage, which may damage the infrastructure. Cleaning and inspection of the sewers, transport pipes and the canal system needs to be done especially before the rainy season.

Model evaluation

Most polder areas are flat and the hydraulic gradient is negligible. Therefore to model the groundwater zone, a conceptualisation as reservoir storage is acceptable.

In case of the Netherlands the discharge to the subsurface drains as a function of the actual storage was applied and it was found that observed and computed data more or less fit. However, some are not fitting well due to unknown components, such as seepage, spatial differences in soil properties, spatial rainfall distribution, estimated open water area, estimated drain depth and distance between the drains, which vary over the area of the polder and vary due to differences in surface height due to topography. The model only represents the average conditions of the polder area. Therefore the observed and computed may be different especially during the dry period, because during this period seepage may have a relatively large influence on the polder area. However, the seepage is not constant due to fluctuations of the water level in the adjacent area, which were not included in this study.

In case of Thailand it has been found that the observed data and the calculation results fitted quite well with each other, especially with respect to the groundwater table. Therefore, the model can be used for simulation with a series of hydrological data and also for variations in the main components of the water management system. However, when the model will be applied in practice, the model can represent only the average conditions for systems.

9 Evaluation

Effect of rainfall and evapotranspiration on the water management in the polder area

In the temperate humid zone the rainfall is distributed over the year round and there is low evaporation during winter. Water management in this area has most of the time only to deal with drainage. Rainfall in the humid tropical zone is unreliable. In the rainy season the amount is more than enough, but in the dry season there is very limited rainfall, while there is high evaporation the whole year round. The water management in polder areas has to deal with these two situations. Water supply for irrigation during the dry season and discharge of excess water during the rainy season are requirements for water management in the humid tropical zone. In the humid tropical zone dams may be constructed for storing water to prevent flooding in the rainy season and to use it for irrigation during the dry season. The polder water area can also be used for storing of water for irrigation during the dry season. Sometimes water scarcity occurs and other measures may be required such as change in crops, change in cropping patterns or reduced area of crops.

Rainfall and runoff relation in the urban areas

The two main functions of the sewer system in the urban area are disposal of storm water and of wastewater from households and industries. In a polder environment the discharge to the polder water management system is coming from paved and unpaved areas. The rainfall runoff relations in the temperate humid and the humid tropical zone are different because of differences in temperature and rainfall pattern. In the case studies on the urban polder in the Netherlands and in Thailand it was found that if the sewer diameter increases during optimalization there will be a more rapid discharge to the canal system and a higher possibility that damage occurs due to too high water levels in the canals in the urban area. However, in the Netherlands rainfall is not severe, so it is difficult to find a good rainfall runoff relation and there is limited damage due to water at the street. Flash floods and water at the street can much more frequently occur in Thailand due to severe rainfall in a short time.

Crops, irrigation and drainage

In the temperate humid zone subsurface drainage by open field drains and subsurface pipe drains is applied to control the groundwater table. The most common crop in the humid tropical zone is rice. Therefore drainage and irrigation generally have to deal with open water, because rice grows in ponded water. In the temperate humid zone there is only drainage and groundwater control, because most of crops are dry food crops, such as potato, winter wheat and grass.

Operation and maintenance

In the temperate humid zone the operation of the water manage system in a rural polder focuses on groundwater control, related to the field drain depth and the polder water level. The operation primarily concerns drainage. The drainage at field level is controlled by subsurface drains. Peat soils are mostly drained by open field drains with a shallow drain depth and a small distance between the drains. Clay soils are mostly drained by corrugated PVC pipes with a deeper drain depth than in peat soil. Groundwater control is seldom done by the farmers. Sometimes farmer can interfere in the water level in the collector drain by constructing a weir or by closing the end of a drain pipe to manage the groundwater table during a rainfall deficit. The maintenance of drainage at field level is very important because it is directly related to damage in the polder area. Therefore, in peat soil the open field drains have to be kept at the required depth, while flushing of drains pipes in a clay polder is needed. The open water level in the temperate humid zone can be kept higher during the summer period to keep the groundwater high and to let the crops benefit of capillary rise.

In the humid tropical zone operation of water management systems in rural polders mostly focuses on open water level control, except for dry food crop areas. Because rice is grown in most of the polder areas, most of the time there has to be a water layer above the ground surface. This water layer needs to be maintained based on the requirements of the rice growth stage. This generally needs to be managed by pumping or gravity flow, which is dependent on the water level in the polder and in the field. The drainage of the field is preferably done by a culvert to control the water level in the field. Control of the polder water level in the humid tropical zone has to deal with irrigation and drainage because in this area there is a high evapotranspiration and unreliable rainfall. The irrigation is done at both the field level and the main system level. Irrigation and drainage at the field level is done by the farmers whereas irrigation and drainage at the main system level is usually done by government officers, together with farmer associations. The irrigation water is normally supplied during the dry season to the polder area through the dikes. Usually at the end of the rainy season around early November water will be stored in the canal systems for use in the dry season. Hence, the canal systems in rural polders in the humid tropical zone may be used for both irrigation and drainage. While the control of the polder water level is different dependent on the season, in the rainy season the polder water level is lower than in the dry season.

Drainage systems in the rural areas of a polder

Due to the differences between crops in the temperate humid zone and the humid tropical zone in a rural polder area, there are different requirements for the drainage systems. In the temperate humid zone the control of the groundwater table is the major function of drainage systems, therefore subsurface drainage is applied in these areas. However, in the humid tropical zone rice is grown in most of the polders, therefore drainage generally has to deal with the water layer above the ground surface. Due to the different purposes of the drainage systems in the temperate humid zone the polder water level should be lower than the depth of the subsurface drains, while the polder water level in the humid tropical zone can be higher than the ground surface to supply water to the plots by gravity or lower than the ground surface for drainage conditions.

Due to the different functions of the drainage systems in polder areas, there are differences in layout and design concepts. Polder areas in the temperate humid zone with different levels of the ground surface, have generally been divided into different areas to control the groundwater table dependent on the ground surface and each area in the polder may have its own weir or pumping station and preferred water level. In such an area the excess water is discharged over the weir or pumped to the main canal, dependent on the preferred water level in the area compared to the polder water level. Finally water is pumped into the collection and transport system and conveyed to the river or the sea. For polders in the humid tropical zone there are usually no different areas but only adaptations of breeding and agricultural practices to the different open water depths. Such as the dry seed method rice for high level land and deep water rice (floating rice) for low lying areas. The canals in the polder areas in the humid tropical zone are used for both drainage and irrigation. During the dry period the irrigation water is supplied to the polder area to be stored in the canal systems, from there farmers can pump or let water flow by gravity into their land, dependent on the water level in the canal system.

Drainage systems in urban areas in polders

The open water in urban areas in polders in the temperate humid zone and humid tropical zone is mostly composed of the canal systems. In urban areas in polders in Thailand some parts of the water area have been converted to roads by filling the urban canals and transport pipes were installed to transport excess water to the open water area. From there the excess water is pumped to a river or the sea. Due to this the water level in the canals rises very fast. Rainfall intensity in the humid tropical zone is higher than in the temperate humid zone, which also makes the canal water level rise faster than in the temperate humid zone. Therefore, the pumping capacity must be very high. The sewer pipe diameter in urban areas in the humid tropical zone should be larger than in the temperate humid zone, because of more intensity of rainfall and more probability of damage due to water at the street. Damage due to too high canal water levels in the humid tropical zone is also more severe than in the temperate humid zone. Moreover, the paved area in the urban area in Thailand is mostly composed of asphalt and concrete, while in the Netherlands it is mostly composed of bricks and tiles, which can absorb water and let it infiltrate into the ground. Hence, less peak water will flow to the canal system at the same rainfall intensity.

Model development

In the temperate humid zone soils in polder areas may be more wet compared to soils in the humid tropical zone, especially in the Netherlands where rainfall is distributed almost over the whole year. Therefore irrigation is rarely needed in the Netherlands. Soil in the humid tropical zone has a higher probability that cracks will develop during dry periods, especially under rainfed rice conditions. In order to simulate such situations the OPOL package has been modified to simulate the behaviour of crops, soil and water in the humid tropical zone, including irrigation and drainage systems. It was found that the OPOL5 package, which has been coupled with the models FLOC and Slit, simulated results that did fit well with the observed data in the experimental area in Thailand. Therefore it can be concluded that the model can be reliable for application in Thailand.

Economic overview

In the rural areas based on the total area of the polder and average cost at the present and optimal conditions the costs for construction of field drainage systems, open water and pumping stations in a polder per unit area were found to be as follows:
    For the rural areas in the Netherlands:
-    6,400 €/ha for the clay polder;
-    8,900 €/ha for the peat polder.
    For the rural areas in Thailand:
-    600 €/ha for a polder under rainfed rice conditions;
-    2,100 €/ha for a polder under irrigated rice conditions;

- 4,200 €/ha for a polder under dry food crop conditions;
- 900 €/ha for the rural polder.

A clay polder may have less construction costs than a peat polder. However, the construction costs are also dependent on the size and topography of the area, which is not taken to account in the model.

The annual equivalent costs in the Netherlands are around 10%, and annual operation and maintenance costs 3 - 4% of the construction costs in the clay polder and 12 - 14% and 6 - 8% respectively for the peat polder. A peat polder usually has higher annual costs due to the fact that of open field drains need more frequent cleaning than subsurface drain pipes in a clay polder and also because workability in a peat polder is more difficult than in a clay polder.

In Thailand the investment costs per ha for a dry food crops polder is the highest. The investment costs per ha for the rural case study is not so high because the cost of construction of fishponds, which cover a large area in this polder, has not been included in this study.

The annual equivalent costs in Thailand are around 15%, and annual operation and maintenance costs about 10% of the construction costs for rainfed rice, 13 - 14% and 8 - 9% for irrigated rice, 11 - 12% and 5 - 6% for the dry food crops polder, 10 - 11% and 5 - 6% for the rural case study area. A rice polder has higher annual equivalent costs than a dry food crops polder, because the costs of earthwork for construction and maintenance involve much higher labour costs, while in a dry food crops polder construction and maintenance costs involve mostly machine costs, which are cheaper.

In the urban areas based on the total area of the polder and average cost at the present and optimal conditions, the costs for construction of sewer systems, open water and pumping stations in the polder per unit area were found to be as follows: the urban areas in the Netherlands of 31,300 €/ha, while in Thailand of 15,100 €/ha. The annual equivalent costs in the Netherlands are around 7%, and annual operation and maintenance costs 1.5% of the construction costs and in Thailand 9% and 2.4% respectively.

The construction costs in Thailand are lower than in the Netherlands due to the lower labour and fuel cost.

The annual equivalent costs and annual operation and maintenance costs in percentages of the total construction costs in the urban areas in Thailand are higher than in the Netherlands because rainfall intensity and temperature in Thailand is higher than in the Netherlands. Consequently there is more erosions of the soil and weeds growth in canal system, which can sediment and obstruct in the canals and sewer systems. Moreover, in the urban polder in Thailand has a higher density of population than in the Netherlands which can produced more garbage, which can deposit in the canal and sewer system. The annual equivalent and annual maintenance costs in percentages of the total construction costs in the rural areas in the Netherlands and in Thailand, may be difficult to compare due to the differences in the field drainage method.

Closing remarks

The OPOL package is a useful tool for determining optimal values for the main components of the water management system in a polder and also for understanding the effects of changes in the values for the main components of the water management system to the costs and damage in a polder. The values for the main components of the water management system as resulting from the simulations in this study are indicative and have to be determined in practice under real conditions such as physical conditions, operation and maintenance practices, land use, agriculture practices, policy, technical aspects, soil conditions, environment and landscape conditions. The particular data have to be based on local conditions.

The annual equivalent costs as found in this study are rather low. Therefore it is recommended to include a safety margin in the optimal values for the main components of the water management system, which are obtained from the OPOL package. The return period, flood risk and risk costs may be included in further studies and analyses on the main components of the water management system.

The annual equivalent costs increase tremendously when the values of the main components of the water management systems are taken smaller than the optimal values. Therefore the risk of under design of the main components has to be seriously considered. The design would also have to take into account future use, lifetime of assets, growth of population, urbanisation, advance of technology, change of environment, economic and social conditions, statistics in hydrology, risk analysis, etc.

The drain depth has most influence on the annual equivalent costs in the rural area in both the clay polder and the peat polder in the Netherlands. In Thailand the polder water level has most influence on the annual equivalent costs in the area of rice for both rainfed and irrigated conditions. In irrigated rice conditions not only the polder water level but also the pumping capacity has most influence on the annual equivalent costs, while the ditch water level has most influence on the annual equivalent costs in the dry food crops polder. The urban area is very sensitive to the canal water level in Thailand, but not so much in the Netherlands.

In this study it is shown that the optimal values for the main components of water management systems in polder areas for the temperate humid and the humid tropical zone can be determined.

References

Acharya, C.L. and Sood, M.C., 1992. Effect of tillage methods on soil physical properties and water expense of rice on an acidic Alfisol, Journal of the Indian Society of Soil Science, 40, 409 - 419

Ali, L., 2002. An integrated approach for the improvement of flood control and drainage schemes in the coastal belt of Bangladesh, PhD Thesis, IHE Delft, Delft, the Netherlands

Asian Engineering Consultants Co. Ltd and Macro Consultants Co. Ltd, 1996. Main report, feasibility report of Bangpakong right bank project, Chacherngsao province, Thailand, 5-19 to 5-20 (in Thai)

Eelaart, A. van den, 1998. A web site by Adriaan van den Eelaart in support of ISDP (Integrated Swamp Development Project), IBRD Loan 3755-IND (http://www.eelaart.com/)

Agro climatic atlas of Alberta, http://www.agraic.gov.ab.ca/agdex/000/710001a.html

Almeya, G., 1979. Mexico's new frontier: the humid tropics, Ceres, 12(5), 27 - 33

Andreae, B., 1981. Farming development and space: a world agricultural geography, Walter de Gruyer, Berlin, Germany

Ayoade, J.O., 1983. Introduction to climatology for the tropics, Department of Geography, University of Ibadan, John Wiley & Sons, Chichester New York Brisbane Toronto Singapore, 189

Baig, M.R., 1997. Country reports C5, Proceedings Volume 1, 7th ICID International Drainage Workshop, Drainage for the 21st Century, 17 - 21 November 1997, Penang, Malaysia

Balek, J., 1977. Hydrology and water resources in tropical Africa, Elsevier, Amsterdam, the Netherlands

Barrow, C., 1986. Water resources and agricultural development in the tropics, Longman, New York, USA

Belinda Fuller, ed., 1996. TDRI quarterly review, Vol 11, No. 4, December 3 - 10, http://www.info.tdri.or.th/library/quarterly/text/d96_1.html

Biswas, A.K., 1972. History of hydrology, 2nd edition, North-Holland Publishing Company, Amsterdam, the Netherlands

Biswas, A.K., 1976. System approach to water management, McGraw Hill, New York e.a., USA

Boekel, P., 1974. De betekenis van de ontwatering voor de bodemstructuur op zavel- en lichte klei-gronden en de financiele consequenties daarvan, Bedrijfsontwikkeling, 5^e jaargang, nr. 10 (in Dutch)

Bouma, J. and Anderson, J.L., 1973. Relationships between soil structure characteristics and hydraulic conductivity. In: R.R. Bruce (Ed.), Field soil moisture regime. Soil Science Society of America, Special Public no. 5, American Society of Agronomy, Madison, 77 - 105

Breaden, J.P., 1973. The generation of flood damage time sequences, University of Kentucky Water Resource Institute, Lexington, Kentucky, USA

Bronswijk, J.J.B., 1988. Modeling of water balance, cracking and subsidence of a clay soils, Journal of Hydrology 97 (1988), 199 - 212

Bronswijk, J.J.B., 1989. Prediction of actual cracking and subsidence in clay soil, Soil Science Society of America, 148(2), 87 - 93

Bronswijk, J.J.B., 1990. Shrinkage geometry of a heavy clay soil at various stresses, Soil Science Society of America, J. 54, 1500 - 1502

Cai Lingen, 1997. Country reports C4, Proceedings Volume 1, 7th ICID International Drainage Workshop, Drainage for the 21st Century, 17 - 21 November 1997, Penang, Malaysia

CIA the world Fact book 2000 Netherlands
http://www.cia.gov/cia/publications/factbook/goes/nl.html

CIA the world Fact book 2000 Thailand
http://www.cia.gov/cia/publications/factbook/goes/th.html

Charnvej, O., 1999. Irrigation water saving for rice production, The symposium on Water saving for irrigation for paddy rice, 10 - 13 October 1999, Peoples Republic of China

Charoenying, S., 1989. Rice farming practice in irrigated areas, Irrigated Agricultural section, Water Allocation Division, Royal Irrigation Department, Bangkok, Thailand, 4-5 (in Thai)

Charoensiri, K., 1991. Diffusion of moisture in section area of soil at rise bed system in southern part of Central Plain of Thailand, MSc thesis, Kasetsart University, Bangkok, Thailand (in Thai)

Chaudhary, T.N., Bhatnagar, V.K. and Prihar, S.S., 1975. Corn yield and nutrient uptake as affected by water-table depth and soil submergence, Agronomy Journal vol. 67, November-December 1975, 745 - 749

Chen Jia-fang and Li Shi-ye, 1981. Some physical properties of paddy profiles in Taihu Lake region, Proceedings of Symposium on Paddy soil, Edited by Institute of Soil Science, Academia Sinca, Science Press Beijing, Springer-Verlag, Berlin Heidelberg New York, 21

Cheng, Yun-sheng, 1984. Effects of drainage on the characteristics of paddy soils in China. (Nanjing). In Organic Matter and Rice. IRRI, Manila, Phillipines

Commissie voor Hydrologisch Onderzoek, 1988. Rapporten en Nota's No 19, Nederlandse Organisatie voor Toegepast, the Netherlands (in Dutch)

Constande, A.K., 1982. From spontaneous settlement to integrated Planning and Development, Paper to be present at 'Polders in the world' October 1982, Lelystad, the Netherlands

Cultuurtechnische Vereniging, 1988. Cultuurtechnishch Vademecum. Werkgroep Herziening Cultuurtechnisch Vademecum (in Dutch)

De Bakker, H. and Van den Berg, M.W., 1982. Proceedings of the symposium on peat lands below sea level, ILRI publication 30, the Netherlands, 130 - 163, 201 - 204

De Bie, K., 1992. Yield constraints for rice in Thailand, International Institute for Aerospace Survey and Earth Sciences (ITC), the Netherlands

De Crecy, J., 1982. Structural behaviors of heavy soils and drainage systems. In Land Drainage. M.J. Gardiner (ed.). Balkema, Rotterdam, the Netherlands, 73 - 84

De Datta, S.K. and Kerim, M.S.A.A.A., 1974. Water and nitrogen economy of rainfed rice as affected by soil puddling. Soil Science Society of America Proceedings, 38, 515 - 518

De Glopper, R.J., 1973. Subsidence after drainage of the deposits in the former Zuyder Zee and in the brackish and marine forelands in the Netherlands, Van Zee tot Land, 50, Staatsuitgeverij, 's-Gravenhage

Depeweg, H.W.Th, 2000. Structures in irrigation networks lecture note, IHE, Delft, the Netherlands, April 2000, 207 - 208

Deinum, B., 1966. Climate nitrogen and grass, Dissertation no 404, Wageningen University, Wageningen, the Netherlands

Delft municipality, 2002. Delft, the Netherlands, Drawing No 9M1818.A0/9320-304 and 4K0140A0/3280-301

De Ron, J. and Van der Werf, R., 2003. Onderzoek naar mogelijk verbeteringen polderwatersystem Delft, Bijlagenraport, Gemeente Delft, the Netherlands, 6 Augustus 2003, Raportage 9M0413 (in Dutch)

Doorenbos, J. and Kassam, A.H., 1986. Yield respond to water, FAO irrigation and drainage paper 33, Rome, Italy

Döring, M., 2000. The drainage tunnel of Lake Fucin - a Roman land reclamation project in Italy, Water & History, submitted paper, Second World Water Forum, 19 March 2000, the Hague, the Netherlands

Delfland, 1990. Peilbesluit polder Schieveen, Hoogheemraadschap van Delfland, Delft the Netherlands (in Dutch)

Department of Drainage and Sewerage, 1998. Sukhumvit survey and design drainage project, Completion report of detail design of flood protection and drainage systems, report No 1/2 prepared by Water development Consultants Co. Ltd, Bangkok Metropolitan Administration, Bangkok, Thailand, 4-20 - 4-22 (in Thai)

Driessen, P.M., 1978. Peat soils, In: Soil and Rice IRRI (Ed.), Los Baoos, Philippines, 763 - 779

Dumm, L.D., 1960. Validity and use of the transient flow concept in subsurface drainage, Paper presented at ASAE meeting, Memphis, December, 4 - 7

ECAFE (Econ Comm., for Asia and far East), 1969. Proceeding of the second symposium on the development of deltaic areas, cited by Jaw-Kai Wang and Wagen R.E., Irrigated Rice Production Systems: Design Procedure, Westview Press, Inc. Colorado, USA

Feddes, R.A., and Bastiaanssen, W.G.M., 1990. Forecasting soil-water plant atmosphere interactions in arid regions, In: Proceedings NATO workshop, Water saving techniques for plant growth, Gent, Belgium (In press)

Fokkens, B., 1970. Berekening van de samendrukking van veenlagen uit het gehalte aan organische stof en water De Ingenieur 82

Follett, R.F., Doering, E.J., Reichman, G.A., Benz, L.C., 1974. Effect of irrigation and water-table depth on crop yields, Agronomy Journal vol. 66, 304 - 308

Forchheimer, P., 1930. Hydralik 3 rd. Teubner, Leipzig-Berlin, Germany

Framji, K.K., Garg, B.C. and Luthra, S.D.L., 1981. Irrigation and drainage in the world, a global review, International Commission on Irrigation and Drainage (ICID), New Delhi, India

Ghosh, A.K., Sadana, S. and Asthana, A.N., 1980. Evaluation of rice culture for submergence tolerance and yield under lowland condition at Tribular, IRRI Newsletter 5 (5), 12

Goudie, A.S., 1981. The human impact: Man's role in environmental change, Basil Blackwel, Oxford, UK

Greacen, E.L. and Gardner, E.A., 1982. Crop behavior on clay soils, Symposium on the management of clay soils held at the University of West Indies, St. Augustine, Trinidad, 15 - 19 September 1980, Tropic. Agric (Trinidad) 59(2), 123 - 132, Butterworth Scientific, Guildford, UK

Greenland, D.J., 1981. Recent progress in studies of soil structure and its relation to properties and management of paddy soils, In: Institute of Soil Science, Academia Sinica ed., Proceedings of Symposium on Paddy Soil. Science Press, Beijing, China, 42 - 58

Grigg, D., 1970. The hash lands: a study in agricultural development, Macmillan Press, London and New York

Grossman, R.B., Brasher, B.R., Franzmeier, D.P., and Walker, J.L., 1968. Linear extensibility as calculated from natural-clod bulk density measurements, Soil Science Society of America, Proc., 32(4), 570 - 573

Harris, D.R., 1980. Human ecology in savanna environments, Academic Press, London and New York

Hendriks, R., F.A., Oostindie, K. and Hamminga, P., 1998. Simulation of bromide tracer and nitrogen transport in a cracked clay soil with the FLOCR/ANIMO model combination, Journal of Hydrology 215 (1999), 94 - 145

Hess, T., 1994. A microcomputer scheduling program for supplementary irrigation, Department of Water Management, Silsoe College, Bedford, UK
http://www.fao.org/docrep/w4367e/w4367e08.htm

Hissink, D.J., 1935. De bodemkundige gesteldheid van de achtereenvolgens ingedijkte Dollar polders Versl Landbk Onderz 41.3, 's-Gravenhage (in Dutch)

Hofwegen, P.J.M. van and Svendsen, M., 2000. A vision of water for food and rural development, the Hague, the Netherlands

Hoogmoed, W.B. and Bouma, J., 1980. A simulation model for predicting infiltration into cracked clay soil, Soil Science Society of America, J., 44, 458 - 461

Houghton, J.T., Jenkins, G.J. and Ephraums, J.J., ed, 1990. Climate change, The IPCC Scientific assessment

Hungsapreg, S., Suthakavatin, R., and Suwanarat, K., 1998. Sustainable development of deltas country position of Thailand, Proceeding International conference at occasion of 200 year Directorate-General for Public Works and Water Management, Amsterdam, the Netherlands, 23 - 27 November 1998, Delft University Press

IHP International Symposium on 'Rivers and People in Southeast Asia and the Pacific - Partnership for the 21st Century'

Inthaiwong, S., 1996. MSc Thesis, Study on rice yield reduction due to flooding, Kasetsart University, Bangkok, Thailand (in Thai)

International Rice Research Institute (IRRI), 1971. Annual report for 1970, Los Banos, Philippines

Government Public Relations Department, Introduction to Thailand,
http://www.thaimain.org/eng/Thailand/history1.html

Institute of Lowland Technology (ISLT), 1998. International Symposium on Lowland Technology, Organized by The, Saga University, Saga, Japan
http://www.groundcontrol.nl/studies/e_power.htm

Jackson, I.J., 1977. Climate, water and agriculture in tropics, Longman, London and New York

Jeurink, N., Gels, J.H.B., Eugelink, A.H., van Duijn, H., 2000. Peilbesluit Duifpolder Midden-Delfland Toelichting, Hoogheemraaschap van Delfland, Project number 3752925 (in Dutch)

Jindasanguan, J., 1995. Irrigation water supply in the Irrigation Office 8, Royal Irrigation Department, Bangkok, Thailand (in Thai)

Jongdee, S., Mitchaell, J.H. and Shu Fukai, 1996. Modelling approach for estimation of rice yield reduction due to drought in Thailand, Department of Agriculture, The University of Queensland, Brisbane, Queensland 4072, Australia

Kaduma, J.D., 1982. Water as constraint on agricultural development in the semi-arid areas of Tanzania', Water Supply and Management, 6 (6)

Kershaw, K.A., 1973. Quantitative and dynamic plant ecology (2nd edn), Edward Arnold, London, UK

Kowal, J.M. and Kassam, A.H., 1978. Agriculture ecology of savanna, Clarendon Press Oxford

Kuester, J.L. and Mize, J.H., 1973. Optimization techniques with FORTRAN, McGraw -Hill, New York, USA, 320 - 330

Kupkanchanakul, T., 1999. Bridging the rice yield gab in Thailand (http://www.fao.org/DOCREP/003/X6905E/x6905e0d.html)

Leeghwater, J. Asz., 1641. Harrlemmer-Meer-boek Dominicus van der Stichel, Amsterdam, the Netherlands

Letey, J., Stolzy, L.H. and Valoras, N., 1965. Relationships between oxygen diffusion rate and corn growth, Agronomy Journal, 57, 91 - 92

Leung, K.W. and Lai, C.Y., 1974. The Characteristics and genesis of paddy soils in the northern part of Taiwan, Agricultural Research, 22, 77 - 97

Lobbrecht, A.H., Sinke, M.D. and Bouma, 1999. Dynamic control of the Netherlands polder and storage basin, the Netherlands, Water Science & Technology, volume 39 number 4, 1999, ISSN 0237-1223, 269 - 279

Lucas, R.E., 1982. Organic soils (Histosols) formation, distribution, physical and chemical properties and management for crop production, Michigan State University, Research Report No. 435 (Farm Science), USA

Luijendijk, J., 1987. The role of civil engineering in lowland development, Symposium ITB, Bandung, Indonesia

Luijendijk, J. and Schultz, E., 1982. Het waterbeheeringsysteem van een polder, PT/Civiele Techniek 37, nr. 9, the Netherlands (in Dutch)

Maaskant, M., Ritzema, H.P. and Wolters, W., 1986. Polders in Egypt, In: Land and Water International no. 58, NEDECO, the Hague, the Netherlands

Mallik, S., C.R.L., Mitra, N.K. and Mandal, B.K., 1998. Breeding for submergence tolerance, International Rice Research Newsletter 13 (4), 12 - 13

Mao Zhi, 2001. Water efficient irrigation and environmentally sustainable irrigated rice production in China, Wuhan University, Wuhan, China

Master Plan for Flood Protection and Drainage in Samut Prakan East, 1988. Prepared by Thailand Institute of Scientific and Technological Research, 30 August 1988,Volume II and appendix 4, Bangkok, Thailand

Matsushima, S., 1962. Ministry of Agriculture. (Malaya), Bull. No. 112

McIntyre, D.S., Loveday, J. and Watson, C.L., 1982. Field studies of water and salt movement in an irrigated swelling clay soil. I: Infiltration during ponding, Australian Journal of Soil Research 20, 81 - 90

Medina, J.E., Martin, D. and Eisenhauer, D., 1998. Infiltration model for furrow irrigation, Journal of Irrigation and Drainage Engineering /March /April 1998/73

Middelkoop (red.), H., 1999. The impact of climate change on the river Rhine and the implications for water management in the Netherlands, summary report of NRP project 952210

Molle, F. and Keawkulaya, J., 1998. Water management and agricultural changes: A case study in the Chao Phraya delta, Southeast Asia Studies 36(1), 32 - 58

Molle, F., Sutti, C., Keawkulaya, J., Korpraditskul, R., 1988. Water management in raised bed systems: a case study from the Chao Phraya delta, Thailand, DORAS Center, Kasetsart University, ORSTOM, Administrative building 10th floor, Bangkhen, 10900 Bangkok, Thailand

Naklang, K., 1996. Direct Seeding for rainfed lowland rice in Thailand, Surin Rice Experiment Station, Surin, Department of Agriculture, Thailand, 28

National Economic and Social Development Board (NESDB), 2001, Gross national product and national income at current market prices by industrial origin (http://www.nesdb.go.th/econSocial/macro/gdp_data/reportagdp.asp?heading_id=9)

Netherlands Organization for Applied Scientific Research TNO, 1989. Water in the Netherlands CHO-TNO, Delft, the Netherlands

Nicholls, R.J. and Mimura, N., 1998. Regional issues raised by sea-level rise and their policy implications, Climate Research, 11, 5 - 18

Notowijoyo, P., 1991. Polder development in Indonesia, Optimization of the main components of drainage system in polders, MSc Thesis H.H. 102, October 1991, IHE, Delft, the Netherlands

Nutalaya, P., Young, R.N., Chumnankit, T. and Buapeng, S., 1996. Land subsidence in Bangkok during 1978 - 1988. in Milliman, J. D. and Haq, B. U (eds.) Sea-Level Rise and Coastal Subsidence, 105 - 130

Pagliai, M. and Painuli, D.K., 1991. The physical properties of paddy soils and their effects on post rice cultivation for upland crops, In: Soil Management for Sustainable Rice Production in the Tropics, IBSRAM Monograph No. 2, International Board for Soil Research and Management, Thailand, 391 - 415

Painuli, D.K., Woodhead, T. and Pagliai, M., 1988. Effective use of energy and water in rice soil puddling, Soil and Tillage Research, 12, 149 - 161

Palada, M.C. and Vergara, B.S., 1972. Environmental effects on the resistance of rice seedling to complete submergence, Crop Science 12, 209 - 212

Poopakdi, A., 1999. Establishing rice legume cropping system in the Central Plain for better productivity and sustainability, International Conference, The Chao Phraya Delta: Historical Development Dynamic and Challenges of Thailand 's Rice Bowl, 12-15 December 2000, Kasetsart University, Bangkok, Thailand

Probert, M.E., Fergus, I.F., Bridge, B.J., McGarry, D., Thomson, C.H. and Russell, J.S., 1987. The properties and management of Vertisols, Commonwealth Agricultural Bureaux International, Oxon., UK

Public Work Department, 1988. Master plan for flood protection and drainage in Samut Prakan East, Bangkok, Thailand, Volume II: Report, August 1988 4 - 13

Purvis, A.C. and Williamson, R.E., 1972. Effects of flooding and gaseous composition of the root environment on growth of corn, Agronomy Journal 66, 674 - 678

Rajatasereekul, S., Sriwisut, S., Porn-uraisanit, P., Rungsook, S., Mitchell and Fukai, S., 1997. Phonology requirement for rainfed lowland rice in Thailand and Loa PDR, Department of Agriculture, University of Queensland, Brisbane, Queensland 4072, Australia

Ratanaphol, P., 1994. MSc Thesis, Case study of drainage system in Plychumphon Irrigation Project, Kasetsart University, Bangkok, Thailand (in Thai)

Rao, M.S. and Murthy, P.S.S., 1986. Effect of complete submergence on plant height, flowering duration and percentage recovery in five rice varieties, IRRI Newsletter 11(6), 13 - 14

Reddy, B.B., Ghosh, B.C. and Panda, M.M., 1986. Flood tolerance of rice at difference crop growth stage as affected by fertilizer application, Rice Abstr.9 (3), 134 - 135

RID (Royal Irrigation Department), 1975. Irrigated Agriculture Branch, Operation and Maintenance Division, Royal Irrigation Department, Bangkok, Thailand (in Thai)

RID (Royal Irrigation Department), 2001. Contract price submit by Uraluksingburi (1994) Co. Ltd for construction pumping station at Srinakarin dam Kanchanaburi province, Office of Water Resource Development 3, Royal Irrigation Department, Bangkok, Thailand (in Thai)

RID (Royal Irrigation Department), 2002. Handbook of unit cost for estimate construction contract price and budget planning approved by Bureau of the Budget Thailand in 2001, Royal Irrigation Department, Bangkok, Thailand (in Thai)

Rice Research Institute 1987a. Progress report of rainfed rice improvement research and development project (1983-1985), Monitoring and Evaluation Section, Rice Research Institute, Department of Agriculture, Bangkok, Thailand, 158

Richards, P., 1964. The tropical rain forest: an Ecological Study (2nd edn), Cambridge University Press, UK

Rijniersce, K., 1983. A simulation model of physical soil ripening Rijksdienst voor de IJsselmeerpolders, Flevobericht 203, Lelystad, the Netherlands, 138

Ritzema, H.P., 1994. Drainage principles and applications, ILRI, Publication 16, Second edition (Completely Revised), Wageningen, the Netherlands, 46

Rosenbrock, H.H, 1960. An automatic method for funding the greatest or least value of a function, Computer Journal 3, 175-184. [341]

Ruthenberg, H., 1980. Farming system in the tropics (3rd edn), Clarendon Press, Oxford, UK

Rycroft, D.W. and Amer, M.H., 1995. Prospects for the drainage of clay soils, FAO, Irrigation and Drainage paper 51, 16 - 25

Safety, accessibility, quality of life and innovation in a dynamic delta, 19 September 2000, Ministry of Transport Publics works and water management http://www.minvenw.nl/cend/dvo/international/english/pressrelease/000919_5929e.htm l

Sabhasri, S. and Suwarnarat, K., 1996. in Milliman, J. D. and Haq, B. U (eds.) Sea-level rise and coastal subsidence, 343 - 356

Saiful Alam; 1994. Optimizing water management in polder Flevoland, the Netherlands, MSc thesis, H.H. 190, IHE, Delft, the Netherlands

Salter, P.J. and Goode, J.E., 1967. Crop response to water at different stages of growth, Comm. Ag. Bureau Res. Rev. No 2

Sanders, Douglas C., 1997. Vegetable crop irrigation, Horticulture Information Leaflet 33-E, Department of Horticultural science, North Carolina State University, College of Agriculture &Life Science

Sanderson, M., 1990. UNESCO sourcebook in climatology, For hydrologist and water resource engineer, UNESCO

Schultz Bart, 1990. Guidelines on the construction of horizontal subsurface drainage system, Working Group on Drainage Construction, ICID, New Delhi, India

Schultz Bart, 2001. Irrigation, drainage and flood protection in a rapidly changing world, ICID

Schultz Bart, Thatte, C.D. and Labhsetwar, V.K., 2005. Irrigation and drainage important contributors to global food production, ICID

Schultz, E.; 1992. Warterbeheersing van de Netherlandse droogmakerijen, PhD thesis, Delft University of technology, Delft, the Netherlands (in Dutch)

Schultz, E., 1982. From natural to reclaimed land, RIJP, Lelystad, the Netherlands

Schultz, E, 1983. From natural to reclaimed land, Land and water management in the polders of the Netherlands, Polder of the world, Final report, International Symposium, Lelystad, the Netherlands, 31-33

Segeren, W.A., 1982. Keynotes international symposium, Polders of the world, International Symposium, Lelystad, the Netherlands

Segeren, W.A., 1983. Polder of the world, Final report, International Symposium, Lelystad, The Netherlands, 15 - 25

Sharma, P.K. and De Datta, S.K., 1985. Puddling influence on soil, rice development and yield, Soil Science Society of America Journal, 49, 1451 - 1457

Sharma, P.K. and De Datta, S.K., 1986. Physical properties and processes of puddled rice soils, Advances in Soil Science, 5, 139 - 178

Shiang Kueen, Hsu, 1997. Country reports C9, proceedings volume 1, 7th ICID International Drainage Workshop,' Drainage for the 21st Century, 17 - 21 November 1997, Penang, Malaysia

Smit, N., 1971. A history of dams, Peter Davies, London, UK

Somboon, J.R.P., 1990. Geomorphology of the Chao Phraya delta Thailand, PhD Thesis, Kyoto University, Japan, 86

Somboon, J.R.P. and Thiramongkol, N., 1993. Effect of sea-level rise on the north coast of the bight of Bangkok, Malaysian Journal of Tropical Geography, 24, 3 - 12

Spatial Planning and the Environmental, 1998. Planning with water, Ten building blocks for policy innovation in spatial planning, the Hague, the Netherlands

Staringcentrum, 1976. Hydrologie waterkwantitiet van Middel West-Nederland Reginaale Studies 9, Wageningen, the Netherlands (in Dutch)

Stoutjesdijk, J.A., 1982. Polders of the world, compendium of polder projects, October 1982, the Netherlands

Sutti, C., 1998. Management water use in rise bed system at Damnernsadeung, MSc thesis, Kasetsart University, Bangkok, Thailand, 78 (in Thai)

Suwannachit, S., 1996. Water management of Klongdan Irrigation Project, B Eng thesis, Irrigation Engineering Faculty, Kasetsart University, Kamphaengsean, Nakhonprathom, Thailand (in Thai)

Sunthorn Ruanglek, Shoombhol Chaveesuk and Maitri Poolsup, 1982. Land reclamation in Thailand, International Symposium, Polders of the world volume1, October 1982, the Netherlands

Stephens, J.C., J.H., Allen, J.R., and Chen, E., 1984. Organic soil subsidence, Geological Society of America, Rev. in Eng Geol., Vol VI

Stephens, J.C. and Stewart, E.H., 1977. Effect of climate on organic soil subsidence, Proc. 2nd Symposium, Land Subsidence (Anaheim) IASH Publication 121

Stevin, H, 1667. Wisconstich filsofisch bedryf Ph. De Cro-Y, Leiden, the Netherlands (in Dutch)

Surajit, K De Datta, 1981. Principle and practices of rice production, John Wiley and Sons, New York, USA

Thailand: Demographics. http://www.boi.go.th/english/Thailand/index

Thailand Malaysian Journal of Tropical Geography, 24, 3 - 12. The power of water (update June 1999), The role of water in physical planning http://www.groundcontrol.nl/studies/e_power.htm

UN Population Bureau, 2003, (http://www.worldpop.org/prbdata.htm)

Undan, R.C., 1977. Lowland rice submergence damage and drainage systems design cited by Jaw-Kai Wang and Wagen R.E., Irrigated Rice Production Systems: Design Procedure, Westview, Inc. Colorado, USA, 300

United Nations, 1994. Statistic Yearbook, 39th issue, United Nations, New York, USA

US Army Corps of engineers, 1996\2000. Department of army, Washington, DC, 20314-1000, Manual No 1110-2-1619, Appendix A How to perform a detailed Evaluation of Flood Proofing Options http://www.usace.army.mil/inet/functions/cw/cecwp/NFPC/fphow/ace8toc.htm

Van De Goor, G.A.W., 1973. Drainage of rice fields, International Institute for Land Reclamation and Improvement, Wageningen, the Netherlands, Pub 16.4, 382 - 384

Van De Ven G.P., ed., 1993\2004. Man-made lowlands, history of water management in the Netherlands, the Netherlands

Van den Berg, J.A., Schultz, E., and de Jong, J.,1977. Some qualitative and quantitative aspects of surface water in an urban area with separate storm water and waste water systems, IHAS-AISH publication No 123, Amsterdam, October, the Netherlands

Van den Bosch, G.H.J.M., 2003. GWW Kosten, bemailingen, grondweken, drainage, 18e editie, the Netherlands (in Dutch)

Van Veen, J., 1962. Dredge drain reclaim, The Art of Nation, Fifth Edition, the Hague, the Netherlands, 54

Ven, G.A., 1980. Runoff from arable land in Flevopolders, In proceedings of the symposium, 'The influence of man on the hydrological regime with special reference to representative and experimental basins', IAHS-AISH publication nr 130, Helsinki, Finland

Vinh Tran Si, 1997. Country reports C11, Proceedings Volume 1, 7th ICID International Drainage Workshop,' Drainage for the 21st Century, 17 - 21 November 1997, Penang, Malaysia

Visser, H., 2000. Bodemgebruik in Nederland. Kwartaalbericht Milieustatistieken 2000/2. CBS, Voorburg/Heerlen, the Netherlands (in Dutch)

Vongvisessomjai, S., Polsi, R., Manotham, C., Srisaengthong, D. and Charulukkana, S., 1996: Coastal erosion in the Gulf of Thailand. In: J.D. Milliman and B.U. Haq, eds., Sea-Level Rise and Coastal Subsidence, Kluwer Academic Publ., Dordrecht, 131 - 150

Wagret, P., 1959. Les polders, Dunod, Paris, France

Walker, B.H.(ed), 1979. Management of semi-arid ecosystems, Elsevier, Amsterdam, the Netherlands (Developments in Agricultural and Managed Forest Ecology, Vol 7)

Wandee; P., 2001. Some examples of optimization of water management in Netherland's Polders, MSc thesis, H.E. 082; IHE, Delft, the Netherlands

Wannasai, C., Nabbheerong, N., Suwanthada, S. and Chairin, S., 1993. Effects of water depths on growth yield and grain quality of KDML 105, Thai Agricultural Research Journal Vol 11, January - April 1993 (in Thai)

Webster, C.C. and Wilson, P.N., 1966. Agriculture in the tropics, Longman, London, UK

Weerakorn, D.W.R., 1997. Country reports C12, Proceedings Volume 1, 7th ICID International Drainage Workshop,' Drainage for the 21st Century, 17 - 21 November 1997, Penang, Malaysia

Weerasinghe, et al., 1996. Climate soil and water problems encountered in Kiralakale at the downstream of Niwala Ganga scheme; OUR Engineering Technology, Vol 2, Number 1, Open University of Sri Lanka, 22 - 27

Wesseling, J., 1979. Saline seepage in the Netherlands, occurrence and magnitude, Research on possible change in the distribution of saline seepage in the Netherlands, Proceedings and information No. 26, 18

Williamson, R.E, 1964. The Effect of root Aeration on plant growth, Soil Science Society Proceedings, 1964, 86 - 90

World Bank, 2000. Beyond the crisis: Strategy for renewing rural development in Thailand, 8

Wosten, J.H.M., Veerman en J. stolte G.J., 2001. Waterretentie-en doorlatendheiskrakteristieken van boven-en ondergronden in Nederland: de Staringreeks, Alterra-rapport 153, ISSN 1566-7197, Wageningen, the Netherlands (in Dutch)

Yingjajaval, S., 1993. A catalogue of water retention Functions of major soil series of series of Thailand, Department of Soil Science, Kasetsart University, Kamphaengsean, Nakhonprathom, Thailand

Yoneo Ishii, ed, Translated by Peter and Stephanie Hawkes, 1975. Thailand: a rice-growing society, the Center for Southeast Asia Studies, Kyoto University, Kyoto, Japan

Yu De-fen, Yao Xian-liang, 1981. Some physical properties of paddy profiles in Taihu Lake region, Proceedings of Symposium on Paddy soil, Edited by Institute of Soil Science, Academia Sinca, Science Press Beijing, Springer-Verlag, Berlin Heidelberg New York, 311

Yun-sheng, C., 1983. Drainage of paddy soils in Taihu lake region and its effects, Institute of Soil Science, Academia Sinica, Nanjing, China, Soil Research Report No. 81, 81

Appendix I

Soil development in polders after reclamation

Soil composition

The soil in the polders has two major epochs, which are the Pleistocene and the Holocene. The Pleistocene developed in the last three ice ages, which was about 400,000 years ago and the Holocene developed about 10,000 years ago. The Pleistocene epoch was sandy and in the Holocene epoch the sea rose and many peat bogs were formed. Holocene can be classified into four units: basal peat, marine, tidal and alluvial. Later the sea deposited marine sediments over the peat. Formation of peat and consequent sedimentation was repeated many times. Therefore Holocene sediment consists of silt and clay and occasionally organic matter.

Peat is formed by the accumulation of vegetable material, which has not decomposed. In a moist and humid environment with moderate temperature decomposition is slowed down. The plants partly change into humus and the net result is that peat is formed. Moreover, a high groundwater table is beneficial for formation of peat. For instance, lakes with a rich vegetation gradually became filled with peat and eventually developed into a peat bog.

Soil formation and subsidence in mineral soils

Soil formation in reclaimed sediments begins when the sediments start to loose part of the water as a result of direct evaporation or transpiration through the vegetation, if present. This soil formation, which possibly begins during sedimentation and takes place more rapidly after reclamation, is called 'soil ripening'.

The loss of water from the soil by evaporation and transpiration causes the capillary potentials to rise, especially in the topmost layer of the sediment. These potentials bring about a contraction of the grain structure, which leads in turn to a reduction in pore space. In Illite type of soils this reduction is largely irreversible. The soil therefore shrinks on all sides. This leads in the first instance in a lowering of the surface level.

Mainly vertical cracks can be formed due to horizontal shrinkage and compaction takes place due to vertical shrinkage. This process results in a further drop of the surface level. In the top layer the soil is mixed due to agricultural activities and cracks disappear, but cracks under the top layer may stay open.

The loss of the water from soft layers at deeper levels can be ascribed entirely to squeezing, as a result of increase in load after emergence.

Therefore the fall of the ground surface or subsidence in a polder after reclamation is due to irreversible reduction in the pore space of the soil, as a result of an increase in the various forces exerted on the grain structure apart from artificial loading.

In order to understand the behaviour of the water storage in the soil in a polder, it is important to consider the following aspects in detail.

Subsidence

As a consequence of drying out of newly reclaimed polder land and consequent soil ripening a significant soil subsidence may occur in the area, generally resulting in the

formation of cracks. The decrease in the surface level is the consequence of two mechanisms:
-   subsidence of the topsoil as a result of ripening;
-   compaction of the subsoil as a result of the increase in granular pressure due to reduced pore water pressure.

The first component can be calculated by using the Hissink's method (Hissink, 1935):

$$d_1 * pb_1 = d_2 * pb_2 \qquad (I\text{-}1)$$

where:
d_1 and d_2 = thickness of the Holocene layer before and after subsidence (cm)
pb_1 and pb_2 = bulk densities of the layer before and after subsidence (g/cm^3)

The bulk density after ripening can be estimated on the basis of reference sites, at which the soil profiles, land use, history and hydrological situation are comparable to the area for which a relation is to be established.

For the second component a soil mechanical approximation involving Terzaghi's formula can be used in a modified form as suggested by Hissink (1935) and De Glopper (1973) as follows:

$$z = h\left[\left(\frac{1}{c_p} + \frac{1}{c_s}\log t\right)\ln\frac{p_b}{p_1} + \left(\frac{1}{c_p'} + \frac{1}{c_s'}\log t\right)\ln\frac{p_2}{p_b}\right] \qquad (I\text{-}2)$$

where:
Z   = subsidence (m)
H   = original soil thickness (m)
T   = time (day)
p_1 = grain stress before loading (Pa)
p_2 = grain stress after loading (Pa)
p_b = boundary stress after loading (Pa)
c_p, c_s, c_p' and c_s' = consolidation constants (-)

In the ripening simulation model, which was developed by Rijniersce (Rijniersce, 1983), compaction is calculated by using the following formula, based on an assumption whether or not it is the result of ripening. The compaction is caused by an increase in the grain stresses. It may occur due to soil moisture suction in the Holocene layer and imposition of load, or increase in stress due to lowering of the groundwater table in deeper Pleistocene layers:

$$\Delta V = \frac{1}{c_r} * \ln\frac{p_2}{p_1} \qquad (I\text{-}3)$$

$$\frac{1}{c_r} = \varepsilon * \frac{1}{\ln\dfrac{p_2}{p_1} + \dfrac{1-\varepsilon}{K_2 * \varepsilon}} \tag{I-4}$$

where:
ΔV = relative compaction (-)
ε = initial pore volume (-)
K_2 = constant (-)

However, if the relative compaction and bulk density before compaction are known the bulk density after compaction may be approximated by using the following formula (Rijniersce, 1983):

$$\rho_2 = \frac{\rho_1}{1 - \Delta V} \tag{I-5}$$

Bulk densities vary with the clay and organic matter contents in the Holocene topsoil layer, which vary from 10% to 30% and 1% to 12% respectively. The thickness of the clay topsoil also varies between 0.9 to 2 m before the ripening.
Therefore lowering of the surface is mainly due to varying thickness and clay content of the Holocene topsoil and its compaction that changed the topography of the land during the ripening process.

Soil storage and cracks

Shrinking does not take place entirely in one direction or uniformly in all three directions, rather it shows a certain distribution of cracking and subsidence. If the distribution of cracks is expressed by 'r_s' the following expression may be used to determine the crack fraction in the soil (Rijniersce, 1983):

$$d_2 = d_1 * (1 - \Delta V)^{\frac{1}{rs}} \tag{I-6}$$

$$\mu_2 = 1 - \frac{(1 - \mu_1) * (1 - \Delta V)}{(1 - \Delta V)^{\frac{1}{rs}}} \tag{I-7}$$

where:
ΔV = relative compaction resulting from shrinkage and over burden load in the soil (-)
d_1, d_2 = layer thickness before and after subsidence (cm)
μ_1, μ_2 = crack fraction before and after subsidence (-)
r_s = 3 if soil shrinks equally in all directions (-)
        = 1 if solely subsidence occurs (-)

Information on the distribution factor (r_s) is available to only a limited extent. From experiments performed during the ripening process it was stated that layers are characterized as soft shrinked when shrinkage occurred only in vertical direction and as firm shrinked when shrinkage occurs in three directions. Experiments in the IJessmeerpolders gave a relation of r_s with overburden pressure and water factor n.

Therefore by assuming a certain distribution factor (r_s) and initial fraction of cracks (μ_1) an approximate value of crack fraction (μ_2) (r_s) in the area may be calculated when the initial and final depths of the Holocene layer are known.

Based on the bulk density as explained above in heavily cracked, heavy soil 30% of its total volume may be occupied by the cracks and it may be as little as 5 - 10% for partially cracked soil.

However, for medium clay and light clay soils in the Netherlands, where almost all clay is Illite, if the soil is composed of less than 8% of the total volume of soil of fine particles, with a diameter less than 2 μm, cracks will not develop during the ripening process.

Consequence of subsidence

The occurrence of subsidence has important consequences for the design and planning of water management in a polder.

Firstly the change in the pattern of contour lines, caused by differences in subsidence, may influence the choice of a site for a pumping station. Allowance will have to be made for subsidence or lowering of the water level, e.g. when installing facilities on the banks of sub-main and main drains and determining the overflow level of weirs.

Secondly in cases where drain pipes are applied the depth of the base of collector drains indirectly depends on subsidence because the depth of a collector drain is closely related to the depth of the subsurface drain outlets. A distance of 0.20 m is often maintained between the subsurface drain outlets and the water level or the bottom (in case of normally dry collector drains) of the collector drain in order to ensure non-submergence of the subsurface drain under normal conditions.

Therefore subsidence, which occurs after reclamation, clearly has several consequences of which some can be of considerable financial significance. While the topmost layers show a rather rapid process of subsidence, in the deeper layers subsidence may be very slowly. Even after a century, some subsidence may still occur in these layers.

Drainage and subsidence in peat soils

Unreclaimed peat soils are generally saturated with water. Their porosity is high, often ranging from 0.8 to 0.9 or even higher. Peat layers can be up to 10 m thick, depending on the rate of supply of organic matter, its decomposition, and the period over which the peat formation took place.

After a peat soil has been reclaimed, a complex process of subsidence starts. The reason is that the supply of organic matter stops while the decomposition increases significantly. Due to a lower groundwater table, air can enter the soil and the oxidation of the organic matter increases. Besides, due to the oxidation in the layer above the groundwater table, this layer will shrink due to increased capillary stress during the evaporation surplus. Because of the high porosity, a relatively small increase of the capillary stress results in major shrinkage. Moreover, the soil layers below the groundwater table are compressed because of an increase in intergranular pressure.

The subsidence depends on many factors such as type of peat, degree of decomposition, climate conditions, type of land use and depth of the groundwater table. The rate of subsidence varies from 1.0 to 2.0 mm/year in cold and temperate air. In the west of the Netherlands, where areas were reclaimed in the Middle Ages, the surface has subsided about 2 m in spite of shallow drainage. Some 85% subsidence can be ascribed to oxidation of organic matter, which will continue at a rate of about 2 mm/year (De Bakker and Van den Berg, 1982). Worldwide subsidence rates vary from less than 1 to more than 8 cm per year. Records on peat subsidence in Southeast Asia (lowland coastal peat) indicated 50 cm to 1 m in the initial year after reclamation with subsequent decrease to less than 6 cm per year (http://www.fao.org/docrep/x5872e/x58 72e09.htm). Table I.1 gives some reported rates of subsidence for sites worldwide.

Table I.1 Subsidence of organic soil measured at specific sites in different areas (Lucas, 1982)

Location of site	Annual subsidence in cm/year	Cumulative subsidence in cm	Time period in years	Average depth to groundwater table in cm-surface
California Delta	2.5 - 8.2	152 - 244	26	
Louisiana	1.0 - 5.0			
Michigan	1.2 - 2.5	7.6 - 15	5	
New York	2.5	150	60	90
Indiana	1.2 - 2.5	7.6 - 15	6	
Florida Everglades	2.7 2.7 - 4.2	147 19 - 29	54 7	90 60
Netherlands	0.7 1.0 - 1.7	70 6 - 10	100 6	10 - 20 50
Ireland	1.8			90
Norway	2.5	152	65	
England	0.5 - 5.0	325 (by 1932) 348 (by 1951)	84 103	
Israel	10			
Russia (Minsk bog)	2.1	100	47	
Malaysia[1]		200	16	

Note: [1] peat land was drained by open ditches one metre deep, distance between the ditches 20 to 40 m to grow rice every year (Driessen, 1978)

The drainage in peat areas requires a strategy. Otherwise the area, which has to be kept constantly at a shallow groundwater table, will subside at considerable magnitude. Unfortunately the subsidence of peat soils is rather difficult to predict. The ultimate subsidence is that all peat will disappear. Subsidence will differ from place to place according to thickness of the peat and the organic matter content. It is known that all peat will disappear by the oxidation process if drainage of the land continues. Due to the high water content, the subsidence will be considerable and this will lead to drainage problems. Especially when the ground surface becomes too low to drain off the excess water by gravity, pumping will be necessary.

The influence of soil temperature was examined by Stephens and Stewart. They found that the rate of oxidation doubles for every 10 °C increase in soil temperature (Stephens and Stewart, 1997). Stephens, et al., 1984 presented an equation based on field experiments and laboratory research in low moor peat soils in the Florida Everglades (Ritzema, 1994). These equations give a relation between the annual subsidences and mean annual soil temperature at a depth of 0.10 m.

$$S = \frac{16.9 \times D_w - 1.04}{1,000} \times 2^{\frac{T-5.0}{10}}$$
(I-8)

where:
S   = annual subsidence (m)
D_w = depth of the groundwater table (m-surface)
T   = mean annual soil temperature at 0.10 m-surface (°C)

The exponent (T-5.0)/10 indicates that the oxidation rate drops at decreasing mean annual soil temperatures, and becomes negligible at 5 °C. Moreover, it was shown that by maintaining a high groundwater table, the subsidence can be controlled. To avoid undue subsidence only shallow groundwater tables are desired, about 0.5 m-surface. A network of narrowly spaced open field drains can maintain this.

The subsidence of the peat layer below the groundwater table can be calculated on the basis of the consolidation theory. Fokkens developed an equation to calculate the compression of the peat layer below the groundwater table (Fokkens, 1970).

$$S_{pc} = \varepsilon \times w \left(25.3 f_0 + w \ln \frac{p_i + \Delta p_i}{p_i} \right)^{-1} \ln \frac{p_i + \Delta p_i}{p_i} \times D_i$$
(I-9)

where:
S_{pc} = subsidence of the peat layer below the groundwater table (m)
D_i = initial thickness of the peat layer (m)
ε = initial porosity (-)
w   = initial water content (-)
f_o = organic matter content expressed as fraction of total dry mass (-)
p_i = intergranular pressure before the groundwater table is lowered (kPa)
Δp_i = increase in the intergranular pressure after the groundwater table is lowered (kPa)

Appendix II

Water control and hydrological conditions in a rice field

Determination of the water layer depth in a rice field

The determination of the water layer depth in rice is dependent on the moisture content in layers I and II, precipitation, interception, evapotranspiration, groundwater table, capacity of the canal and pumping capacity. The water level on the surface is determined as follows:

If P (Δt)> 0 and V_{vcs1} >θ_1(t) >= θ_1 at pF = 2.53 and h_1(t) + h_2(t) < D_1(t) + D_2(t)

   If E_i (t+Δt)>E_{imax} and E_i (t+Δt)>V_{max} (Δt)

$$W_L(t + \Delta t) = W_L(t) + V_{al}(\Delta t) - 1/3 * f_t * E_0(\Delta t)$$ (II-1)

   If E_i (t+Δt)<E_{imax}

$$W_L(t + \Delta t) = W_L(t) - S_{i1}(\Delta t) - S_{i2}(\Delta t) - 1/3 * f_t * E_0(\Delta t)$$ (II-2)

If P (Δt) > 0 and V_{vcs1} >θ_1(t) >= θ_1 at pF =2.53 and h_1(t) + h_2(t) > D_1(t) + D_2(t)

   If E_i (t+Δt)>E_{imax} and E_i (t+Δt)>V_{max} (Δt)

$$W_L(t + \Delta t) = W_L(t) + V_{al}(\Delta t) - 1/3 * f_t * E_0(\Delta t) + W_{eg}(\Delta t)$$ (II-3)

where:
P(Δt)	= precipitation (mm/time step)
W_{eg}	= volume above the ground surface that remains in the rice field (mm)
W_L(t+Δt)	= actual water layer depth in rice field at time step (mm)
E_i	= interception (mm)
f_t	= crops factors for rice (-)
E_0(Δt)	= open water evaporation (mm/time step)
Val	= surplus to be added to water depth at field surface (mm)
h_1	= groundwater depth above the bottom of layer I (m)
h_2	= groundwater depth above the bottom of layer II (m)
D_1	= depth of layer I (m)
D_2	= depth of layer II (m)
V_{vcs1}	= soil moisture at saturation in layer I (mm/m)

   If E_i (t+Δt)<E_{imax}

$$W_L(t + \Delta t) = W_L(t) - S_{i1}(\Delta t) - S_{i2}(\Delta t) - 1/3 * f_t * E_0(\Delta t) + W_{eg}(\Delta t)$$ (II-4)

If P (Δt)> 0 and θ_1(t) >V $_{vcs1}$ and θ_2(t) >V $_{vcs2}$

If E$_i$ (t+Δt)>E$_{imax}$ and E$_i$ (t+Δt)>V$_{max}$ (Δt)

$$W_L(t + \Delta t) = W_L(t) + V_{al}(\Delta t) - 1/3 * f_t * E_0(\Delta t) - Q_2(\Delta t) \qquad (\text{II-5})$$

where:

V_{vcs2} = soil moisture at saturation in layer II (mm/m)
E_{imax} = maximum interception (mm)
$Q_2(\Delta t)$ = discharge due to different water level in the field and the lateral canal (mm/time step)

If E$_i$(t+Δt)<E$_{imax}$

$$W_L(t + \Delta t) = W_L(t) - S_{i1}(\Delta t) - S_{i2}(\Delta t) - 1/3 * f_t * E_0(\Delta t) - Q_2(\Delta t) \quad (\text{II-6})$$

where:
$S_{i1}(\Delta t)$ = maximum amount of moisture that can be stored in layer I at any instant (mm)
$S_{i2}(\Delta t)$ = maximum amount of moisture that can be stored in layer II at any instant (mm)

If P(Δt) = 0 and $W_L(t) = 0$ then $W_L(t + \Delta t) = 0$

If P(Δt) = 0 and V_{vcs1} >θ_1(t) >= θ_1 at pF =2.53 and h$_1$(t) + h$_2$(t) > D$_1$(t) + D$_2$(t)

$$W_L(t + \Delta t) = W_L(t) - S_{i1}(\Delta t) - S_{i2}(\Delta t) - 1/3 * f_t * E_0(\Delta t) \qquad (\text{II-7})$$

If P(Δt) = 0 and V_{vcs1} >θ_1(t) >= θ_1 at pF =2.53 and h$_1$(t) + h$_2$(t) > D$_1$(t) + D$_2$(t)

$$W_L(t + \Delta t) = W_L(t) - S_{i1}(\Delta t) - S_{i2}(\Delta t) - 1/3 * f_t * E_0(\Delta t) + W_{eg}(\Delta t) \quad (\text{II-8})$$

If P(Δt)> 0 and θ_1(t) >V $_{vcs1}$ and θ_2(t) >V $_{vcs2}$

$$W_L(t + \Delta t) = W_L(t) - S_{i1}(\Delta t) - S_{i2}(\Delta t) - f_t * E_0(\Delta t) - Q_2(\Delta t) \qquad (\text{II-9})$$

For all case:

If $W_L(t + \Delta t) < 0$ then $W_L(t + \Delta t) = 0$ \qquad (II-10)

Determination of the water layer depth in a rice field in rainfed conditions

In case of rainfed conditions it is assumed that the farmer will only drain water from the field when the water level exceeds the maximum allowable depth at each stage of rice growth. In rainfed conditions after harvesting between November and December the field is dry. The soil is also dry and cracks may occur in the soil surface. Also during the crop season the soil can be dry and cracks may occur because of drought and unreliable rainfall. So determination of the water layer depth is as follows:

IF $W_L(t + \Delta t) > W_{LM}(t + \Delta t)$ then

If $W_L(t + \Delta t) - W_{LM}(t + \Delta t) <= D_{cT}(\Delta t)$ then

$$W_L(t + \Delta t) = W_{LM}(t + \Delta t) \qquad \text{(II-11)}$$

If $W_L(t + \Delta t) - W_{LM}(t + \Delta t) > D_{cT}(\Delta t)$ then

$$W_L(t + \Delta t) = W_L(t + \Delta t) - D_{cT}(\Delta t) \qquad \text{(II-12)}$$

where:

$W_{LM}(t+\Delta t)$ = maximum allowable water layer depth for rice at different growth stages (mm)

$D_{cT}(\Delta t)$ = total discharge capacity of the surface drainage system for rice (mm/time step)

In rainfed conditions it is assumed that only gravity discharge through a culvert into a rice field occurs. The discharge capacity is dependent on the water level in the field and the polder water level and that there is no pumping for irrigation.

The water layer depth under rainfed conditions is calculated as follows:

IF $W_L(t + \Delta t) < W_{LS}(t)$ then

If $W_L(t + \Delta t) - W_{LS}(t) > I_{rf}(\Delta t)$ then

$$W_L(t + \Delta t) = W_L(t + \Delta t) + I_{rf}(t + \Delta t) \qquad \text{(II-13)}$$

If $W_L(t + \Delta t) - W_{LS}(t + \Delta t) < I_{rf}(\Delta t)$ then

$$W_L(t + \Delta t) = W_{LS}(t) \qquad \text{(II-14)}$$

where:

$W_{LS}(t+\Delta t)$ = preferred water layer depth for rice at different growth stages (mm)

$I_{rf}(\Delta t)$ = discharge capacity of water into a rice field through a culvert (mm/time step)

Determination of the water layer depth in irrigated conditions

In case of irrigated conditions the water can be fully controlled. The water level is calculated as follows.

Water layer depth for drainage conditions:

IF $W_L(t+\Delta t) > W_{LM}(t+\Delta t)$ then

If $W_L(t+\Delta t) - W_{LM}(t+\Delta t) <= D_{cT}(\Delta t)$ then
$$W_L(t+\Delta t) = W_{LM}(t+\Delta t) \qquad\qquad (II\text{-}15)$$

If $W_L(t+\Delta t) - W_{LM}(t+\Delta t) > D_{cT}(\Delta t)$ then
$$W_L(t+\Delta t) = W_L(t+\Delta t) - D_{cT}(\Delta t) \qquad\qquad (II\text{-}16)$$

Water layer depth for irrigation conditions:

IF $W_L(t+\Delta t) < W_{LS}(t)$ then

If $W_L(t+\Delta t) - W_{LS}(t) > I_{cT}(\Delta t)$ then
$$W_L(t+\Delta t) = W_L(t+\Delta t) + I_{cT}(t+\Delta t) \qquad\qquad (II\text{-}17)$$

If $W_L(t+\Delta t) - W_{LS}(t+\Delta t) < I_{cT}(\Delta t)$ then
$$W_L(t+\Delta t) = W_{LS}(t) \qquad\qquad (II\text{-}18)$$

where:
$I_{cT}(\Delta t)$ = total discharge capacity of the irrigation system (mm/time step)

Determination of irrigation and drainage discharge for irrigated rice condition

Determination of drainage discharge

In case of heavy rainfall the farmers will keep a water layer in the rice field at maximum allowable water depth (W_{LM}) at each crop stage. The drainage capacity is limited by the storage capacity of the lateral, the main canal and the pumping capacity in the lateral and the main drain.

If $W_L(t+\Delta t) <= W_{LM}(t+\Delta t)$ then $Q_{1n}(\Delta t) = 0$, $Q_1(\Delta t) = 0$ $\qquad (II\text{-}19)$

If $W_L(t+\Delta t) > W_{LM}(t+\Delta t)$ then $Q_{1n}(\Delta t) = W_{LM}(t+\Delta t) - W_L(t+\Delta t)$ (II-20)

If $Q_{1n}(\Delta t) > D_{cT}(\Delta t)$ then $Q_1(\Delta t) = D_{cT}$ $\qquad\qquad (II\text{-}21)$

If $Q_{1n}(\Delta t) < D_{cT}(\Delta t)$ then $Q_1(\Delta t) = Q_{1n}(\Delta t)$ $\qquad\qquad (II\text{-}22)$

The total discharge to the canal during each time step is:

$$Q_T(\Delta t) = Q_1(\Delta t) + Q_2(\Delta t) \tag{II-23}$$

where:

Q_T = total discharge into the canal during each time step (mm/time step)
Q_{1n} = discharge that needs to be drained from the ground surface (mm/time step)
Q_1 = actual surface drainage discharge during each time step (mm/time step)
Q_2 = discharge due to different head in the field and the lateral canal (mm/time step)

Determination of the surface drainage capacity in a rice field for irrigated conditions

The surface drainage system in an irrigated rice field is assumed to be an open pipe culvert, which is used in the surface field drain as shown in Figure II.1. The drainage capacity is based on the maximum allowable water depth, the polder water level and the pumping capacity as follows.

The discharge through the culvert is calculated based on the design drainage modulus in mm/day. The maximum discharge through the culvert cannot exceed the drainage modulus at the maximum allowable water depth. The maximum allowable water depth is the depth which rice can tolerate without adversely affecting grain yield. Experiments suggest that continual flooding is not essential for a high grain yield but modern rice can tolerate at least 15 cm of water depth without adversely affecting grain yield (De Datta, 1981 and Wannasai, et al., 1993).

In the rainy season the water level in the lateral canal is usually kept low and it is assumed that the polder water level is the same as or lower than the water level. So in normal conditions the water level in the lateral canal is less than the level outside. The farmer can discharge the excess water through the pipe culvert, which varies with the water depth as follows.

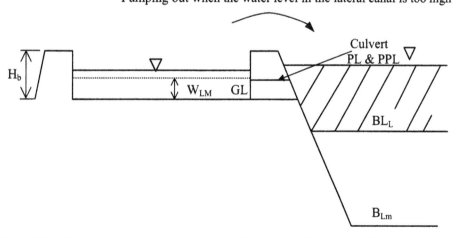

Pumping out when the water level in the lateral canal is too high

Figure II.1 Schematic presentation of the drainage from a rice field in a rice polder

If PL(t+Δt) < GL and W_L(t+Δt) > W_{LM}(t+Δt), PL(t+Δt)> = PPL(t+Δt)

The discharge capacity from the pipe is calculated as follows (after Depeweg, 2000):

$$D_{c1}(\Delta t) = \frac{8,640}{N} * C * \frac{\pi}{4} * (D/1,000)^2 * N_P *$$
$$\sqrt{2g(W_L(t+\Delta t) - D/2)/1,000}$$

$$\tag{II-24}$$

$$D_{c1}(\Delta t) = MIN(D_{c1}, W_L(t+\Delta t) - W_{LM}(t+\Delta t)) \tag{II-25}$$

$$D_{c2}(\Delta t) = W_L(t+\Delta t) - W_{LM}(t+\Delta t) - D_{c1}(\Delta t) \tag{II-26}$$

If $D_{c2}(\Delta t) <= 0$ then $D_{c2}(\Delta t) = 0$ $\hspace{2cm}$ (II-27)

If $D_{c2}(\Delta t) > 0$ then $D_{c2}(\Delta t) = MIN(D_{c2}(\Delta t), P_{ca}(\Delta t))$ $\hspace{1cm}$ (II-28)

where:

D_m	= design drainage modulus for the culvert (mm/time step)
D_{c1}	= discharge capacity of the culvert (mm/time step)
D_{c2}	= required pumping capacity (mm/time step)
D	= internal diameter of the culvert (mm)
C	= discharge coefficient for a circular pipe = 0.90 (-)
G	= acceleration of gravity (m^2/s)
N	= number of time steps per day (-)
N_P	= number of culverts per ha (-)
P_{ca}	= maximum pumping capacity (mm/time step)
PL	= water level in the lateral canal (m±MSL)
PPL	= polder water level (m±MSL)

If the water level in the lateral canal is above the bottom of the pipe culvert but less than water level in the field the discharge capacity will decrease. In this case the farmer tries to increase the discharge capacity by pumping out of excess water. In this case the farmer will open the gate of the culvert to drain by gravity. After the water level in the field is equal to the water level in the lateral canal, the gate will be closed. Then the farmer will pump the excess water out of the field until the water depth is equal or lower than the maximum allowable water depth.

If GL(t+Δt) + D <= PL(t+Δt) < GL(t+Δt) + W$_L$(t+Δt) and W$_L$(t+Δt) > W$_{LM}$(t+Δt), PL(t+Δt)> =PPL(t+Δt)

$$D_{c1}(\Delta t) = \frac{8,640}{N} C \frac{\pi}{4} (D/1,000)^2 N_P \sqrt{2g(W_L(t+\Delta t)/1,000 + GL - PL(t+\Delta t))}$$

$$\tag{II-29}$$

$$D_{c1}(\Delta t) = MIN(D_{c1}, W_L(t+\Delta t) - W_{LM}(t+\Delta t)) \tag{II-30}$$

$$D_{c2}(\Delta t) = W_L(t+\Delta t) - W_{LM}(t+\Delta t) - D_{c1}(\Delta t) \tag{II-31}$$

If $D_{c2}(\Delta t) <= 0$ then $D_{c2}(\Delta t) = 0$ $\qquad\qquad$ (II-32)

If the water level in the lateral canal is higher than the water level in the field the discharge capacity is completely determined by the pump. The farmer will close the pipe. The pumping requirement is at least equal to the capacity of the pump.

If PL(Δt) > GL + W$_L$(t+Δt) and W$_L$(t+Δt) > W$_{LM}$(t+Δt), PL(t+Δt) = PPL(t+Δt)

If PL(Δt) > GL + W$_L$(t+Δt) and W$_L$(t+Δt) > W$_{LM}$(t+Δt), PL(t+Δt) = PPL(t+Δt) Then

$$D_{c1}(\Delta t) = 0, \ D_{c2}(\Delta t) = W_L(t + \Delta t) - W_{LM}(t + \Delta t) \qquad\qquad \text{(II-33)}$$

If the water level in the lateral canal is higher than the level of the bunt the discharge capacity will become zero. No water can be pumped out and the pumping will be stopped.

If PL(t+Δt) >= GL(t+Δt) + H$_b$ and W$_L$(t+Δt) > W$_{LM}$(t+Δt), PL(t+Δt) = PPL(t+Δt) Then

$$D_{c1}(\Delta t) = 0, \ D_{c2}(\Delta t) = 0 \qquad\qquad\qquad \text{(II-34)}$$

where:
H$_b$ $\qquad$ = height of the bund around the rice field (m+surface)
GL $\qquad$ = ground level of the rice field (m-MSL)

If $D_{c2}(\Delta t) > 0$ then $D_{c2}(\Delta t) = MIN(D_{c2}(\Delta t), P_{ca}(\Delta t))$ $\qquad\qquad$ (II-35)

Hence the total discharge capacity is calculated as follows:

$$D_{cT}(\Delta t) = D_{c1}(\Delta t) + D_{c2}(\Delta t) \qquad\qquad\qquad \text{(II-36)}$$

where:
D$_{cT}$ $\qquad$ = total discharge capacity of the surface drainage system (mm/time step)

Determination of irrigation discharge capacity in a rice field for irrigated conditions

In irrigated rice conditions the canals will be used as irrigation and drainage canals. So there are two main functions of the canals in polder areas. The irrigation requirement is computed on a weekly basis based on the assumption that in case of a dry spell the farmer will keep the water layer in the rice field at the required water depth (W$_{LS}$) for each of the crop stages. Irrigation water supply in rice fields is limited by the storage capacity of the lateral and main canals in the polders.

In dry periods the irrigation capacity is calculated based on available water in the lateral and main drain. Related to the agricultural practice the irrigation supply is calculated on a weekly basis.

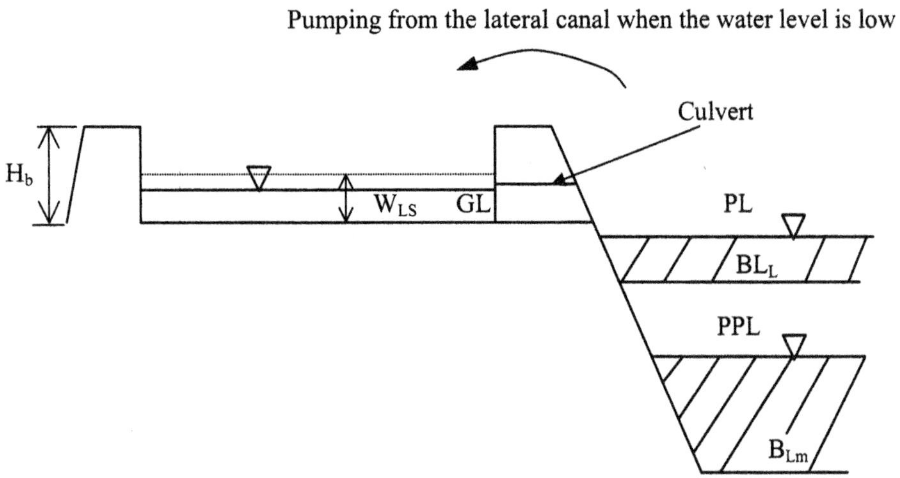

Pumping from the lateral canal when the water level is low

Figure II.2 Schematic presentation of irrigation in the rice field in rice polders

Before or at the end of rainy season between November and December, the rainwater within polder is stored in the main and lateral canals in polder areas. The open pipe culvert can be used for supply of irrigation water as shown in Figure II.2, if the water level in the lateral canal is high enough. The irrigation capacity is calculated based on the polder water level and the pumping capacity as follows.

If the water level in the main canal is higher than the water level required at the rice growth stages and the water depth in the field is lower than the required water depth, the farmer will open the gate of the culvert, and the lateral inlet gate. Then the water will flow into the lateral canal, and pumping to the field is required which can be calculated as follows:

If $GL + W_{LS}(t+\Delta t)/1,000 < PL(t+\Delta t)$ and $W_L(t+\Delta t) < W_{LS}(t+\Delta t)$, $PPL(t+\Delta t)$, $> = PL(t+\Delta t)$

$$I_{c1}(\Delta t) = \frac{8,640}{N} * C * \frac{\pi}{4} * (D/1,000)^2 * N_p *$$
$$\sqrt{2g(PL(t+\Delta t) - GL - W_L(t+\Delta t)/1,000)}$$
(II-37)

If $I_{c1}(\Delta t)-I_m(\Delta t) > 0$ then $I_{c2}(\Delta t) = MIN(I_{rn}(\Delta t) - I_{c1}(\Delta t), P_{ca}(\Delta t))$ (II-38)

If $I_{c1}(\Delta t)-I_m(\Delta t) <= 0$ then $I_{c1}(\Delta t) = 0$ (II-39)

where:
$W_{LS}(t+\Delta t)$	= water depth required for rice growth stage at time step (mm)
$W_L(t+\Delta t)$	= actual water depth at time step (mm)
$W_{LM}(t+\Delta t)$	= maximum allowable of water depth in the field for rice growth stage (mm)

$I_{c1}(\Delta t)$ = supply capacity of the irrigation water through the culvert (mm/time step)

$I_{c2}(\Delta t)$ = supply capacity required for irrigation water by pumping (mm/time step)

$I_m(\Delta t)$ = amount of water that needs to be supplied (mm/time step)

If the water level in the main canal is lower than the water level in the field but still higher than the bed level of the lateral canal plus 0.10 m, while the water depth in the field is lower than the water depth required, the farmer will let the water flow to the lateral canal. After that the water will be pumped from the lateral until the water depth is equal to the required water depth.

If GL + $W_L(t+\Delta t)$/1,000 > PL($t+\Delta t$) and $W_L(t+\Delta t)$ < $W_{LS}(t+\Delta t)$, PPL($t+\Delta t$) > = PL($t+\Delta t$) > = BL_L + 0.10 then

$$I_{c1}(\Delta t) = 0,\ I_{c2}(\Delta t) = MIN(I_{rn}(\Delta t), P_{ca}(\Delta t)) \tag{II-40}$$

where:
BL_L = bed level of the lateral canal (m+MSL)

If the polder water level is lower than the bed level of the lateral canal plus 0.10 m and the water depth in the field is lower than the required water depth. It is assumed that the farmer will close the gate at the inlet of the lateral canal and pump the water from the main drain to the lateral canal. Finally the water is pumped from lateral canal to the rice field. In this case it is assumed that the farmer will not allow unnecessary storing of water in the lateral canal.

If GL + $W_L(t+\Delta t)$/1,000 > PPL($t+\Delta t$) and $W_L(t+\Delta t)$ < $W_{LS}(t+\Delta t)$, PPL($t+\Delta t$) < BL_L + 0.10 then

$$I_{c1}(\Delta t) = 0,\ I_{c2}(\Delta t) = MIN(I_{rn}(\Delta t), P_{ca}(\Delta t)/2) \tag{II-41}$$

If the water level in the main drain is lower than its bed level plus 0.30 m, the farmer will not pump any water from the main drain due to safety of the canal embankment.

If $W_L(t+\Delta t)$ < $W_{LS}(t+\Delta t)$, PPL($t+\Delta t$) <= BL_M + 0.30 then

$$I_{c1}(\Delta t) = 0,\ I_{c2}(\Delta t) = 0 \tag{II-42}$$

where:
BL_M = bed level of main canal (m+MSL)

Hence the total supply capacity of the irrigation system is calculated as follows:

$$I_{rs}(\Delta t) = I_{c1}(\Delta t) + I_{c2}(\Delta t) \tag{II-43}$$

where:
$I_{rs}(\Delta t)$ = irrigation water that is supplied to the plot (mm/time step)
$I_{c1}(\Delta t)$ = supply capacity of the irrigation water through the culvert (mm/time step)
$I_{c2}(\Delta t)$ = supply capacity of the irrigation water by pumping (mm/time step)

Calculation of the groundwater table

The above computation is done for each time step (hours) and cumulated for presentation and averaged for each day. The computation of the groundwater table due to the surplus in layer II is based on the following schematisation of layer II and the rise in the water level in the crack (Figure 6.7).

$$V_{ad} = \sum_{i=1} \frac{V_a}{O_1} \tag{II-44}$$

$$A_2 = \frac{(100 - V_{s2}) * (\theta_{ve2} - \theta_{v2})}{200 * D_2} \tag{II-45}$$

$$A_2 * h_2(t)^2 + V_{s2}(t) * h_2(t) - \frac{V_{ad}}{10} = 0 \tag{II-46}$$

where:
O_1 = number of steps in a day (-)
V_{ad} = surplus in layer II cumulated (mm/day)

The moisture condition in layer II is calculated based on the average moisture conditions of the groundwater depth above bottom of layer II as follows:

If $h_2(t) >= D_2(t)$ then $\theta_2(t) = \theta_{ve2}$ \hfill (II-47)

If $D_2(t) > h_2(t) >= 0$ then $\theta_2(t) = \dfrac{\theta_{ve2} * h_2(t) + \theta_{v2} * (D_2 - h_2(t))}{D_2}$ \hfill (II-48)

where:
θ_2 = moisture content of soil layer II (%)
θ_{v2} = moisture content at field capacity in layer II (%)
θ_{ve2} = moisture content at saturation in layer II (%)

The moisture deficit to 80% of saturation in layer II is calculated as follows:

If $\theta_2 < \theta_{ve2}$ then $S_{v2R} = (0.80 * \theta_{ve2} - \theta_2) * 10$ \hfill (II-49)

If $\theta_2 >= \theta_{ve2}$ then $S_{v2R} = 0$ (II-50)

where:

$S_{v2R}(\Delta t)$ = soil moisture deficit in layer II related to 0.8 saturation (mm/m)

If the depth of the groundwater is above the top of layer II the groundwater table is calculated as follows:

If $h_2(t) > D_2(t)$

$$V_{ad2} = 10 * \left(A_2 * D_2(t)^2 + V_{s2}(t) * D_2(t) \right)$$ (II-51)

$$A_1 = \frac{(100 - V_{s1}) * (\theta_{ve1} - \theta_{v1})}{200 * D_1}$$ (II-52)

$$A_1 * h_1(t)^2 + V_{s1}(t) * h_1(t) - \frac{(V_{ad} - V_{ad2})}{10} = 0$$ (II-53)

where:

θ_1 = moisture content of soil layer I (%)
θ_{v1} = moisture content at field capacity in layer I (%)
θ_{ve1} = moisture content at saturation in layer I (%)
V_{ad2} = volume of water in the cracks in layer II at full submergence depth (mm)

The moisture condition in layer I is calculated based on the average of the water depth above layer I as follows:

If $h_1(t) >= D_1(t)$ then $\theta_1(t) = \theta_{ve1}$ (II-54)

If $D_1(t) > h_1(t) >= 0$

If $D_1(t) > h_1(t) >= 0$ then $\theta_1(t) = \dfrac{\theta_{ve1} * h_1(t) + \theta_{v1} * (D_1 - h_1(t))}{D_1}$ (II-55)

The moisture deficit to 80% of saturation in layer I is calculated as follows:

If $\theta_1 < \theta_{ve1}$ then $S_{v1R}(\Delta t) = (0.80 * \theta_{ve1} - \theta_1(t)) * 10$ (II-56)

If $\theta_1 >= \theta_{ve1}$ then $S_{v1R}(\Delta t) = 0$ (II-57)

where:

$S_{v1R}(\Delta t)$ = soil moisture deficit in layer I related to 0.8 saturation (mm/m)

If the groundwater depth is greater than the thickness of layers I and II it is important that there is a water layer on the field. The water depth is calculated as follows:

If $h_1(t) + h_2(t) > D_1(t) + D_2(t)$

$$V_{ad1} = 10 * \left(A_1 * D_1(t)^2 + V_{s1}(t) * D_1(t) \right)$$ (II-58)

$$W_{ge}(t) = V_{ad} - V_{ad1} - V_{ad2}$$ (II-59)

where:
V_{ad1} = volume of water in the cracks in layer I at full submergence depth (mm)
W_{ge} = volume of water that is added to the water layer above the surface (mm)

Solving the quadratic equation for $h_1(t)$ and $h_2(t)$ gives the change in the groundwater table above the bottom of layers I and II and the water depth above the surface.

Groundwater table under buildings and infrastructure

The surface level for buildings and infrastructure in rice polders is determined by the ground level plus landfill. The landfill height for buildings is usually higher than for infrastructure in rice polders because farmers like to protect themselves and their properties. The groundwater below buildings and infrastructure is calculated based on the height of the landfill as follows.

In case there is no water layer in the field.

$$SOW1 = D_1 + D_2 - h_1 - h_2 + Hbd$$ (II-60)

$$SOW2 = D_1 + D_2 - h_1 - h_2 + Hif$$ (II-61)

In case of there is a water layer in the field; the groundwater table is assumed to be the same as the water level in the field.

$$SOW1 = Hbd - WL/1,000$$ (II-62)

$$SOW2 = Hif - WL/1,000$$ (II-63)

where:
SOW1   = depth of the groundwater table from the base level of the buildings (m)
SOW2   = depth of groundwater table from the base level of the infrastructure (m)
Hbd   = height of the landfill for buildings (m)
Hif   = height of the landfill for infrastructure (m)

Appendix III

Damage formula for yield reduction due to flooding of rice

Formula for the simulation of yield reduction due to flooding of rice

Based on Figure 6.23 the equation can be written as follows:

If $R_{WLV} >= 1.05$ then $1-(Y_a/Y_m)_V = 0.0766 * R_{WLV} + 0.0805$ (III-1)

If $R_{WLH} >= 1.05$ then $1-(Y_a/Y_m)_H = 0.0830 * R_{WLH} + 0.0871$ (III-2)

If $R_{WLF} >= 1.05$ then $1-(Y_a/Y_m)_F = 0.0664 * R_{WLF} + 0.0698$ (III-3)

If $R_{WLY} >= 1.05$ then $1-(Y_a/Y_m)_Y = 0.0139 * R_{WLY} + 0.0146$ (III-4)

However, when there is less water depth at the different stages of growth, it is assumed that yield would will occur in the same amount as yield formation stage:

If $R_{WLV}, R_{WLH}, R_{WLF}, R_{WLY} < 1.05$ then

$$1-(Y_a/Y_m)_Y = 0.0139 * R_{WLV}, R_{WLH}, R_{WLF}, R_{WLY} + 0.0146$$ (III-5)

where:

R_{WLV} = (Σ Actual water layer)/(Σ Water layer preferred) at the vegetative stage

R_{WLH} = (Σ Actual water layer)/(Σ Water layer preferred) at the head development stage

R_{WLF} = (Σ Actual water layer)/(Σ Water layer preferred) at the flowering stage

R_{WLY} = (Σ Actual water layer)/(Σ Water layer preferred) at the yield formation stage

Appendix IV

Cost calculation formula for field pumping for rice

The costs for field pumping are calculated based on the quantity of water that was pumped for drainage and irrigation dependent on the water level in the field. The quantity of water to be pumped for drainage is calculated as follows:

If $D_{c2}(\Delta t) > 0$ then $Q_{Di} = D_{c2} * \Delta t$ (IV-1)

If $D_{c2}(\Delta t) = 0$ then $Q_{Di} = 0$ (IV-2)

Therefore the total amount of water to be pumped for drainage is:

$$Q_{DT} = 10 * N_b * B_g * \sum_{i=1}^{n} Q_{Di}$$ (IV-3)

where:
Q_{Di} = amount of water to be pumped for drainage during the time step (mm)
n = total number of time steps per year (-)
Q_{DT} = total amount of water to be pumped for drainage (m^3)

Hence the costs for field pumping for surface drainage is calculated as follows:

$$Costpd = Q_{DT} * Cp_d$$ (IV-4)

where:
Costpd = total annual costs for pumping by field pumps for surface drainage (€)
Cp_d = cost for pumping by small pumps for surface drainage (€/m^3)

The quantity of water to be pumped for irrigation depends much on the water level in the field, in the lateral canal and in the main canal and also on the surface level, the lateral canal bed level and the main canal bed level. The quantity of water to be pumped for irrigation is calculated as follows:

If $GL + W_L(\Delta t) < PL(\Delta t) < GL + W_{LS}(\Delta t)$ and $W_L(\Delta t) < W_{LS}(\Delta t)$, $PPL(\Delta t) > PL(\Delta t)$ then

$$Q_{IRi} = I_{c2} * \Delta t$$ (IV-5)

If $GL + W_L(\Delta t) > PL(\Delta t)$ and $W_L(\Delta t) < W_{LS}(\Delta t)$, $PPL(\Delta t) > = PL(\Delta t) > = BL_L + 0.10$ then

$$Q_{IRi} = I_{c2} * \Delta t$$ (IV-6)

If GL + $W_L(\Delta t)$ > PPL(Δt) and $W_L(\Delta t)$ < $W_{LS}(\Delta t)$, PPL(Δt) <= $BL_L(\Delta t)$ + 0.10 then

$$Q_{IRi} = 2 * I_{c2} * \Delta t \tag{IV-7}$$

Therefore the total amount of water to be pumped for irrigation is:

$$Q_{IRT} = 10 * N_b * B_g * \sum_{i=1}^{n} Q_{IRi} \tag{IV-8}$$

where:

Q_{IRi} = amount of water to be pumped for irrigation during the time step (mm)
n   = total number of time steps per year (-)
Q_{IRT} = total amount of water to be pumped for irrigation (m^3)

Hence the costs for field pumping for irrigation are calculated as follows:

$$Costpdi = Q_{IRT} * Cp_i \tag{IV-9}$$

where:

Costpdi   = total annual costs for pumping by small pumps for irrigation (€)
Cp_i = cost for pumping by small pumps for irrigation (€/m^3)

Appendix V

Water control conditions in a dry food crops polder

Drainage capacity and water control

The surface drainage system in the dry food crops field is assumed to be an open pipe culvert as shown in Figure V.1. The drainage capacity is based on the allowable water level in the plot and the lateral water level or polder water level and the pumping capacity as follows.

The discharge through the culvert is calculated based on the design drainage modulus in mm/day. The maximum discharge throughout the culvert cannot exceed the drainage modulus at the maximum allowable water level. The maximum allowable water level is the level in the ditches, which crops can tolerate without adversely affecting yield due to a too high groundwater table.

The discharge through the culvert is calculated based on the design drainage modulus in mm/day. The maximum discharge throughout the culvert cannot exceed the drainage modulus at the maximum allowable water layer depth. The maximum allowable water level in the ditch, which crops can tolerate without adversely affecting yield, is dependent on the type of crops.

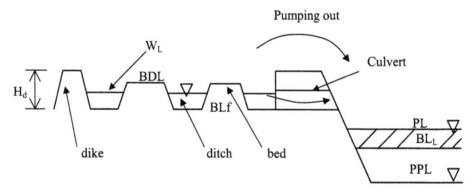

Figure V.1 Schematic presentation of the drainage for a dry food crops polder

In the rainy season the lateral canal water level is usually kept low and it is assumed, that the polder water level is the same as or lower than the lateral water level. So in normal conditions the water level in the lateral is lower than the outlet level. The farmer can discharge the excess water through the pipe culvert, which is varied with the water level as follows.

If $PL(t+\Delta t) < BLf$ and $W_L(t+\Delta t) > W_{LA}(t+\Delta t)$, $PL(t+\Delta t) >= PPL(t+\Delta t)$

where:
W_L	= water level in the ditch in a dry food crops polder (m-surface)
W_{LA}	= maximum allowable water level in the ditch in a dry food crops polder (m-surface)
BLf	= bottom level of the ditches in a dry food crops polder (m-surface)

BDL   = bed surface level of a raised bed systems in a dry food crops polder
         (m+surface)
PL   = water level in the lateral canal (m-surface)
PPL   = polder water level (m-surface)
BL_L = bed level of the lateral canal (m-surface)
H_d = height of the surrounding dike (m+surface)

The discharge capacity from the pipe is calculated as follows (after Depeweg, 2000):

$$D_{c1}(\Delta t) = \frac{86,400}{N} * C * \frac{\pi}{4} * (D/1,000)^2 * N_P *$$
$$\sqrt{2g(W_L(t+\Delta t) - BLf - D/2,000)}$$
(V-1)

$$D_{c1}(\Delta t) = MIN(D_{c1}(\Delta t), Vaf(t+\Delta t))$$
(V-2)

$$D_{c2}(\Delta t) = Vaf(t+\Delta t) - D_{c1}(\Delta t)$$
(V-3)

If $D_{c2}(\Delta t) <= 0$ then $D_{c2}(\Delta t) = 0$
(V-4)

If $D_{c2}(\Delta t) > 0$ then $D_{c2}(\Delta t) = MIN(D_{c2}(\Delta t), P_{CA} * \Delta t)$
(V-5)

If the water level in the lateral canal is lower than the water level in the plot but higher than the bottom of the culvert the discharge capacity will decrease. In this case it is assumed that the farmers try to increase the discharge capacity by pumping out excess water. In this case farmers will open the gate of the culvert. After the water level in the field is equal to the water level in the lateral canal, the gate will be closed. Then the farmers will pump the excess water out of the field until the water depth is equal or less than the maximum allowable water depth.

If $BLf + D <= PL(t+\Delta t) < W_L(t+\Delta t)$ and $W_L(t+\Delta t) > W_{LA}(t+\Delta t)$, $PL(t+\Delta t) > =PPL(t+\Delta t)$

$$D_{c1}(\Delta t) = \frac{86,400}{N} * C * \frac{\pi}{4} * (D/1,000)^2 * N_P * \sqrt{2g(W_L(t+\Delta t) - PL(t+\Delta t))}$$
(V-6)

$$D_{c1}(\Delta t) = MIN(D_{c1}(\Delta t), Vaf(t+\Delta t))$$
(V-7)

If $D_{c2}(\Delta t) <= 0$ then $D_{c2}(\Delta t) = 0$
(V-8)

If $D_{c2}(\Delta t) > 0$ then $D_{c2}(\Delta t) = MIN(D_{c2}(\Delta t), P_{CA} * \Delta t)$
(V-9)

$$D_{c2}(\Delta t) = Vaf(t+\Delta t) - D_{c1}(\Delta t)$$
(V-10)

If $D_{c2}(\Delta t) <= 0$ then $D_{c2}(\Delta t) = 0$
(V-11)

If $D_{c2}(\Delta t) > 0$ then $D_{c2}(\Delta t) = MIN(D_{c2}(\Delta t), P_{CA} * \Delta t)$ (V-12)

If the water level in the lateral canal is higher than the water level in the ditch the discharge capacity is totally determined by the pump. The farmer will close the culvert. The pumping capacity requirement is at least equal to the pumping capacity of the pump.

If PL(Δt) > W$_L$(t+Δt) and W$_L$(t+Δt) > W$_{LA}$(t+Δt), PL(t+Δt) = PPL(t+Δt)

$$D_{c1}(\Delta t) = 0$$ (V-13)

$$D_{c2}(\Delta t) = Vaf(t + \Delta t)$$ (V-14)

If $D_{c2}(\Delta t) > 0$ then $D_{c2}(\Delta t) = MIN(D_{c2}(\Delta t), P_{CA} * \Delta t)$ (V-15)

If the water level in the lateral canal is higher than the dike level the discharge capacity will become zero. No water can be pumped out and the pumping will be stopped due to flooding.

If PL(t+Δt) >= BDL(t+Δt) and W$_L$(t+Δt) > W$_{LA}$(t+Δt), PL(t+Δt) = PPL(t+Δt) then

$$D_{c1}(\Delta t) = 0 , \ D_{c2}(\Delta t) = 0$$ (V-16)

Hence the total discharge capacity is calculated as follows:

$$D_{cT}(\Delta t) = D_{c1}(\Delta t) + D_{c2}(\Delta t)$$ (V-17)

where:
$D_{cT}(\Delta t)$ = total discharge capacity of the surface drainage system (m^3/ time step)

Irrigation capacity and water control

The irrigation supply is dependent on the water level in the plot as follows:

If $\theta_1 \leq \theta_{1b}$ and $WL(\Delta t) \geq BLf + 0.15$

$$I_{rn}(t + \Delta t) = \left(\frac{1,000 * P_{cb} * \Delta t_p}{A_p * opp} \right)$$ (V-18)

$$I_{rs}(t + \Delta t) = I_{rn}(t + \Delta t)$$ (V-19)

If $\theta_1 \leq \theta_{1b}$ and $WL(\Delta t) < BLf + 0.15$ then I$_{rn}$(t+Δt) is calculated by Equation 6.186 and 6.187.

$$I_{rs}(t+\Delta t) = 0 \tag{V-20}$$

If $\theta_1 \geq \theta_{1f}$ then $I_{rn}(t+\Delta t) = 0$, $I_{rs}(t+\Delta t) = 0$ (V-21)

where:
$I_{rs}(t+\Delta t)$ = irrigation that is supplied to the plot (mm/time step)
P_{cb} = pumping capacity for spraying from the small boat (m³/hr)
Δt_p = time of spraying of water in hour (usually less than 2 hours)
A_p = area of the plot (ha)
Q_{nd} = discharge that needs to be drained (mm/time step)
Q_c = capacity of the canal system (mm/time step)
θ_{1b} = moisture content of the soil in layer I at –0.25 bar (pF=3.017) (%)
θ_{1f} = moisture content of the soil in layer I at field capacity (%)
opp   = percentage of open water in the plot (%)

In case the lateral canal water level is higher water level in the ditch, while the water level in the ditch is lower than the maximum allowable water level, the farmer will open the gate and let water flow through the pipe culvert as shown in Figure V.2. The control of the water level in the plot is done based on the main canal water level and the pumping capacity as follows.

The farmer will let the water flow to the field by opening the gate of the culvert. After the water level in the field is the same as the water level in the main canal the farmer will close the gate of the culvert and pump the water from the lateral canal until the water depth is equal to the required water depth. The water level in the ditch in the next time step will be calculated dependent on required discharge capacity and the pumping capacity.

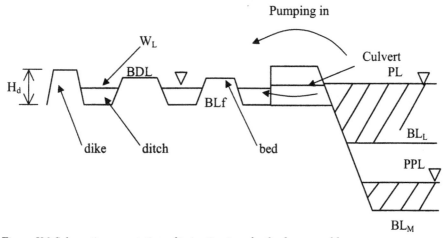

Figure V.2 Schematic presentation of irrigation in a dry food crops polder

If the water level in the lateral canal is higher than the water level in the plot and the water level in the plot is lower than the maximum allowable water level. The farmer will open the gate of the culvert and let the water flow into the ditch until the water level in the ditch is equal to the maximum allowable water level. It is assumed that the farmer will not pump while the water level in the lateral canal is higher than the maximum water level in the plot, unless the discharge of the culvert is not enough for the spraying water to the crops in plot.

If $PL(\Delta t) > W_L(\Delta t)$ and $W_L(\Delta t) < W_{LA}(\Delta t)$, $PPL(\Delta t) >= PL(\Delta t)$

$$I_{wc1}(\Delta t) = \frac{8,6400}{N} * C * \frac{\pi}{4} * (D/1,000)^2 * N_P *$$

$$\sqrt{2g(PL(t + \Delta t) - W_L(t + \Delta t)}$$

$$\text{(V-22)}$$

If $I_{wc1}(\Delta t) > I_{rn}(\Delta t)$ then $I_{wc2}(\Delta t) = 0$ $\qquad$ (V-23)

If $I_{wc1}(\Delta t) < I_{rn}(\Delta t)$ then

$$I_{wc2}(\Delta t) = \min((I_{rn}(\Delta t) - I_{WC1}(\Delta t)), P_{CA} * \Delta t) \qquad \text{(V-24)}$$

If $I_{wc1}(\Delta t) \geq V_{ao}(t)$ then $I_{wc1}(\Delta t) = V_{ao}(t)$, $I_{wc2}(\Delta t) = 0$ $\qquad$ (V-25)

where:
$I_{wc1}(\Delta t)$ = discharge capacity of the irrigation water through a culvert (m³/time step)
$I_{wc2}(\Delta t)$ = discharge capacity of the irrigation water by pumping from the lateral canal (m³/time step)
$V_{ao}(t)$ = volume above the actual water level to the maximum allowable water level in the plot at any instant (m³)
P_{CA} = pumping capacity for pumping water to the plot (m³/hour)

In case of the water level in the ditch is higher than or equal to the water level in the lateral canal, but lower than the maximum allowable water level. The farmer will close the gate and pump water to the plot.

If $W_L(\Delta t) >= PL(\Delta t)$ and $W_L(\Delta t) < W_{LA}(\Delta t)$, $PPL(\Delta t) >= PL(\Delta t)$ then

$$I_{wc1}(\Delta t) = 0, I_{wc2}(\Delta t) = \min(V_{ao}(t), p_{CA} * \Delta t) \qquad \text{(V-26)}$$

If the polder water level is lower than the water level in the plot but still higher than the bed level of the lateral canal plus 0.10 m and the water level in the plot is lower than the allowable water level. The farmer will let the water flow to the lateral canal. After that the water will be pumped from the lateral until the water level in the plot is equal to the allowable water level.

If $W_L(\Delta t) > PL(\Delta t)$ and $W_L(\Delta t) < W_{LA}(\Delta t)$, $PPL(\Delta t) >= PL(\Delta t) > = BL_L + 0.10$ then $I_{wc1}(\Delta t)$ and $I_{wc2}(\Delta t)$ are calculated by formulas as given in Equation V-26.

If the polder water level is lower than the bed level of the lateral canal plus 0.10 m and the water level in the plot is lower than the allowable water level, the farmer will close the gate at the inlet of the lateral canal and then pump the water from the main canal to the lateral canal. Finally the water will be pumped from the lateral canal to the field. In this case it is assumed that the farmer will not allow unnecessary storing of water in the lateral canal.

If PL(Δt) < W$_{LA}$(Δt) and W$_L$(Δt) < W$_{LA}$(Δt), PPL(Δt) <= BL$_L$(Δt) + 0.10 then

$$I_{wc1}(\Delta t) = 0 \, , I_{wc2}(\Delta t) = MIN(V_{ao}(t), P_{CA} / 2 * \Delta t) \tag{V-27}$$

where:

BL$_L$ = bed level of the lateral canal (m-surface)

If the water level in the main canal is lower than its bed level plus 0.30 m, the farmer will not allow pumping any water from the main drain due to safety of the canal structure.

If W$_L$(Δt) < W$_{LA}$(Δt), PPL(Δt) <= BL$_M$(Δt) + 0.30 then

$$I_{wc1}(\Delta t) = 0 \, , I_{wc2}(\Delta t) = 0 \tag{V-28}$$

where:

BL$_M$ = bed level of the main canal (m-surface)

Therefore the irrigation supply from the canal is calculated as follows:

$$I_{rsd}(\Delta t) = I_{wc1}(\Delta t) + I_{wc2}(\Delta t) \tag{V-29}$$

where:

I$_{rsd}$ (Δt)   = irrigation that is supplied to the dry food plot (m^3/time step)

Determination of the water level and control condition in the ditch for irrigation

The water level in the ditch is calculated based on the storage volume of water in the ditch by water balance with rainfall, open water evaporation, seepage from field, surface runoff and volume available from the drain condition, in case the groundwater is above the water level in the ditch.

Appendix VI

Determination of parameters for the soil layers

Program for the determination of the constant a, b and c in the formula:

$$pF = a + b*LN((V/V0)^c - 1) \qquad \text{(VI-1)}$$

Where:
pF   = moisture tension (cm)
a,b,c   = constant of soil type (m-surface)
V   = soil moisture (%)
V0   = porosity of soil (%)

Soil layer I for the clay polder Schieveen

Location: Schesoil

Layer I

There are 8 pairs of observations.

pF-observed	V-observed
1.0	74.1
1.3	71.3
1.7	67.8
2.0	63.8
2.4	57.2
3.0	47.1
3.4	41.8
4.2	30.9

The pore volume is 75.0%.
The start values for the constants are: a = 4.000 b = 1.000 c = -0.300.
The start value for the iteration step is S = 0.02.
The computation is terminated when the total difference is smaller than 0.02 times the total of the observed pF-values or at the 50th iteration.
Total difference of the values = 2.48
Average difference   = 0.31

Observed pF	Computed pF	Difference
1.00	-0.29	1.29
1.30	1.05	0.25
1.70	1.70	0.00
2.00	2.15	-0.15
2.40	2.64	-0.24
3.00	3.17	-0.17
3.40	3.40	-0.00
4.20	3.83	0.37

The values for the constants are: a = 5.085 b = 0.934 c = -0.261.

Values to be used:

Moisture content at saturation   73.8%
Moisture content at field capacity 65.3%
Moisture content at wilting point 21.3%

Soil layers II and III for the clay polder Schieveen

Location: Schesoi2

Layers II and III

There are 8 pairs of observations.

PF-observed	V-observed
1.0	56.0
1.3	54.6
1.7	52.9
2.0	50.7
2.4	47.1
3.0	41.0
3.4	37.2
4.2	30.6

The pore volume is 57.0%.
The start values for the constants are: a = 4.000 b = 1.000 c = -0.300.
The start value for the iteration step is S = 0.02.
The computation is terminated when the total difference is smaller than 0.02 times the total of the observed pF-values or at the 50th iteration.
Total difference of the values = 1.70
Average difference   = 0.21

Observed pF	Computed pF	Difference
1.00	0.33	0.67
1.30	1.17	0.13
1.70	1.70	-0.00
2.00	2.13	-0.13
2.40	2.61	-0.21
3.00	3.14	-0.14
3.40	3.40	-0.00
4.20	3.78	0.42

The values for the constants are: a = 5.433 b = 0.950 c = -0.261.

Values to be used:

Moisture content at saturation   56.3%
Moisture content at field capacity 51.5%
Moisture content at wilting point 22.6%

Soil layer I for the peat polder Duifpolder

Location: Duifsoil

Layer I

There are 8 pairs of observations.

pF-observed	V-observed
1.0	71.6
1.3	68.9
1.7	65.7
2.0	62.0
2.4	56.0
3.0	46.7
3.4	41.1
4.2	31.7

The pore volume is 72.0%.
The start values for the constants are: a = 4.000 b = 1.000 c = -0.300.
The start value for the iteration step is S = 0.02.
The computation is terminated when the total difference is smaller than 0.02 times the total of the observed pF-values, or at the 50th iteration.
Total difference of the values = 3.09
Average difference   = 0.39

Observed pF	Computed pF	Difference
1.00	-0.85	1.85
1.30	1.03	0.27
1.70	1.70	0.00
2.00	2.15	-0.15
2.40	2.64	-0.24
3.00	3.15	-0.15
3.40	3.40	-0.00
4.20	3.78	0.42

The values for the constants are: a = 5.117 b = 0.908 c = -0.251.

Values to be used:

Moisture content at saturation   71.0%
Moisture content at field capacity 63.4%
Moisture content at wilting point 20.9%

Soil layers II and III for the peat polder Duifpolder

Location: Duifsoi2

Layers II and III

There are 8 pairs of observations.

pF-observed	V-observed
1.0	54.2
1.3	52.2
1.7	48.6
2.0	44.1
2.4	36.4
3.0	25.7
3.4	20.3
4.2	12.7

The pore volume is 55.0%.
The start values for the constants are: a = 4.000 b = 1.000 c = -0.300.
The start value for the iteration step is S = 0.02.
The computation is terminated when the total difference is smaller than 0.02 times the total of the observed pF-values or at the 50th iteration.
Total difference of the values = 2.31
Average difference   = 0.29

Observed pF	Computed pF	Difference
1.00	0.06	0.94
1.30	1.05	0.25
1.70	1.72	-0.02
2.00	2.18	-0.18
2.40	2.68	-0.28
3.00	3.17	-0.17
3.40	3.40	0.00
4.20	3.73	0.47

The values for the constants are: a = 4.718 b = 0.775 c = -0.168.

Values to be used:

Moisture content at saturation   54.3%
Moisture content at field capacity 46.1%
Moisture content at wilting point   4.7%

Soil layer I for rice polder under rainfed conditions

Location: rsupun1

Layer I

There are 4 pairs of observations.

pF-observed	V-observed
2.0	54.2
2.5	43.1
3.5	31.2
4.2	24.2

The pore volume is 55.0%.
The start values for the constants are: a = 4.000 b = 1.000 c = -0.300.
The start value for the iteration step is S = 0.02.
The computation is terminated when the total difference is smaller than 0.01 times the total of the observed pF-values or at the 50th iteration.
Total difference of the values = 2.08
Average difference   = 0.52

Observed pF	Computed pF	Difference
2.01	0.91	1.10
2.53	3.14	-0.61
3.49	3.85	-0.36
4.18	4.18	0.00

The values for the constants are: a = 5.200 b = 0.798 c = -0.299.

Values to be used:

Moisture content at saturation   54.7%
Moisture content at field capacity 51.8%
Moisture content at wilting point 23.8%

Soil layer II for rice polder under rainfed conditions

Location: rsupun2

Layer II

There are 4 pairs of observations.

pF-observed	V-observed
2.0	55.3
2.5	44.8
3.5	32.4
4.2	24.8

The pore volume is 56.0%.

The start values for the constants are: a = 4.000 b = 1.000 c = -0.300.
The start value for the iteration step is S = 0.02.
The computation is terminated when the total difference is smaller than 0.01 times the total of the observed pF-values or at the 50th iteration.
Total difference of the values = 1.90
Average difference = 0.48

Observed pF	Computed pF	Difference
2.01	1.14	0.87
2.53	3.19	-0.66
3.49	3.87	-0.38
4.18	4.18	0.00

The values for the constants are: a = 5.240 b = 0.722 c = -0.255.

Values to be used:

Moisture content at saturation   55.8%
Moisture content at field capacity 53.6%
Moisture content at wilting point 24.3%

Soil layer III for rice polder under rainfed conditions

Location: rsupun3

Layer III

There are 4 pairs of observations.

pF-observed   V-observed
   2.0   52.4
   2.5   41.9
   3.5   26.4
   4.2   21.0

The pore volume is 53.0%.

The start values for the constants are: a = 4.000 b = 1.000 c = -0.300.

The start value for the iteration step is S = 0.02.

The computation is terminated when the total difference is smaller than 0.01 times the total of the observed pF-values, or at the 50th iteration.
Total difference of the values = 1.68
Average difference = 0.42

Observed pF   Computed pF   Difference
   2.01   1.70   0.31
   2.53   3.38   -0.85
   3.49   4.01   -0.52
   4.18   4.18   0.00

The values for the constants are: a = 5.180 b = 0.561 c = -0.169.

Values to be used:

Moisture content at saturation   53.0%
Moisture content at field capacity 51.9%
Moisture content at wilting point 20.4%

Soil layer I for rice polder under irrigated conditions and a dry food crops polder

Location: drsupan1

Layer I

There are 3 pairs of observations.

pF-observed   V-observed
   2.0   44.1
   3.0   23.4
   4.2   10.4

The pore volume is 45.0%.

The start values for the constants are: a = 6.000 b = 1.000 c = -0.300.

The start value for the iteration step is S = 0.02.

The computation is terminated when the total difference is smaller than 0.01 times the
total of the observed pF-values or at the 50th iteration.
Total difference of the values = 0.74
Average difference   = 0.25

Observed pF   Computed pF   Difference
   2.01   2.01   0.00
   3.01   3.75   -0.74
   4.18   4.18   -0.00

The values for the constants are: a = 4.760 b = 0.488 c = -0.181.

Values to be used:

Moisture content at saturation   45.0%
Moisture content at field capacity 44.1%
Moisture content at wilting point   9.8%

Soil layer II for rice polder under irrigated conditions and a dry food crops polder

Location :drsupan2

Layer II

There are 4 pairs of observations.

pF-observed	V-observed
2.5	31.9
3.0	23.8
3.5	19.9
4.2	12.9

The pore volume is 37.3%.

The start values for the constants are: a = 5.000 b = 1.000 c = -0.300.

The start value for the iteration step is S = 0.02.

The computation is terminated when the total difference is smaller than 0.01 times the total of the observed pF-values or at the 50th iteration.
Total difference of the values = 0.64
Average difference   = 0.16

Observed pF	Computed pF	Difference
2.53	2.43	0.10
3.01	3.37	-0.36
3.49	3.67	-0.18
4.18	4.18	0.00

The values for the constants are: a = 5.000 b = 0.850 c = -0.304.

Values to be used:

Moisture content at saturation   36.9%
Moisture content at field capacity 33.9%
Moisture content at wilting point 12.6%

Soil layer III for rice polder under irrigated conditions and a food crops polder

Location :drsupan3

Layer III

There are 4 pairs of observations.

pF-observed	V-observed
2.5	31.6
3.0	25.2
3.5	20.5
4.2	12.9

The pore volume is 37.3%.

The start values for the constants are: a = 5.000 b = 1.000 c = -0.300.

The start value for the iteration step is S = 0.02.

The computation is terminated when the total difference is smaller than 0.01 times the total of the observed pF-values or at the 50th iteration.
Total difference of the values = 0.46
Average difference   = 0.12

Observed pF	Computed pF	Difference
2.53	2.21	0.32
3.01	3.10	-0.09
3.49	3.54	-0.05
4.18	4.18	0.00

The values for the constants are: a = 5.140 b = 0.979 c = -0.300.

Values to be used:

Moisture content at saturation   36.6%
Moisture content at field capacity 32.6%
Moisture content at wilting point 12.6%

Soil layer I for a rural polder in Thailand

Location: samutav8

Layer I

There are 8 pairs of observations.

pF-observed	V-observed
2.0	59.9
2.3	57.8
2.5	55.2
3.0	50.3
3.5	44.1
3.7	39.0
4.0	34.5
4.2	33.5

The pore volume is 65.0%.
The start values for the constants are: a = 3.500 b = 1.000 c = -1.500.
The start value for the iteration step is S = 0.05.
The computation is terminated when the total difference is smaller than 0.03 times the total of the observed pF-values or at the 50th iteration.
Number of iteration steps 1
Total difference of the values = 0.41
Average difference   = 0.05

Observed pF	Computed pF	Difference
2.00	1.89	0.11
2.30	2.23	0.07
2.52	2.54	-0.02
3.00	3.00	-0.00
3.48	3.44	0.04
3.70	3.76	-0.06
4.00	4.04	-0.04
4.18	4.10	0.08

The values for the constants are: a = 3.750 b = 0.875 c = -1.375.

Values to be used:

Moisture content at saturation   64.4%
Moisture content at field capacity 59.3%
Moisture content at wilting point 31.8%

Soil layers II and III for a rural polder in Thailand

Location: samut2-38

Layers II and III

There are 8 pairs of observations.

pF-observed	V-observed
2.0	62.8
2.3	62.4
2.5	61.0
3.0	55.0
3.5	29.9
3.7	26.2
4.0	24.1
4.2	23.5

The pore volume is 65.0%.
The start values for the constants are: a = 3.500 b = 1.000 c = -1.500.
The start value for the iteration step is S = 0.05.
The computation is terminated when the total difference is smaller than 0.03 times the total of the observed pF-values or at the 50th iteration.
Total difference of the values = 1.02
Average difference = 0.13

Observed pF	Computed pF	Difference
2.00	2.23	-0.23
2.30	2.30	-0.00
2.52	2.50	0.02
3.00	2.96	0.04
3.48	3.82	-0.34
3.70	3.93	-0.23
4.00	4.00	0.00
4.18	4.02	0.16

The values for the constants are: a = 3.600 b = 0.442 c = -1.254.

Values to be used:

Moisture content at saturation   65.0%
Moisture content at field capacity 63.6%
Moisture content at wilting point 18.3%

Abbreviations and acronyms

BMA	Bangkok Metropolitan Administration
CPU	Central Processing Unit
Dia.	Diameter
ECAFE	Economic Commission for the Asia and the Far East
FAO	Food and Agriculture Organization of the United Nations
FC	Field capacity
GIS	Geographic Information System
GDP	Gross Domestic Product
HYVrice	High Yielding Variety rice
ICID	International Commission on Irrigation and Drainage
I.C.W.	Instituut voor Cultuurtechniek en Waterhuishouding (Institute for culture technique and water management in the Netherlands)
IRRI	International Rice Research Institute
KDML	Kao Dok Mali
Kp	Crop factor for class A pan evaporation
Max.	Maximum
MSL	Mean Sea Level
m-s.	Metre below ground surface
NAP	Normal Amsterdam Level (approximate Mean Sea Level)
NESDB	National Economic and Social Development Board (Thailand)
NRCS	Natural Resources Conservation Service of the United States of America
RID	Royal Irrigation Department (Thailand)
Sat	Saturation
TNO	Nederlandse Organisatie voor Toegepast onder zoke (Netherlands Institute of Applied Geoscience-National Geological Survey)
USDA	United States Department of Agriculture
WMO	World Meteorological Organization
WP	Wilting point

List of symbols

A_{c1}	= relative area of the cracks in layer I (-)
A_{dr}	= cost of subsurface drain or open field drain (€/m)
A_g	= total investment costs for pumping station (10^6 €)
a_{ke}	= drainage modulus (mm/day)
A_{m1}	= relative area of the soil matrix in layer I (-)
A_p	= area of the plot (ha)
Ar	= total area in the polder where rice is grown (ha)
B	= bed width (m)
B_d	= bed width of ditches (m)
BDL	= bed surface level of a raised bed systems in a dry food crops polder (m+surface)
B_g	= size of the farm (ha)
B_p	= size of the plot (ha)
BLf	= bottom level of the ditches in a dry food crops polder (m-surface)
BL_L	= bed level of the lateral canal (m-surface)
BL_M	= bed level of the main canal (m-surface)
C	= discharge coefficient for a circular pipe = 0.90 (-)
C	= runoff coefficient (-)
c	= hydraulic resistance of the confined layer (day)
C_i	= coefficient of linear extension or linear extensibility (-)
C_{i1}	= coefficient of linear extension of soil layer I (-)
C_{i2}	= coefficient of linear extension of soil layer II (-)
Cp_d	= cost for pumping by small pumps for surface drainage (€/m³)
costb	= total costs for bund construction (€/year)
costbm	= total maintenance costs for the bund (€/year)
costd	= total costs of subsurface drains or open field drains (€)
costdk	= total costs for the surrounding dike construction (€)
costdkm	= total maintenance costs for the surrounding dike (€/year)
costdm	= total maintenance costs for subsurface drains or open field drains (€ /year)
costdrf	= total costs for the surface ditches system for a dry food crops polder (€)
costdrfm	= total costs for maintenance of the surface ditches system for a dry food crops polder (€)
costds	= total costs for construction of the PVC pipe culvert (€)
costdsm	= maintenance costs for the PVC pipe culvert (€/year)
costfd	= total costs for construction of the ditches (€)
costfdm	= maintenance costs for the ditches (€/year)
Costpd	= total annual costs for pumping by small pumps for surface drainage (€)
Costpdi	= total annual costs for pumping by small pumps for irrigation (€)
$c_p, c_s,$	= consolidation constants (-)
$c_p', c_s',$	= consolidation constants (-)
C_{pi}	= cost for pumping by small pumps for irrigation (€/m³)
C_u	= earth compaction cost (€/m³)
D	= thickness of the crack layer (m)
D	= aquifer thickness (m)
D	= thickness of the aquifer below the drain level (m)
D	= depth of the aquifer below the water level in the lateral canal (m)

D	= internal diameter of the culvert (mm)
d	= equivalent thickness of the aquifer below the drain level (m)
d	= days from the last limiting date (-)
d	= number of days of delay from the last date of sowing (-)
D'	= thickness of the confined layer (m)
D_1	= depth of layer I (m)
D_1	= depth of the soil in layer I where there is no shrinkage (m)
d_1	= thickness of the Holocene layer before subsidence (cm)
d_1	= layer thickness before subsidence (cm)
D_2	= depth of layer II (m)
D_2	= depth of layer II at no shrinkage (m)
d_2	= thickness of the Holocene layer after subsidence (cm)
d_2	= layer thickness after subsidence (cm)
D_3	= thickness of layer III (m)
d_a	= diameter of aggregates (m)
D_{c1}	= discharge capacity of the culvert (mm/time step)
D_{c2}	= required pumping capacity (mm/time step)
D_{cT}	= total discharge capacity of the surface drainage system for rice (mm/time step)
D_{cT}	= total discharge capacity of the surface drainage system (m^3/ time step)
D_d	= drain depth of subsurface drain or open field drain (m)
D_d	= subsurface drain depth (m-surface)
D_{hr}	= diameter of the main sewer (m)
Di	= initial thickness of the peat layer (m)
di	= sum of distances move in the i direction since last rotation of axes (-)
D_m	= design drainage modulus for the culvert (mm/time step)
D_m	= drainage modulus (mm/day)
D_o	= dewatering depth (m-surface)
D_{pi}	= increase in the intergranular pressure after the groundwater table is lowered (kPa)
D_r	= drain depth or minimum water level from the surface of the bed (m)
D_r	= root depth of the crops (m)
D_r	= ripening depth (m)
D_S	= distance between the ditches (m)
D_w	= depth of the groundwater table (m-surface)
E'	= evapotranspiration from layer II (mm/time step)
Ea	= actual evapotranspiration (mm/time step)
E_{as}	= actual evapotranspiration from the root zone (mm/time step)
E_C	= cost for earth movement for the ditches (€/m^3)
E_i	= interception (mm)
E_{ir}	= irrigation efficiency (-)
E_{imax}	= maximum interception (mm)
E_{m}	= potential evapotranspiration of crops (mm)
E_{ng}	= operation power for the pumping station (kWh)
E_o	= open water evaporation (mm/time step)
E_p	= potential evapotranspiration (mm/time step)
E_{ps}	= potential evapotranspiration from the root zone (mm/time step)
ET_a	= actual evapotranspiration of the crops (mm)
ET_m	= potential evapotranspiration of the crops (mm)
f	= crop factor (-)

f_o	= organic matter content expressed as fraction of total dry mass (-)
f_r	= empirical reduction factor accounting for tortuousity and necking of the cracks (-)
G	= acceleration of gravity (m^2/s)
G_c	= pumping capacity (m^3/s)
G_c	= pumping capacity (m^3/time step)
G_c	= operation of pumping capacity (mm/day)
G_{cmax}	= maximum pumping capacity (mm/day)
G_{cap}	= power required for the pumping station (KW)
GL	= ground level of the rice field (m-MSL)
G_{oi}	= percentage of infrastructure in the net area (%)
G_{ow}	= excavation (m^3)
G_{W0}	= groundwater table (m-surface)
H	= original soil thickness (m)
h	= groundwater table in layer II (m)
h	= water depth above the bottom of layer II (m)
h_0	= initial height of the groundwater table (m)
h_1	= groundwater depth above the bottom of layer I (m)
h_2	= groundwater depth above the bottom of layer II (m)
h_3	= height of the water level in the rice field above the equivalent of the aquifer level (m)
h_4	= height of the canal water level above the equivalent of the aquifer level (m)
h_a	= start level (m+polder water level)
H_b	= height of the bund around the rice field (m+surface)
H_b	= height of timbering (m)
Hbd	= height of the landfill for buildings (m)
H_d	= height of the surrounding dike (m+surface)
h_g	= average groundwater table (m-surface)
Hif	= height of the landfill for infrastructure (m)
H_{is}	= start level below the required water level in the plot (m)
h_l	= headloss (m)
h_p	= stop level (m-polder water level)
h_p	= lowering of the water level from the polder water level (m)
h_p	= stop level above the required water level in the plot (m)
h_p	= total operating head of pump (m)
h_s	= pressure head (m)
h_{s2}	= depth of the canal water level above the bottom of layer II (m)
i	= variable index = 1, 2, 3,........., N (-)
I_{c1}	= supply capacity of the irrigation water through the culvert (mm/time step)
I_{c2}	= supply capacity of the irrigation water by pumping (mm/time step)
I_{cT}	= total discharge capacity of the irrigation system (mm/time step)
I_{rf}	= discharge capacity of water into a rice field through a culvert (mm/time step)
I_m	= amount of water that needs to be supplied (mm/time step)
I_m	= amount of water that needs to be supplied by spraying from the small boat (mm/time step)
I_{rs}	= irrigation water that is supplied to the plot (mm/time step)
I_{rsd}	= irrigation water that is supplied to the dry food crop plot (m^3/time step)
I_w	= indicator to start pumping (-)
I_w	= water depth related to preferred polder water level (m)
I_{wc}	= indicator to start pumping into the plot (-)

I_{wc1}	= discharge capacity of the irrigation water through a culvert (m^3/time step)
I_{wc2}	= discharge capacity of the irrigation water by pumping from the lateral canal (m^3/time step)
j	= direction index = 1, 2, 3,.........., N (-)
K	= hydraulic conductivity of the aquifer (mm/day)
K	= saturated hydraulic conductivity of the soil (mm/time step)
K	= seepage from the bottom (mm/time step)
K	= average seepage (mm/year)
k	= stage index (-)
K	= soil hydraulic conductivity (mm/time step)
k, n	= constants (-)
$K1$	= permeability of layer I (m/day)
$K2$	= constant (-)
$K2$	= permeability of layer II (m/day)
K_a'	= hydraulic conductivity of the aquifer (mm/day)
K_{avg}	= composite permeability of the two layers (m/day)
K_{beg}	= annual maintenance costs for the pumping station (10^6 €)
K_e	= energy costs for pumping (€/year)
K_{gw}	= cost of maintenance in earth work (€/m^3)
K_o	= cost of maintenance in the sub-main drains (€/m)
K_{ob}	= cost of maintenance in the main drains (€/m)
K_{obt}	= cost of sub-main drains (€/m)
K_{obv}	= cost of main drains (€/m)
K_{og}	= cost of excavation (€/m^3)
k_{ot}	= unsaturated hydraulic conductivity (m/day)
K_{sc}	= saturated hydraulic conductivity of the cracks (m/day)
L	= distance between the drains (m)
L	= leakage factor (m)
L	= lifetime of the project components (years)
L	= length of the collector drains (m)
L	= distance between the canals (m)
L_1	= clay content in layer I (%)
L_2	= clay content in layer II (%)
L_b	= horizontal distance between the berm and the lateral canal (m)
L_g	= distance between canals (m)
L_{hr}	= length of the main sewer (variable diameter) (m)
L_p	= air content in the soil (%)
L_s	= distance between the drains (m)
Ls	= distance between the lateral canals (m)
L_{s1}	= linear shrinkage of soil layer I (-)
L_{s2}	= linear shrinkage of soil layer II (-)
L_{wr}	= length of secondary sewers (0.3 m diameter)
m_1	= crack fraction before subsidence (-)
m_2	= crack fraction after subsidence (-)
$M_{i,j}$	= direction vector component (normalized)
N	= number of time steps per day (-)
n	= porosity (%)
n	= total number of time steps per year (-)
n_1	= porosity of soil in layer I (%)
n_2	= porosity of soil in layer II (%)

n_3	= porosity of soil in layer III (%)
N_b	= total number of plots in the polder area (-)
N_b	= number of farms (-)
N_P	= number of culverts per ha
N_{pt}	= total number of plots (-)
N_R	= number of rows of the ditches in the plot (-)
O_{af}	= urban area for the rainfall volume (m^2)
O_{bb}	= area of building/farm (ha)
O_d	= total land area (ha)
O_d	= total area in the polder (ha)
ODR	= average oxygen diffusion rate during a rainy period (g*10^{-8}/cm^2/min)
O_g	= gross area (ha)
O_1	= number of steps in a day (-)
O_1	= total land area (ha)
O_1	= total land for agriculture (ha)
opp	= percentage of open water in the plot (%)
O_{si}	= sub-areas like housing, shops, etc. (ha)
O_v	= facilities within the farm (%)
O_w	= open water area (ha)
O_{wm}	= area of open water at instant t (m^2)
p	= soil moisture tension (cm)
P	= precipitation (mm/time step)
p_1	= grain stress before loading (Pa)
p_1	= pressure in layer I (cm)
p1	= moisture tension or suction pressure in layer I (cm)
p_2	= grain stress after loading (Pa)
p_2	= pressure in layer II (cm)
p^2	= suction pressure in the middle of layer II (cm)
p^3	= suction pressure at the top of layer III (cm)
p^4	= average suction pressure in layer III (cm)
p_b	= boundary stress after loading (Pa)
pb_1	= bulk densities of the layer before subsidence (g/cm^3)
pb_2	= bulk densities of the layer after subsidence (g/cm^3)
P_{CA}	= pumping capacity for pumping water to the plot (m^3/hour)
P_{ca}	= maximum pumping capacity (mm/time step)
P_{cb}	= pumping capacity for spraying from the small boat (m^3/hr)
P_d	= water depth in the ditch required for pumping by a removable boat (m)
pi	= intergranular pressure before the groundwater table is lowered (kPa)
PL	= water level in the lateral canal (m-surface)
PL	= polder water level (m-surface)
P_L	= length of the plot (m)
P_1	= parcel length (m)
P_n	= rainfall surplus over the open water area (m^3/time step)
P_o	= average of excess precipitation (mm/year)
P_o	= pressure halfway between layer II and layer III (cm)
P_{ow}	= open water at polder water level (%)
P_{ow}	= open water up to water level (%)
P_p	= cost for construction of the PVC pipe culvert (€/m)
PP	= polder water depth from the surface (m)
PP	= polder water level (m-surface)

PPL	= polder water level (m-surface)
P_W	= width of the plot (m)
Q	= discharge to the lateral canal (mm/time step)
Q	= discharge into a drain from the land (mm/time step)
q	= moisture (%)
$Q_{(j)}$	= different type of land use in the rural area (ha)
Q_1	= actual surface drainage discharge during each time step (mm/time step)
Q_{1n}	= discharge that needs to be drained from the ground surface (mm/time step)
Q_2	= discharge due to different head in the field and the lateral canal (mm/time step)
Q_3	= total discharge during each time step (mm/time step)
Q_{ave}	= average discharge (mm/day)
Q_c	= capacity of the canal system (mm/time step)
Q_{comi}	= computed discharge (mm/day)
Q_d	= design discharge of the pipe (m³/s)
Q_{Di}	= amount of water to be pumped for drainage during the time step (mm)
Q_{d1}	= crop yield reduction due to late date of sowing (%)
Q_{d2}	= damage due to late date of harvesting (%)
Q_{DT}	= total amount of water to be pumped for drainage (m³)
Q_g	= discharge during each time step if the remaining volume is > 0 (mm)
Q_{nd}	= discharge that needs to be drained (mm/time step)
Q_{obsi}	= observed discharge (mm/day)
Q_{opp}	= discharge available for surface runoff (mm)
Q_r	= total irrigation water supply taken from the canal system (m³/time step)
Q_{rd}	= drainage discharge requirement related to the drainage modulus (m³/s/ha)
Q_{ra}	= canal discharge (mm/time step)
Q_{ri}	= sewer discharge (mm/time step)
Q_{IRi}	= amount of water to be pumped for irrigation during the time step (mm for rice or m³ for dry food crops)
Q_{IRT}	= total amount of water to be pumped for irrigation (m³)
Q_s	= cumulative discharge (m³/time step)
Q_T	= total discharge into the canal during each time step (mm/time step)
R	= model efficiency (-)
R_o	= relative yield (%)
R_o	= relative yield reduction (%)
r	= annual interest rate (%)
r_s	= distribution direction of cracks (-)
S	= annual subsidence (m)
S_{pc}	= subsidence of the peat layer below the groundwater table (m)
S_c	= cracks (%)
S_{i1}	= maximum amount of moisture that can be stored in layer I (mm)
S_{i1}	= moistening rate of layer I (mm)
S_{i2}	= maximum amount of moisture that can be stored in layer II (mm)
S_{i3}	= maximum moistening capacity of layer III (mm/time step)
S_L	= side slope of ditches (-)
SOW1	= depth of the groundwater table from the base level of the buildings (m)
SOW2	= depth of groundwater table from the base level of the infrastructure (m)
SP	= depth of the water level (m-surface)
S_{v1}	= soil moisture deficit in layer I related to field capacity (mm/m)
S_{v1}	= increase in storage at instant t in layer I up to field capacity (mm/m)

S_{v1}	= moisture added in layer I (mm/m)
S_{v1}	= moisture content deficit related to field capacity in layer I (mm/m)
S_{v1R}	= soil moisture deficit in layer I related to 0.8 saturation (mm/m)
S_{v2}	= increase in storage at instant t in layer II up to field capacity (mm/m)
S_{v2}	= soil moisture deficit in layer II related to field capacity (mm/m)
S_{v2R}	= soil moisture deficit in layer II related to 0.8 saturation (mm/m)
T	= time (day)
T	= mean annual soil temperature at 0.10 m-surface (°C)
t_{dk}	= side slope of the surrounding dike embankment (-)
T_{gb}	= horizontal component of the side slope above the water level (-)
T_{go}	= horizontal component of the side slope below the water level (-)
T_s	= side slope of the drains (-)
T_s	= side slope for the lateral canals (-)
T_s	= side slopes of collector drains (-)
T_t	= side slopes of sub-main drains (-)
T_v	= side slopes of main drains (-)
T_t	= side slope of the collector drains (-)
T_v	= side slope of the canals (-)
u	= wetted perimeter of the lateral canal (m)
V_1	= value of building/farm (€)
V_a	= surplus of water in layer II cumulated (mm/day)
V_a	= saturated volume of groundwater in layer II (mm)
V_{a2}	= volume available to layer III or to be drained (mm/time step)
V_{ad}	= surplus in layer II cumulated (mm/day)
V_{ad}	= volume of water in cracks (mm)
V_{ad1}	= volume of water in the cracks in layer I at full submergence depth (mm)
V_{ad2}	= volume of water in the cracks in layer II at full submergence depth (mm)
V_{af}	= surplus to be added to the water depth at field surface (mm)
V_{al}	= surplus to be discharged to the drain (mm)
V_{ald}	= volume of water that is added to the ditch (mm)
V_{als}	= volume of water that is stored in the soil matrix (mm)
V_{ao}	= volume above the actual water level to the maximum allowable water level in the plot (m³)
V_c	= relative volume of cracks (m³/m²)
V_{ep}	= soil moisture at which evapotranspiration is at potential rate (mm/m)
V_{ia}	= accumulated precipitation (mm)
V_{imax}	= maximum capacity of surface retention (mm)
V_1	= length of main drains (m)
V_{max}	= maximum volume of water that can be absorbed at any instant (mm)
V_{max1}	= maximum rate of moistening in layer I (mm/m/hour)
V_{max2}	= maximum rate of moistening in layer II (mm/m/hour)
V_{min}	= soil moisture at which evapotranspiration is zero (mm/m)
V_o	= open water volume above the polder water level (m³)
V_{ob}	= depth above the dewatering level (m)
V_{op}	= difference of water depth between the polder water level and the dewatering level (m)
V_{opp}	= maximum volume available for sewer discharge (m³)
V_{ow}	= minimum volume in the open water that will be pumped into the plot (m³)
V_{ow}	= minimum volume in the open water that can be bailed out (m³)
V_{ri}	= maximum capacity of sewer retention (mm)

V_s	= volume change in the soil (%)
V_{s1}	= volume change of soil layer I (%)
V_{s2}	= volume change of soil layer II (%)
V_t	= annual recovery factor (-)
V_t	= volume to be replenished in layer III (mm)
V_{t1}	= annual distribution factor for excavation (-)
V_{t2}	= annual distribution factor for timbering (-)
V_{t4}	= the annual distribution factor of pump cost of life span 't4' years (-)
V_{vc}	= soil moisture at field capacity (mm/m)
V_{vcs}	= soil moisture at saturation (mm/m)
V_{vcs1}	= soil moisture at saturation in layer I (mm/m)
V_{vcs2}	= soil moisture at saturation in layer II (mm/m)
V_{z1}	= capillary rise from layer II to layer I (mm/time step)
w	= initial water content (-)
w_c	= crack width (m)
W_{cr}	= width of crest of the dike (m)
W_{ge}	= volume of water that is added to the water layer above the surface (mm)
W_{ge}	= volume above the ground surface that remains in the rice field (mm)
W_{IP}	= irrigation need to be pumped by a removable boat (mm/time step)
W_L	= actual water layer depth in rice field at time step (mm)
W_L	= water level in the ditch in a dry food crops polder (m-surface)
W_{LA}	= maximum allowable water level in the ditch in a dry food crops polder (m-surface)
$wl1$	= value of building and facility in the farm (€/m^2)
$wl2$	= value of infrastructure in the farm (€/m^2)
W_{LM}	= maximum allowable water layer depth for rice at different growth stages (mm)
W_{LMR}	= maximum allowable water depth in the rice field (mm)
w_{lmax}	= maximum investment costs for i crop (€/ha)
w_{lmin}	= minimum investment costs for i crop (€/ha)
W_{LS}	= preferred water layer depth for rice at different growth stages (mm)
y	= depth of water in the drain (m)
Y_a	= actual yield of the crops (ton/ha)
Ya	= actual rice yield (ton/ha)
Y_m	= potential yield of the crops (ton/ha)
Ym	= potential rice yield (ton/ha)
Y_p	= water level in the plot (m-surface)
Y_r	= required water level in the plot (m-surface)
y_s	= depth of water from the polder water level (m)
Y_s	= water level (m+polder water level)
Z	= subsidence (m)
ΔD_1	= change in depth of the soil in layer I due to swelling and shrinking (m)
ΔD_1	= change in depth of layer I (m)
ΔD_2	= change depth of the soil in layer II due to swell and shrinkage (m)
ΔD_2	= change in depth of layer II (m)
Δp_i	= increase in the intergranular pressure after the groundwater table is lowered (kPa)
Δt_p	= time of spraying of water (hour)
ΔV	= relative compaction (-)

ΔV	= relative compaction resulting from shrinkage and over burden load in the soil (-)
Δt	= time step (hours)
€	= Euro currency
ε	= initial porosity (-)
ε	= initial pore volume (-)
θ_{1b}	= moisture content of the soil in layer I at -0.25 bar (pF=3.017) (%)
θ_{1F}	= moisture content of the soil in layer I at field capacity (%)
$\propto$	= reaction factor (-)
μ	= drainable pore space (%)
μ_1	= crack fraction before subsidence (-)
μ_2	= crack fraction after subsidence (-)
θ	= moisture (%)
θ_1	= soil moisture in layer I (%)
$\theta_{11/3}$	= water content in the soil layer I at a potential of -33 kPa (pF = 2.53) (%)
θ_{115}	= water content in soil layer I at a potential of -1,500 kPa (pF = 4.18) (%)
$\theta_{1/3}$	= water content in soil at a potential of -33 kPa (pF = 2.53) (%)
θ_{15}	= water content in the soil at a potential of -1,500 kPa (pF = 4.18) (%)
θ_{1fc}	= field capacity of soil layer I (%)
θ_2	= soil moisture at the top of layer II (%)
θ_2	= water content of soil layer II (%)
$\theta_{21/3}$	= water content in soil layer II at a potential of -33 kPa (pF = 2.53) (%)
θ_{215}	= water content in soil layer II at a potential of -1,500 kPa (pF = 4.18) (%)
θ_{2fc}	= field capacity of soil layer II (%)
θ_3	= soil moisture at bottom of layer II (%)
θ_4	= soil moisture in layer III (%)
θ_{v1}	= moisture content at field capacity in layer I (%)
θ_{v2}	= moisture content at field capacity in layer II (%)
θ_{ve2}	= moisture content at saturation in layer II (%)
θ_{ve3}	= moisture content at saturation in layer III (%)
ρ_d	= bulk density of a dry clod at -1,500 kPa (pF = 4.18) moisture potential (kg/m^3)
ρ_m	= bulk density of a moist clod at -33 kPa (pF = 2.53) moisture potential (kg/m^3)

Samenvatting

Het weer en het klimaat bepalen welke gewassen in een regio verbouwd kunnen worden en zijn met name bepalend voor de jaarlijkse variatie in de oogsten. Met betrekking tot de landbouwwaterhuishouding kan het klimaat in de wereld in hoofdlijnen verdeeld worden in drie klimaatzones, te weten: gematigd, humide tropen en aride en semi-aride.

In Nederland heerst een gematigd klimaat, met een min of meer evenredige verdeling van de neerslag over het jaar. De gemiddelde jaarlijkse neerslag is ongeveer 785 mm. De verdamping van open water ligt tussen de 0 mm/dag 's-winters en 3 - 4 mm/dag 's-Zomers. 's-Zomers is er een neerslagtekort. Anderzijds is er 's-winters een overmaat aan neerslag die afgevoerd moet worden.

Thailand heeft een humide tropisch klimaat. Het is een moesson klimaat, dat gekenmerkt wordt door een duidelijk regenseizoen van ongeveer mei tot september en een betrekkelijk droog seizoen gedurende de rest van het jaar. De jaarlijkse neerslag in het centrale gedeelte van Thailand is ongeveer 1.200 mm. De verdamping van open water ligt gedurende het gehele jaar tussen de 4 en 5 mm/dag. In het regenseizoen is de neerslag in het algemeen meer dan voldoende om rijst of andere gewassen te verbouwen. In het droge seizoen is de hoeveelheid neerslag echter gering. Vandaar dat de voornaamste functie van waterbeheer bestaat uit de combinatie van irrigatie en drainage.

De behoefte aan land voor voedselproductie, stedelijke ontwikkeling, industrie, infrastructuur, alsmede voor recreatie is gedurende de geschiedenis van de mensheid voortdurend toegenomen. Dit heeft onder andere tot gevolg gehad dat moerassen, overstromingsgebieden langs rivieren, getijde gebieden en zelfs meren zijn ingepolderd. Een polder is een stuk land dat is omringd door waterkeringen en waarin tot op zekere hoogte een van de omgeving onafhankelijke waterstand kan worden gerealiseerd. De hoofdelementen van de waterbeheersingsystemen in landelijke en stedelijke polders zijn tegenwoordig als volgt:

- *voor landelijke gebieden:*
  - afstand tussen de ontwateringsmiddelen;
  - diepte van de ontwateringsmiddelen;
  - afvoercapaciteit van het ontwateringsysteem;
  - percentage open water;
  - polderpeil ten opzichte van het maaiveld;
  - bemalingscapaciteit;
- *voor stedelijke gebieden:*
  - doorsnede van de rioolbuizen;
  - afstand tussen de stadsgrachten of transportleidingen;
  - percentage open water;
  - grachtpeil ten opzichte van het maaiveld;
  - afvoer of bemalingscapaciteit.

De belangrijkste doelstelling van deze studie was het evalueren van de huidige ontwerp praktijk voor waterbeheersingsystemen van polders in de gematigde klimaatzone - Nederland - en de humide tropen - Thailand - en het formuleren van aanbevelingen voor een nieuw waterbeheersingsbeleid voor huidige en toekomstige omstandigheden, zoals aanbevelingen voor ontwerp, beheer en onderhoud.

Om inzicht te verkrijgen in het gedrag van systemen bij verschillende vormen van grondgebruik en bodemsamenstelling in gematigde en humide tropische omstandigheden kunnen modelsimulaties worden uitgevoerd. Het bestaande programmapakket OPOL, dat is gebaseerd op een niet stationair model, is verder ontwikkeld tot de versie OPOL5 om de hydrologische condities te simuleren en de hoofdelementen van waterbeheersingsystemen in poldergebieden te optimaliseren, voor condities in de gematigde en humide tropische klimaatzones. Het model laat het gedrag van een systeem zien, evenals de effecten van veranderingen in de hoofdelementen van het systeem op de totale kosten. Op deze manier kunnen de ontwerpen voor waterbeheersingsystemen van poldergebieden geoptimaliseerd worden door de hoofdelementen te variëren totdat de equivalente jaarlijkse kosten minimaal zijn. Een GIS pakket is ter aanvulling van OPOL5 toegepast om de werkelijke toestand van de gebieden te simuleren, zoals topografie, grondsoort, grondgebruik en schade.

Simulatie voor het landelijke gebied

Het waterbeheersingsysteem in landelijke gebieden heeft tot doel het scheppen van goede condities voor de groei van de gewassen en goede ontwatering voor bebouwing en infrastructuur. Om te komen tot optimale waarden voor de hoofdelementen van waterbeheersingsystemen voor de landbouw zijn de voor het ontwerp maatgevende situaties gebruikt om op basis van de transformatie van neerslag in verdamping en afvoer het volgende te bepalen:
-       veranderingen in het vochtgehalte in de onverzadigde zone;
-       verloop van de grondwaterstand;
-       verloop van de open waterstand;
-       fluctuaties in de inundatiediepte van een rijstveld;
-       irrigatiewater voorziening onder humide tropische condities.
Door gebruik te maken van de relatie tussen bovenstaande condities en landbouwkundige criteria kunnen de optimale waarden voor de hoofdelementen van de waterbeheersingsystemen worden bepaald.

Simulatie voor het stedelijke gebied

Een stedelijk gebied vormt in het algemeen een geheel of is een deelgebied van het waterbeheersingsysteem van een polder. Het waterbeheersingsysteem heeft tot doel te voorzien in goede ontwatering en afvoer. Om te komen tot optimale waarden voor de hoofdelementen van waterbeheersingsystemen voor stedelijke gebieden zijn de voor het ontwerp maatgevende situaties gebruikt om op basis van de transformatie van neerslag in afvoer het volgende te bepalen:
-       het voorkomen van water op straat;
-       stijging van de grondwaterstand onder invloed van regenval en peiloverschrijdingen in de stadsgrachten;
-       fluctuaties van het waterpeil in de stadsgrachten.
Bij afvoer over een stuw worden eerst de waarden van de hoofdelementen van het waterbeheersingsysteem in het stedelijke gebied bepaald onder aanname van vrije stroming over de stuw of het gemaal. Hierdoor worden optimale waarden voor de hoofdelementen van het waterbeheersingsysteem voor de stedelijke gebieden gevonden. Afhankelijk van de situatie moet worden gearalyseed of deze waarden nog optimaal zijn als het hele systeem wordt beschouwd.

Case-studies in Nederland en Thailand

Voor Nederland betrof het de volgende case-studies:
- Schieveen, een kleipolder;
- Duifpolder, een veenpolder;
- Hoge en Lage Abtswoudse polder, een stedelijke polder.

Voor Thailand betrof het de volgende case-studies:
- een rijstpolder zonder irrigatie;
- een rijstpolder onder irrigatie;
- een akkerbouw polder;
- een polder tussen de mondingen van de Chao Phraya rivier en de Bangpakong rivier in oostelijk Bangkok, een landelijke polder;
- Sukhumvit polder, een stedelijke polder.

In deze gebieden zijn gegevens verzameld voor het kalibreren van de parameters die de meeste invloed hebben op het grondwater, het open water en de afvoer.

De polders voor de case-studies in Nederland zijn gesitueerd in het gebied van het Hoogheemraadschap van Delfland. Delfland ligt in het westen van Nederland. Het gebied wordt in het westen begrensd door de Noordzee en in het zuiden door de Nieuwe Maas en de Nieuwe Waterweg. Delfland is een typisch poldergebied met een groot aantal verschillende polderpeilen. Het heeft een totale oppervlakte van ongeveer 40.000 ha en bestaat uit 60 polders.

Voor de case-studies in Thailand is eerst een proefgebied in de Centrale Vlakte uitgezocht met zowel niet geïrrigeerde en geïrrigeerde rijst, alsmede akkerbouw. Het proefgebied was gesitueerd in de provincie Suphanburi, ongeveer 208 km ten noordwesten van Bangkok. Het gebied van de provincie Suphanburi bestaat voor het merendeel uit lage riviervlakten met kleine bergketens in het noorden en westen. In het zuidoostelijke gedeelte van de provincie ligt de zeer lage vlakte van de Tha Chin rivier waar rijst wordt verbouwd.

Het gebied voor de case-studie voor de landelijke polder was gesitueerd in het zuidoostelijke gedeelte van Bangkok en het centrale gedeelte van de provincie Samut Prakan. Het wordt in het noorden begrensd door het Klongpravetburirom kanaal, in het westen door de dijk van de Chao Praya rivier, in het oosten door de King Initiated dijk en in het zuiden door het Klonchythale kanaal. Dit is een vlak, laag gebied gesitueerd tussen de mondingen van de Chao Phraya en Bangpakong rivier. Het gebied loopt af in de richting van de Golf van Thailand.

Het gebied voor de case-studie voor de stedelijke polder was gesitueerd in het oostelijke gedeelte van Bangkok. Het oostelijke gedeelte van Bangkok is verdeeld in 10 polders en de polder Sukhumvit, een van de inliggende stedelijke polders van Groot Bangkok, was geselecteerd voor deze studie.

Resultaten van de case-studies

De kleipolder Schieveen in Nederland

De polder Schieveen ligt in de gemeente Rotterdam. Het oppervlak van deze polder is 584 ha. Het grootste gedeelte is weiland (81,3%). De bodem bestaat uit kleiachtig veen voor de toplaag en zware klei voor de ondergrond. De waarden voor de huidige componenten van het waterbeheersingsysteem zijn als volgt: winterpeil is 0,71 m-maaiveld; oppervlakte open water is 5,6%; diepte van de drainbuizen is ongeveer 0,50 m-maaiveld; afstand tussen de drainbuizen is ongeveer 20,00 m; bemalingscapaciteit is 13,4 mm/dag. Als resultaat van deze studie zouden de optimale waarden voor de componenten van het waterbeheersingsysteem als volgt zijn: winterpeil 1,51 m-maaiveld; oppervlakte open water 1,9%; diepte van de drainbuizen ongeveer 1,30 m-maaiveld; afstand tussen de drainbuizen 23,19 m; bemalingscapaciteit 16,2 mm/dag. In dit gebied heeft de diepte van de drainbuizen de grootste invloed op de equivalente jaarlijkse kosten. Overwogen kan worden om de waterstand in de polder 's zomers hoger te houden en de draindiepte lager te kiezen dan nu het geval is.

De veenpolder Duifpolder in Nederland

De Duifpolder is een gedeelte van de regio Midden Delfland en ligt in de gemeenten Schipluiden en Maasland. De oppervlakte van de Duifpolder is ongeveer 370 ha. Het merendeel van het gebied is weiland (88,8%). De bodem bestaat uit kleiachtig veen aan de oppervlakte en veen met niet verteerde plantenresten in de ondergrond. De waarden voor de huidige componenten van het waterbeheersingsysteem zijn als volgt: winterpeil is 0,78 m-maaiveld; oppervlakte open water is 4,4%; de diepte van de drainbuizen is ongeveer 0,35 m-maaiveld; de afstand tussen de drainbuizen is ongeveer 10,00 m; de bemalingscapaciteit is 14,9 mm/dag. Als resultaat van deze studie zouden de optimale waarden voor de componenten van het waterbeheersingsysteem als volgt zijn: winterpeil 1,35 m-maaiveld; oppervlakte open water 1,9%; diepte van de drainbuizen ongeveer 1,15 m-maaiveld; afstand tussen de drainbuizen 26,50 m; bemalingscapaciteit 4,9 mm/dag. De draindiepte heeft de grootste invloed op de equivalente jaarlijkse kosten. De bemalingscapaciteit is momenteel aan de hoge kant. Overwogen kan worden om de waterstand in de polder 's zomers hoger te houden en de draindiepte lager te kiezen dan nu het geval is.

De stedelijke polder Hoge en Lage Abtswoudse polder in Nederland

Het gebied voor deze case-studie heeft een oppervlak van 713 ha en wordt gekenmerkt door veel bebouwing. Het gebied omvat de Hoge Abtswoudse polder (216 ha) en het bebouwde deel van de Lage Abtswoudse polder (497 ha). Deze polders worden beide bemalen door het gemaal Het Voorhof. In de uitgevoerde modelsimulatie wordt er van uitgegaan dat de afvoer van de Hoge Abtswoudse polder over een stuw plaatsvindt naar de Lage Abtswoudse polder. De waarden voor de huidige componenten van het waterbeheersingsysteem zijn als volgt: diameter van de rioolbuizen 0,30 m; afstand tussen de stadsgrachten is 1.870 m; grachtpeil 0,87 m-maaiveld; oppervlakte open water is 0,2% en de bemalingscapaciteit is 5,0 mm/dag. De uitgevoerde simulatie bestond uit een optimalisatie voor de Lage Abtswoudse polder en leverde de volgende resultaten op: diameter van de rioolbuizen 0,30 m, afstand tussen de stadsgrachten tussen de 175 en 350 m, grachtpeil 0,55 m-maaiveld. De oppervlakte open water 5,0% en de bemalingscapaciteit 8,5 mm/dag. De componenten die het meeste van invloed zijn op de equivalente jaarlijkse kosten zijn de diameter van de rioolbuizen en de bemalingcapaciteit. Deze componenten zijn vooral van invloed omdat bij de optimalisatie is uitgegaan van niet extreme dagelijkse neerslag. Dit uitgangspunt heeft een grote invloed op de nauwkeurigheid van het resultaat van de simulatie. Verder is gevonden dat het verlagen van het grachtpeil van 0,55 naar 0,87 m-maaiveld overwogen zou kunnen worden en dat het huidige percentage open water en de bemalingscapaciteit ruim voldoende zijn voor een goed waterbeheer. Echter, het berekende percentage open water bij een optimale conditie is klein omdat alleen dagcijfers beschikbaar waren voor de simulatie. Dit is ook de reden waarom een piek afvoer ten gevolge van een 'flashflood' is niet meegenomen in deze simulatie.

Een rijstpolder zonder irrigatie in Thailand

In de praktijk zijn er in Thailand eigenlijk geen rijstpolders zonder irrigatie, boeren zullen altijd water vanuit het kanaal pompen om te irrigeren. Het optimale waterpeil in de kanalen in een polder is 0,09 m+maaiveld. Met het water op deze hoogte kan het rijstveld optimaal profiteren van irrigatie onder zwaartekracht. Tijdens een periode van droogte zal het water naar het veld stromen, terwijl in een periode met overmatige regenval het overtollige water door de zwaartekracht gedraineerd wordt. Het oppervlak water in de polder is idealiter 1,1 %, de drainage capaciteit 6,7 mm/dag en de bemalingscapaciteit is 7,5 mm/dag. (Note: this text deals with irrigation and has to be changed, also in the English version). Het waterpeil heeft de meeste invloed op de jaarlijkse kosten en opbrengsten van het gebied. Hoe dieper het waterpeil in een, voornamelijk door regen gevoede, rijstpolder hoe meer het water door scheuren vanuit het rijstveld naar de nabij liggende kanalen stroomt. Dit is vooral nadelig in perioden van droogte en heeft gewasschade en daardoor indirect oogstderving tot gevolg. In perioden met overmatige regenval kan dit de boer ten goede komen omdat minder water achterblijft op het rijstveld. In het geval van hoge waterstanden in de polder zal de kans op overstroming van gewassen en van de ruggen rond de rijstakkers toenemen wat grote schade tot gevolg kan hebben.

Een geïrrigeerde rijstpolder in Thailand

De waarden van de belangrijkste componenten van het waterbeheersingsysteem voor een standaard geïrrigeerde rijstpolder zijn als volgt: het waterpeil in de kanalen is 0,80 m-maaiveld; de oppervlakte open water in de polder is 2,4%; de veld drainage capaciteit is 46,0 mm/dag en de bemalingscapaciteit is 26,0 mm/dag. Als resultaat van de simulaties zouden de optimale waarden voor de componenten van het waterbeheersingsysteem als volgt zijn: het waterpeil in de kanalen 0,84 m-maaiveld; oppervlakte open water 2,5 %; veld drainage capaciteit 44,8 mm/dag en de bemalingscapaciteit 32,9 mm/dag. Bij optimale condities moet in droge perioden water in de rijstvelden gepompt worden omdat het waterpeil in de polder dan onder het maaiveld ligt. In het geïrrigeerde polder systeem wordt het meeste irrigatiewater van buiten de polderdijk aangevoerd. Daarom is in het model aangenomen dat het waterpeil in de polder het hele jaar op dit niveau wordt gehouden. De bemalingscapaciteit is groter dan de huidige capaciteit, terwijl de andere componenten in dezelfde orde van grootte liggen. De bemalingscapaciteit en het polderpeil hebben de grootste invloed op de equivalente jaarlijkse kosten. Vergroten van de bamalingscapaciteit zou overwogen kunnen worden.

Een akkerbouw polder in Thailand

De waarden van de belangrijkste componenten van het waterbeheersingsysteem voor een typische akkerbouw polder zijn momenteel als volgt: het streefpeil in de sloten is 0,60 m-maaiveld; de afstand tussen de sloten is tussen de 6 en 8 m; de grondwaterstand is 0,50 m-maaiveld; de oppervlakte open water is 1,1%; de afvoercapaciteit van de velddrainage is 46,0 mm/dag en de bemalingscapaciteit is 26,0 mm/dag. De optimale waarden die op basis van de simulaties zijn gevonden zijn als volgt: het streefpeil in de sloten 0,74 m-maaiveld; afstand tussen de sloten 8,45 m; polderpeil 0,06 m+maaiveld; oppervlakte open water 2,4%; afvoercapaciteit van de velddrainage 34,8 mm/dag en de bemalingscapaciteit 36,5 mm/dag. Bij de gesimuleerde optimale condities ligt het polderpeil boven het maaiveld hetgeen betekent dat het water voor irrigatie onder zwaartekracht naar de kavel kan stromen. Het slootpeil en de afstand tussen de sloten hebben de meeste invloed op de equivalente jaarlijkse kosten. Overwogen zou kunnen worden het slootpeil voor bestaande polders in eerste instantie te verlagen van 0,60 tot 0,75 m-maaiveld. Voor nieuw te ontwikkelen polders zou de slootafstand vergroot kunnen worden van 6,0 tot 8,5 m. Ook kan worden overwogen de bemalingscapaciteit te vergroten. Het polderpeil kan, met name in het droge seizoen, hoger gehouden worden dan de slootbodem, zodat water onder zwaartekracht naar de kavel kan stromen.

De landelijke polder in Thailand

De oppervlakte van de landelijke polder was 18.760 ha, bestaande uit 7.390 ha stedelijk gebied en 11.370 landbouwgebied. Het gebied voor rijstteelt besloeg 12,8% en dat voor viskwekerijen 87,2%. De waterbeheersing in dit gebied wordt 'het behouden van water' genoemd. Het water wordt ten behoeve van irrigatie en andere doeleinden voor het einde van het regenseizoen opgeslagen in de hoofd- en zijkanalen. Normaal gesproken pompen de boeren water uit de hoofdkanalen naar de zijkanalen om het vandaar door te pompen naar de gewassen. De waarden voor de componenten van het waterbeheersingsysteem in een polder met landelijk gebied zijn als volgt: het polderpeil in droge perioden is 0,80 m-maaiveld; de oppervlakte open water beslaat 3,3%; de afvoercapaciteit van de velddrainage is 46,0 mm/dag en de bemalingscapaciteit is 5,3 mm/dag. De gesimuleerde optimale condities zijn als volgt: polderpeil 0,24 m-maaiveld; oppervlakte open water 2,0%; afvoercapaciteit van de velddrainage 62,9 mm/dag en bemalingscapaciteit 6,3 mm/dag. Het polderpeil en de bemalingscapaciteit hebben de meeste invloed op de equivalente jaarlijkse kosten. Overwogen zou kunnen worden om de bemalingscapaciteit te vergroten. De waterstand zou gedurende de bergingsperiode hoger gehouden kunnen worden teneinde meer water vast te kunnen houden in het kanaalsysteem, alsmede om de kosten om water naar de rijstvelden en de viskwekerijen te pompen te verlagen. Het blijkt dat visvijvers tevens kunnen dienen als berging van overtollig water. Aanpassing van de agrarische praktijk aan de tijdschema's van de kweekvijvers teneinde water tijdens de natte periode te kunnen bergen zullen leiden tot verbeterde optimale omstandigheden. Dit vereist echter nadere studie naar de technische en sociaal-economische aspecten.

De stedelijke polder Sukhumvit in Thailand

Het studiegebied voor de stedelijke polder was gesitueerd in het oostelijke gedeelte van Bangkok. Het deelgebied DF van de Sukhumvit polder, met een duidelijke begrenzing en met beschikbare data, was geselecteerd. Dit deelgebied van de polder heeft een oppervlakte van 368 ha, samengesteld uit 248 ha verhard oppervlak en 120 ha onverhard gebied. De huidige waarden voor de belangrijkste componenten van het waterbeheersingsysteem van de stedelijke sub-polder DF zijn als volgt: de diameter van de rioolbuizen is 0,60 m; de afstand tussen de transportleidingen bedraagt 600 m; de waterstand in de kanalen is 2,08 m-maaiveld; de oppervlakte aan open water is 0,6% en de bemalingscapaciteit is 137 mm/dag. De waarden behorend bij de gesimuleerde optimale condities zijn als volgt: diameter rioolbuizen 1,00 m; afstand tussen de transportleidingen 1.480 m; waterstand in de kanalen 2,56 m-maaiveld; oppervlakte open water 1,4% en bemalingscapaciteit 117 mm/dag. De waterstand in de kanalen heeft de meeste invloed op de equivalente jaarlijkse kosten. Verlaging van de waterstanden gedurende de natte tijd en vergroting van de diameter van de rioolbuizen kan de schade door water op straat verminderen. Ook zou overwogen kunnen worden om het oppervlak van het open water te vergroten teneinde meer berging te verkrijgen en de snelle stijging van het waterpeil af te vlakken.

Evaluatie van de resultaten

Capillaire stroming en bodemvochtgehalte in de onverzadigde zone

Na inpoldering kunnen zich scheuren in de bodem vormen ten gevolge van horizontale krimp. Daarnaast kan compactie optreden ten gevolge van verticale krimp van de bodem. Meestal is de toplaag van bodem gemengd door landbouwkundige activiteiten maar blijven de scheuren onder de toplaag open. In de simulatie is de bodem dan ook als volgt geschematiseerd: laag I toplaag, laag II de onderliggende laag met scheuren en daaronder laag III een onderdringbare of semi doorlaterde laag.

De capillaire stroming in de onverzadigde zone in of nabij de wortelzone is gesimuleerd in het model dat gebaseerd is op de bodemvocht spanning bij de randvoorwaarden, die afhangen van de grondwaterstand, de bodemeigenschappen, enz. In de gematigde zone ligt het bodemvochtgehalte boven het niveau van de drainbuizen in de winter boven de veldcapaciteit, derhalve is er geen capillaire opstijging. In deze periode is het bodemvochtgehalte in laag II hoger dan het vochtgehalte in laag I, omdat het model de bodemvochtbalans in laag I berekent met verdamping terwijl er slechts geringe capillaire opstijging vanuit laag II is. 's-Zomers ligt het bodemvochtgehalte beneden veldcapaciteit en daarom speelt de capillaire opstijging een belangrijke rol. In de zomer periode neemt het vochtgehalte in laag I in de klei en de veenpolder in Nederland sneller af dan het vochtgehalte in laag II. Dit kan worden veroorzaakt door de capillaire opstijging van laag II naar laag I en de infiltratie naar laag I verminderd met de wateropname door de gewassen, die groter zijn dan de percolatie van laag I naar laag II vermeerderd met de capillaire opstijging van laag III naar laag II. De bodem in de humide tropen kan droger zijn dan in de gematigde zone vanwege de hoge verdamping over het gehele jaar. Onder droge bodem condities zal klei in de humide tropen scheuren en indien er plotseling regenval optreedt zal het water verloren gaan van het veld naar het systeem van afvoerkanalen, speciaal onder de condities die gelden voor natte rijstteelt zonder irrigatie. Wanneer er geen irrigatie plaatsvindt is de capillaire stroming naar de toplaag ook klein onder natte omstandigheden tijdens de droge periode, vanwege de diepe grondwaterstand. Bij geïrrigeerde rijst is het bodemvochtgehalte bijna verzadigd, waardoor er geen capillaire opstijging is. Diepe percolatie kan een rol spelen in de waterbeheersing van geïrrigeerde systemen. Bij akkerbouw was het bodemvochtgehalte bijna op veldcapaciteit door het irrigatiewater dat over de bedden gesproeid is; de bodem was alleen droog tijdens een korte periode na de oogst. De capillaire opstijging onder akkerbouw condities kan voordelig zijn voor de gewassen en de boeren, omdat er minder gepompt hoeft te worden om te sproeien tijdens droge perioden. In natte perioden kan de capillaire opstijging klein zijn omdat het bodemvochtgehalte dan bijna boven veldcapaciteit is.

Waterberging in de onverzadigde zone

Waterberging in de onverzadigde zone kan een belangrijke rol spelen met betrekking tot fluctuaties van de waterstanden in het afwateringssysteem van een polder, omdat deze waterberging, met name in veengrond, zeer groot kan zijn. Waterberging in de onverzadigde zone is afhankelijk van het bodemvochtgehalte toestand en de grondeigenschappen. Waterberging in de onverzadigde zone speelt in de humide tropen, speciaal bij akkerbouw, een belangrijke rol in de waterstand op de kavel omdat hoe lager deze berging is hoe groter de mogelijkheid van overstroming van de gewassen in de bedden is. Waterberging in de onverzadigde zone speelt ook een belangrijke rol in de natte rijstteelt zonder irrigatie, omdat de bodem bijna droog kan zijn en het vergt behoorlijk veel neerslag om deze berging te vullen. Boeren kunnen rijst telen en het vocht kan in de bodem blijven.

Fluctuaties in de grondwaterstanden

De meeste poldergebieden zijn vlak en de verhanglijnen zijn te verwaarlozen. Daardoor kan bij het modelleren van de grondwaterzone een conceptualisatie van reservoir berging en afvoer naar de draineerbuizen als een functie van de werkelijke berging worden toegepast. Daardoor kan voor de gematigde zone de afvoer als niet-lineaire relatie met de berging in laag II worden gesimuleerd. De berekende grondwaterstanden in zowel de kleipolder als de veenpolder schommelen boven de draindiepte. In werkelijkheid kan het grondwater, speciaal in veengrond, in de zomer onder het niveau van de drains zakken. Daardoor kan het tekort aan bodemvocht bij grasland in de zomer vanwege verminderde capillaire opstijging groter zijn dan in de simulatie is gevonden. In de humide tropen betreft de waterstand zowel het grondwater als het open water. Het waargenomen grondwaterpeil of oppen waterpeil is nodig om het te vergelijken met het gesimuleerde grondwaterpeil of open waterpeil. Bij de case-studies in Thailand komen de resultaten goed overeen met de waargenomen grondwaterpeilen of open waterpeilen, speciaal in het geval van de akkerbouw polder en de geïrrigeerde rijstpolder. De gesimuleerde grondwaterpeilen in de situatie zonder irrigatie weken behoorlijk af van de waarnemingen. Dit werd veroorzaakt door fluctuaties van het waterpeil in de rivier in de nabijheid van de kavel In het proefgebied in Thailand reageert het grondwater zeer snel op de neerslag, omdat de bodem een grotere doorlatendheid heeft. Bovendien werd het grondwater in het geval van geïrrigeerde rijst en akkerbouw eveneens beïnvloed door het peil in het laterale kanaal. Bij geïrrigeerde rijst is niet alleen de fluctuatie in het grondwaterpeil maar ook de fluctuatie in het open waterpeil zeer gevoelig voor het aangrenzende gebied wegens verlies van water door de kaden en diepe percolatie.

Neerslag afvoer relatie in het stedelijke gebied

De twee belangrijkste functies van het rioolsysteem in stedelijke gebieden zijn het verwijderen van regenwater en huishoudelijk en industrieel afvalwater. In een poldergebied komt de afvoer naar het hoofdsysteem van zowel bestrate als onbestrate gebieden. De verhouding tussen de neerslag en de oppervlakteafvoer is verschillend voor de gematigde zone en de humide tropen vanwege verschillen in temperatuur en neerslagpatroon. Bij de case-studies voor de stedelijke polders in zowel Nederland als Thailand is er meer afvoer naar het hoofdsysteem en is de kans op schade groter door hoge waterpeilen in de stedelijke gebieden bij toenemende diameters van rioolbuizen door optimalisatie. Er is in Nederland echter eigenlijk geen extreme neerslag en is het lastig om een goede relatie te vinden tussen neerslag, afvoer en schade door water in de straten. Kort durende overstromingen en water op straat komen met grote regelmaat voor in Thailand vanwege de hevige neerslag in een korte tijd.

Polderpeil

De aanname in het model van een bergingsreservoir voor het hoofdsysteem bij de berekening van het waterpeil is aanvaardbaar voor een vlak poldergebied. In de gematigde zone kan het peil in het hoofdafvoerkanaal de stroming in het drainage systeem onder de grond beïnvloeden in het geval dat het water hoger staat dan de draindiepte. Het kan zelfs de stroming van bodemvocht in de toplaag beïnvloeden. Het peil boven de draindiepte heeft echter weinig effect als de duur van die periode kort is. In de humide tropen wordt het peil in het hoofdsysteem beïnvloed door de afvoercapaciteit van de rijstkavels en de akkerbouw polders. Tijdens hoge waterstanden in de hoofdsystemen is het noodzakelijk om te pompen omdat het stilstaande water in de velden niet door de duikers kan stromen en daardoor schade aan de gewassen kan ontstaan. Bovendien is de mogelijkheid groot dat dijken of kaden overstromen als het peil in het hoofdsysteem hoog is. Tijdens de droge tijd is een hoog peil in de hoofdsystemen goed voor irrigatie als er onder vrij verval water door de duiker ingelaten wordt of als er minder hoog hoeft worden opgepompt. Het waterpeil in de hoofdsystemen in polders in de humide tropen kan aan het eind van het regenseizoen hoog gehouden worden om water te kunnen bergen voor irrigatie in de droge tijd.

Percentage open water

De oppervlakte aan open water is van direct belang voor de waterberging in de polder. Hoe kleiner het open wateroppervlak des te minder bergingsvolume zal er beschikbaar zijn. Normaal gesproken zullen kleinere bergingsvolumes bij gelijke drainage condities hoge waterpeilen in de polder veroorzaken. De kans op schade aan stedelijke eigendommen neemt toe bij een kleiner open wateroppervlak. Ook zal er meer gepompt moeten worden voor zowel rijst als akkerbouw gewassen vanwege de hogere waterpeilen onder overeenkomstige drainage condities. Bovendien neemt de mogelijkheid van overstroming van dijken en kaden toe naarmate de oppervlakte open water in gebieden met rijst en akkerbouw kleiner is.

Afvoercapaciteit

De afvoer van polders geschiedt door middel van bemaling, door een duiker of over een overlaat. Hoe groter de afvoercapaciteit hoe minder het waterpeil in de polder stijgt. In Nederland wordt de afvoercapaciteit doorgaans gerealiseerd door middel van bemaling. De afvoercapaciteit in de humide tropen kan worden onderscheiden in de categorieën oppervlakteafvoer van het veld en afvoer door het kanalen stelsel. Lagere oppervlakteafvoer capaciteit van rijstvelden is schadelijker vanwege hoge waterpeilen in het veld. Lagere oppervlakteafvoer capaciteit van akkerbouw kavels verhoogt de kans op overstroming van de akkers in de kavel met grote schade als gevolg. Bovendien kan een hoog peil in de sloten van een polder voor akkerbouw hoge grondwaterstanden veroorzaken die schade zullen veroorzaken aan de gewassen op de kavel.

Slot opmerkingen

Het OPOL-pakket is een bruikbaar instrumentarium om de optimale waarden voor de hoofdcomponenten van het waterbeheering systeem in een polder te bepalen. Ook voor de begripsvorming over veranderingen in en het gedrag van hoofdcomponenten van het waterbeheering systeem en de effecten op de kosten en schade aan de polder. De gekwantificeerde waarden van de hoofdcomponenten resulterend uit de simulaties die zijn uitgevoerd in deze studie zijn indicatief en moeten in de praktijk nader bepaald worden. Van invloed zijn onder andere de spraktijk van beheer en onderhoud, landgebruik, landbouwkundige exploitatie, vigerend beleid, bodemgesteldheid, milieu- en landschap condities. Daarnaast moeter de hoofdcomponenten resulterend uit de simulaties getoetst worden aan de lokale omstandigheden.

De jaarlijkse kosten die zijn gevonden in deze studie zijn relatief laag. Het wordt aangeraden om een veiligheidsmarge in te bouwen in de optimale waarden van de hoofdcomponenten van het waterbeheering systeem bepaald met het OPOL-pakket.

De jaarlijkse kosten sterk nemen toenemen als de waarden voor de componenten kleiner worden genomen dan de optimale waarden. Hiermee moet in de ook rekening worden gehouden bepaling van de hoofdcomponenten.

Bij het ontwerp moeten daarnaast de volgende aspecten in beschouwing genomen worden: toekomstig gebruik, levensduur van de onderdelen, groei van de bevolking en verstedelijking, vooruitgang in technologie, wijzigingen in de omgevings, economische- en sociale condities, de hydrologische statistiek, risico analyse e.d.

De drainagediepte heeft de grootste invloed op de jaarlijkse kosten in zowel de Nederlandse klei als veenpolder. In Thailand heeft het niveau van het polderwater het meeste effect op de jaarlijkse kosten. Dit is het geval in het gebied met rijstbouw zonder irrigatie maar ook voor geïrrigeerde rijstbouw. Bij geïrrigeerde rijstbouw is naast het het niveau van het polderwater ook de bemalings capaciteit van invloed op de kosten, bij landbouw zonder irrigatie is dit het waterniveau in de sloten en greppels. In Thailand is het waterniveau in de kanalen heel gevoelig voor het waterniveau in de sloten en greppels, in Nederland is dit niet zo zeer het geval.

In deze studie is aangetoond dat de optimale waarden voor de componenten van het waterbeheering systeem condities in een polder in zowel gematigde streken als in de humide tropen kan worden bepaald.

สรุปความโดยย่อ

สภาพอากาศเป็นตัวกำหนดชนิดพืชที่สามารถปลูกได้โดยอาจเปลี่ยนแปลงได้ตลอดปีความสัม-
พันธ์ระหว่างการ จัดการน้ำกับสภาพอากาศในโลกสามารถแบ่งออกได้เป็น๓กลุ่มหลักคือสภาพ
อากาศแบบปานกลาง,สภาพอากาศแบบร้อนชื้นและสภาพอากาศแบบแห้งแล้ง-กึ่งแห้งแล้ง
ประเทศเนเธอร์แลนด์มีสภาพอากาศแบบปานกลางและมีฝนตกโดยสม่ำเสมอตลอดปีฝนโดย
เฉลี่ยประมาณ ๗๘๕ มิลลิเมตรต่อปีการระเหยพื้นผิวน้ำเปลี่ยนแปลงจาก ๐ ในฤดูหนาว เป็น๓
ถึง๔ มิลลิเมตร ระหว่างฤดูร้อนมีการการระเหยคายน้ำมากกว่าฝนตกแต่ในช่วงฤดูหนาวมีฝนตก
เกินความต้องการจึง ต้องทำการระบายน้ำ

ประเทศไทยอยู่ในเขตร้อนชื้นมีสภาพอากาศแบบมรสุมฤดูฝนเริ่มตั้งแต่พฤษภาคมถึงกันยา-
ยนและค่อนข้างแห้งในช่วงเวลาที่เหลือฝนโดยเฉลี่ยประมาณ๑,๒๐๐มิลลิเมตรต่อปีการระเหย
จากพื้นผิวน้ำประมาณ๔ถึง๕มิลลิเมตรตลอดทั้งปีโดยทั่วไปในฤดูฝนมีฝนตกเกินความต้องการ
แต่ในช่วงฤดูแล้งมีฝนตกน้อยมากดังนั้นการจัดการน้ำโดยเบื้องต้นเกี่ยวข้องกับการชลประทาน
และ การระบายน้ำ

ความต้องการพื้นที่สำหรับผลิตอาหาร และชุมชนเมืองได้แก่ ที่อยู่อาศัย อุตสาหกรรมพานิช-
ยกรรม พื้นที่โครงสร้างพื้นฐาน และพื้นที่พักผ่อนหย่อนใจเพิ่มขึ้นตลอดในประวัติศาสตร์ของ
มนุษยชาติที่ผ่านมา ดังนั้นจึงทำให้มีการปรับปรุงพื้นที่หนองน้ำ, พื้นที่น้ำท่วมถึงพื้นที่น้ำท่วม
เนื่องอิทธิพลจากกระแสน้ำและแม้กระทั่งทะเลสาบโดยทำการปิดล้อมพื้นที่เพื่อป้องกันน้ำท่วม

'พื้นที่ปิดล้อมหมายถึงพื้นที่ราบลุ่มซึ่งดั้งเดิมถูกน้ำท่วมหรือมีระดับน้ำใต้ดินสูงบางฤดูกาล
หรือตลอดปีและพื้นที่นี้ได้ถูกแยกระบบอุทกจากพื้นที่โดยรอบเพื่อที่จะสามารถควบคุมระดับน้ำ
ใต้ดินหรือน้ำผิวดินในพื้นที่ดังกล่าว'

องค์ประกอบหลักของจัดการน้ำในพื้นที่ปิดล้อมในเขตชนบทและเขตเมือง ณ ปัจจุบันมีดัง
ต่อไปนี้

- สำหรับเขตชนบท

  • ระยะห่างระหว่างท่อระบายน้ำใต้ดินหรือร่องระบายน้ำ;

  • ความลึกของท่อระบายน้ำใต้ดินหรือร่องระบายน้ำ;

  • ความสามารถในการระบายน้ำออกจากสนาม;

  • เปอร์เซ็นต์ของพื้นที่ผิวน้ำ;

  • ระดับน้ำจากผิวดินของพื้นที่ผิวน้ำ;

- ● ความสามารถในการระบายน้ำหรือความสามารถในการสูบน้ำ

- สำหรับเขตเมือง

  - ● พื้นที่หน้าตัดของท่อระบายน้ำ;
  - ● ระยะห่างระหว่างคลองระบายน้ำหรือท่อส่งน้ำ;
  - ● เปอร์เซ็นต์ของพื้นที่ผิวน้ำ;
  - ● ระดับน้ำจากผิวดินของคลอง;
  - ● ความสามารถในการระบายน้ำหรือความสามารถในการสูบน้ำ

วัตถุประสงค์หลักของการศึกษานี้คือเพื่อที่จะประเมินการออกแบบ,การปฏิบัติงานสำหรับจัด
การน้ำในพื้นที่ ปิดล้อมในสภาพอากาศแบบปานกลาง(เนเธอร์แลนด์)ในสภาพอากาศแบบเขต-
ร้อนชื้น(ไทย)และเพื่อที่จะกำหนดแนวทางจัดทำแผนและนโยบายสำหรับการออกแบบการปฏิ-
บัติงานและบำรุงรักษาเพื่อที่จะมีความเข้าใจอย่างถ่องแท้ถึงพฤติกรรมของการใช้พื้นที่และส่วน-
ประกอบของดินภายใต้ในสภาพอากาศแบบปานกลางและสภาพอากาศแบบเขตร้อนชื้น แบบ-
จำลองทางคณิตศาสตร์สามารถใช้ในการนี้แบบจำลองทางคณิตศาสตร์สามารถแสดงพฤติกรรม
ของระบบและยังสามารถแสดงผลกระทบของการเปลี่ยนแปลงขนาดขององค์ประกอบหลักของ
จัดการน้ำเปรียบเทียบกับมูลค่าเทียบเท่าของระบบทั้งหมด ดังนั้นในแนวทางนี้การออกแบบ
ระบบในพื้นที่ปิดล้อมสามารถหาความเหมาะสมที่สุดโดยการเปลี่ยนแปลงขนาด ขององค์ประ-
กอบดังกล่าวจนกระทั่งมูลค่าเทียบเท่าของระบบทั้งหมดต่อปีมีค่าต่ำสุด แบบจำลองโอปอล๕
ซึ่งเป็นแบบจำลองที่มีมิติเป็นศูนย์ได้ถูกพัฒนาขึ้นมาเพื่อจำลองสภาวะอุทกและหาความเหมาะ
สมที่สุดของขนาดขององค์ประกอบหลักของจัดการน้ำในพื้นที่ปิดล้อม ระบบภูมิศาสตร์สารสน-
เทศได้นำมาใช้ร่วมกับแบบจำลองโอปอล๕เพื่อจำลองสภาพจริงในพื้นที่ได้แก่ การใช้พื้นที่,
ความเสียหาย, ภูมิประเทศ และชนิดของดิน

การจำลองในเขตชนบท

วัตถุประสงค์ของการจัดการน้ำในเขตชนบทก็เพื่อที่จะทำให้เกิดสภาวะที่เหมาะสมสำหรับการเพ
าะปลูกพืชและการระบายน้ำที่เหมาะสมสำหรับอาคารและ โครงสร้างพื้นฐานในการหาผลลัพธ์ที่
เหมาะสมที่สุดขององค์ประกอบหลักของการจัดการน้ำสำหรับการเพาะปลูกพืชคาบของการออก
แบบสามารถใช้ในการเปลี่ยนแปลงฝนและการระเหยการคายน้ำของพืชและปริมาณน้ำส่วนเกิน
เป็น

- การเปลี่ยนแปลงความชื้นในมวลดิน;
- การเปลี่ยนแปลงของระดับน้ำใต้ดิน;
- การเปลี่ยนแปลงของระดับน้ำในพื้นที่;
- การเปลี่ยนแปลงของระดับน้ำในแปลงปลูกข้าว;
- การให้น้ำเพื่อการชลประทานในสภาพอากาศแบบเขตร้อนชื้น

ใช้ความสัมพันธ์ระหว่างเงื่อนไขดังกล่าวและข้อจำกัดทางการเกษตรจะสามารถหาผลลัพธ์ที่เหมาะสมที่สุดขององค์ประกอบหลักของการจัดการน้ำได้

การจำลองในเขตเมือง

การจำลองในเขตเมืองโดยทั่วไปเขตเมืองเป็นส่วนหนึ่งของการจัดการน้ำในพื้นที่ปิดล้อมวัตถุประสงค์ของการจัดการน้ำในเขตเมืองก็เพื่อที่จะทำให้เกิดสภาวะการระบายน้ำที่เหมาะสมในการหาผลลัพธ์ที่ดีที่สุดขององค์ประกอบหลักของการจัดการน้ำสำหรับในเขตเมืองคาบของการออกแบบสามารถใช้ในการเปลี่ยนแปลงฝนและปริมาณน้ำส่วนเกินเป็น

- การเกิดน้ำนองบนถนน;
- การเปลี่ยนแปลงของระดับน้ำใต้ดินภายใต้อิทธิพลของฝนและการเปลี่ยนแปลงของระดับน้ำในคลอง;
- การเปลี่ยนแปลงของระดับน้ำในพื้นที่

ในครั้งแรกขององค์ประกอบหลักของการจัดการน้ำสำหรับในเขตเมืองสามารถหาได้โดยสมมุติเกิดการล้นจากฝายน้ำล้นโดยอิสระหรือโดยสูบน้ำผ่านสถานีสูบน้ำซึ่งวิธีนี้สามารถหาผลลัพธ์ที่ดีที่สุดขององค์ประกอบหลักของการจัดการน้ำสำหรับในเขตเมืองได้

กรณีศึกษาในประเทศเนเธอร์แลนด์และประเทศไทย

สำหรับประเทศเนเธอร์แลนด์กรณีศึกษามีดังนี้
- สคีแฟน พื้นที่ปิดล้อมที่เป็นดินเหนียวเป็นส่วนใหญ่;
- เดาฟ์โพลเดอร์ พื้นที่ปิดล้อมที่มีเป็นดินพรุเป็นส่วนใหญ่;
- โฮกและลากอับสวัดโพลเดอร์ พื้นที่ปิดล้อมเมือง

สำหรับประเทศไทยกรณีศึกษามีดังนี้
- พื้นที่ปิดล้อมที่จำลองปลูกข้าวนาน้ำฝน;
- พื้นที่ปิดล้อมจำลองปลูกข้าวนาชลประทาน;
- พื้นที่ปิดล้อมที่จำลองปลูกผักแบบยกร่อง;

- พื้นที่ปิดล้อมบริเวณระหว่างปากแม่น้ำเจ้าพระยาและปากแม่แม่น้ำบางปะกงพื้นที่ปิดล้อม
  ชนบท;
- พื้นที่ปิดล้อมสุขุมวิทพื้นที่ปิดล้อมเมือง

ข้อมูลของกรณีศึกษาที่ได้เก็บจากพื้นที่ใช้สำหรับสอบเทียบเพื่อที่จะหาค่าตัวแปรซึ่งเป็นปัจจัย
สำคัญต่อระดับน้ำใต้ดิน, น้ำผิวดินและการไหลของน้ำจากมวลดิน

พื้นที่ปิดล้อมสำหรับกรณีศึกษาในประเทศเนเธอร์แลนด์อยู่ในความรับผิดชอบของคณะ-
กรรมการน้ำของเดลฟแลนด์ เดลฟแลนด์ตั้งอยู่ทางแถบตะวันตกของประเทศเนเธอร์แลนด์
พื้นที่ของเมืองนี้มีเขตติดต่อกับทะเลเหนือทางทิศตะวันตกและนิวมาสส์และนิววาเตอร์เวกทาง
ทิศใต้ เมืองเดลเลฟท์ประกอบด้วย พื้นที่ปิดล้อมทั้งหมด มีระดับน้ำในแต่ละพื้นที่แตกต่างกัน
เมืองเดลเลฟท์มีพื้นที่ประมาณ ๔๐,๐๐๐ เฮคแตร์ ซึ่งประกอบด้วยพื้นที่ปิดล้อมจำนวน ๖๐ แห่ง
สำหรับกรณีศึกษาในประเทศไทยก่อนอื่นได้กำหนดพื้นที่สำหรับทำการทดลองในเมืองไทย
สำหรับข้าวนาน้ำฝน,ข้าวนาชลประทานและผักแบบยกร่องบริเวณภาคกลางของประเทศไทย
ที่ตำบลสามชุกในจังหวัดสุพรรณบุรีสถานที่ทดลองอยู่ห่างจากกรุงเทพไปทางทิศตะวันตกเฉียง
เหนือ๒๘๐กิโลเมตรสภาพพื้นที่ส่วนใหญ่ของจังหวัดประกอบด้วยพื้นที่ราบลุ่มแม่น้ำและมีภูเขา
ขนาดเล็กตั้งอยู่บริเวณทางเหนือและตะวันตกของจังหวัดบริเวณตะวันออกเฉียงใต้เป็นที่ราบลุ่ม
แม่น้ำท่าจีนซึ่งเป็นพื้นที่นาข้าวกรณีศึกษาในประเทศไทย
สำหรับพื้นที่ปิดล้อมชนบทตั้งอยู่ในบริเวณตะวันออกเฉียงใต้ของกรุงเทพมหานครและตอน
กลางของจังหวัดสมุทรปราการซึ่งขอบเขตพื้นที่ล้อมด้วยคลองประเวศน์บุรีรมย์ทางทิศเหนือ
คันกั้นแม่น้ำเจ้าพระยาทางทิศตะวันตกคันกั้นน้ำพระราชดำริทางทิศตะวันออกและคลองชาย-
ทะเลทางทิศใต้ลักษณะพื้นที่เป็นที่ราบลุ่มระหว่างปากแม่น้ำเจ้าพระยาและปากแม่น้ำบางปะกง
ลาดลงสู่อ่าวไทย
กรณีศึกษาพื้นที่ปิดล้อมเมือง ได้แก่ พื้นที่ปิดล้อมสุขุมวิทซึ่งตั้งอยู่ในบริเวณแถบตะวันออก
ของกรุงเทพมหานคร ในบริเวณนี้ประกอบด้วยพื้นที่ปิดล้อม ๑๐ แห่ง พื้นที่ปิดล้อมสุขุมวิท
เป็นพื้นที่ปิดล้อมชั้นในของกรุงเทพมหานคร

ผลการศึกษามีดังต่อไปนี้

พื้นที่ปิดล้อมที่เป็นดินเหนียวสคีแฟนในประเทศเนเธอร์แลนด์

พื้นที่ปิดล้อมสคีแฟนตั้งอยู่ในเขตเทศบาลของเมืองรอตเตอร์ดัม สคีแฟนมีพื้นที่ประมาณ ๕๘๔ เฮกแตร์ พื้นที่ส่วนใหญ่ปลูกหญ้า (ร้อยละ๘๑.๓) ดินชั้นบนเป็นดินพรุปนดินเหนียวและดินชั้น-ล่างเป็นดินเหนียวมาก องค์ประกอบหลักของการจัดการน้ำโดยประมาณในปัจจุบันมีดังนี้ ระดับน้ำในพื้นที่ในช่วงฤดูหนาวเท่ากับ ๐.๗๑ เมตรจากผิวดิน, เปอร์เซ็นต์ของพื้นที่ผิวน้ำเท่ากับ ๕.๖, ความลึกของท่อระบายน้ำใต้ดินเท่ากับ ๐.๕๐ เมตรจากผิวดิน, ระยะห่างระหว่างท่อระบาย-น้ำใต้ดินเท่ากับ ๕๐ เมตร และความสามารถในการสูบน้ำของสถานีสูบน้ำเท่ากับ ๑๓.๔ มิลลิเมตรต่อวัน องค์ประกอบหลักของการจัดการน้ำที่เหมาะสมที่สุดมีดังนี้ ระดับน้ำในพื้นที่ในช่วงฤดูหนาวเท่ากับ๑.๕๑เมตรจากผิวดิน,เปอร์เซ็นต์ของพื้นที่ผิวน้ำเท่ากับ๑.๘๘,ความลึกของท่อระบายน้ำใต้ดินเท่ากับ๑.๓๐เมตรจากผิวดิน,ระยะห่างระหว่างท่อระบายน้ำใต้ดินเท่ากับ ๒๓.๑๘ เมตรและความสามารถในการสูบน้ำของสถานีสูบน้ำเท่ากับ ๑๖.๒ มิลลิเมตรต่อวัน ความลึกของท่อระบายน้ำใต้ดินเป็นองค์ประกอบที่มีอิทธิพลต่อมูลค่าเทียบเท่าทั้งหมดของระบบต่อปี สามารถสรุปว่ารักษาระดับน้ำในฤดูร้อนสูงกว่าระดับปัจจุบันและความลึกของท่อระบายน้ำใต้ดินอาจต่ำกว่าปัจจุบัน

พื้นที่ปิดล้อมที่เป็นดินพรุเดาฟ์โพลเดอร์ ในประเทศเนเธอร์แลนด์

เดาฟ์โพลเดอร์ เป็นส่วนหนึ่งของมิดเดิน-เดลฟ์แลนด์ ตั้งอยู่ในเขตเทศบาลของสคิปูลเดนและมาสแลนด์ มีพื้นที่ประมาณ ๓๗๐.๓๘ เฮกแตร์ พื้นที่ส่วนใหญ่ปลูกหญ้า(ร้อยละ๘๘.๘) ดินชั้น-บนเป็นดินพรุปนดินเหนียวและดินชั้นล่างเป็นดินพรุ องค์ประกอบหลักของการจัดการน้ำ โดยประมาณในปัจจุบันมีดังนี้ ระดับน้ำในพื้นที่ในช่วงฤดูหนาวเท่ากับ ๐.๗๘ เมตรจากผิวดิน, เปอร์-เซ็นต์ของพื้นที่ผิวน้ำเท่ากับ ๔.๔, ความลึกของร่องระบายน้ำใต้ดินเท่ากับ ๐.๓๕ เมตรจากผิวดิน, ระยะห่างระหว่างร่องระบายน้ำใต้ดินเท่ากับ ๑๐ เมตรและความสามารถในการสูบน้ำของสถานีสูบน้ำเท่ากับ ๑๔.๕ มิลลิเมตรต่อวัน องค์ประกอบหลักของการจัดการน้ำที่เหมาะสมที่สุดมีดังนี้ ระดับน้ำในพื้นที่ในช่วงฤดูหนาวเท่ากับ ๑.๓๕ เมตรจากผิวดิน, เปอร์เซ็นต์ของพื้นที่ผิวน้ำเท่ากับ ๑.๘๔ ความลึกของร่องระบายน้ำใต้ดินเท่ากับ ๑.๑๕ เมตรจากผิวดิน, ระยะห่างระหว่างร่องระบายน้ำใต้ดินเท่ากับ ๒๖.๕๐ เมตรและความสามารถในการสูบน้ำของสถานีสูบน้ำเท่ากับ ๔.๕ มิลลิเมตรต่อวัน ความลึกของร่องระบายน้ำใต้ดินเป็นองค์ประกอบที่มีอิทธิพลต่อมูลค่าเทียบเท่าทั้งหมดของระบบต่อปีสามารถสรุปว่ารักษาระดับน้ำในฤดูร้อนสูงกว่าระดับปัจจุ-บัน และความลึกของร่องระบายน้ำใต้ดินอาจต่ำกว่าปัจจุบัน

พื้นที่ปิดล้อมเมืองโฮกลากอับสวัดโพลเดอร์ในประเทศเนเธอร์แลนด์

พื้นที่กรณีศึกษาครอบครองพื้นที่ทั้งหมดของโฮกอับสวัดโพลเดอร์และบางส่วนของลากอับสวัด
โพลเดอร์ พื้นที่ปิดล้อมนี้ระบายโดยสถานีสูบน้ำอวฮอฟ มีพื้นที่ทั้งหมดประมาณ ๗๑๓ เฮคแตร์
ซึ่งประกอบด้วย พื้นที่ของโฮกอับสวัดโพลเดอร์ ๒๑๖ เฮคแตร์ และพื้นที่ของลากอับสวัด
โพลเดอร์ ๔๕๗ เฮคแตร์ เนื่องจากสภาพการระบายน้ำจึงทำการประยุกต์แบบจำลองการระบาย
น้ำล้นฝายจากโฮกอับสวัดโพลเดอร์ลงสู่ ลากอับสวัดโพลเดอร์ซึ่งต่ำกว่า องค์ประกอบหลักของ
การจัดการน้ำโดยประมาณในปัจจุบันมีดังนี้ ขนาดของท่อระบายน้ำเท่ากับ 0.30 เมตร ระยะห่าง
ระหว่างคลองระบายน้ำเท่ากับ ๑๗๕ ถึง ๓๕๐ เมตร ระดับน้ำในคลองระบายน้ำเท่ากับ ๐.๕๕
เมตรจากผิวดิน, เปอร์เซ็นต์ของพื้นที่ผิวน้ำเท่ากับ ๕.๐ และความสามารถใน การสูบน้ำของ
สถานีสูบน้ำเท่ากับ ๘.๕ มิลลิเมตรต่อวัน องค์ประกอบหลักของการจัดการน้ำที่เหมาะสม
ที่สุดได้ทำการ คำนวณหา ที่พื้นที่ของลากอับสวัดโพลเดอร์ มีดังนี้ ขนาดของท่อระบายน้ำเท่ากับ
๐.๓๐ เมตร ระยะห่างระหว่าง คลองระบายน้ำเท่ากับ ๑,๘๗๐ เมตร ระดับน้ำในคลองระบาย
น้ำเท่ากับ ๐.๘๗ เมตรจากผิวดิน, เปอร์เซ็นต์ของพื้นที่ผิวน้ำเท่ากับ ๐.๒ และความสามารถใน
การสูบน้ำของสถานีสูบน้ำเท่ากับ ๕.๐ มิลลิเมตรต่อวัน การเปลี่ยนแปลงค่าขนาดของท่อระบาย-
น้ำและความสามารถในการสูบน้ำของสถานีสูบน้ำเป็นองค์ประกอบที่มีอิทธิพลมากต่อกับมูลค่า
เทียบเท่าทั้งหมดของระบบต่อปี องค์ประกอบดังกล่าวมีมีอิทธิพลมากเพราะว่า ปริมาณฝนตก
กระจายสม่ำเสมอไม่มีความเข้ม ไม่มากและใช้ค่าเฉลี่ยของฝนรายวันมาใช้ในการคำนวนทำให้ไม่
ได้การเกิดการขึ้นจริงของน้ำ ลดระดับน้ำในคลองระบายจาก0.55 เป็น ๐.๘๗ เมตรจากผิวดิน
สามารถนำมาปฏิบัติได้และพบว่าพื้นที่ผิวน้ำและพบว่าความสามารถในการสูบน้ำของสถานีสูบ-
น้ำในปัจจุบันและเปอร์เซ็นต์ของพื้นที่ผิวน้ำมีมีค่ามากกว่าความจำเป็น

พื้นที่ปิดล้อมที่จำลองปลูกข้าวนาน้ำฝนในประเทศไทย

ในทางปฏิบัติอาจจะไม่มีพื้นที่ปิดล้อมที่ปลูกข้าวนาน้ำฝนในประเทศไทยเพราะชาวนาส่วนมาก
ในพื้นที่ปิดล้อมจะทำการสูบน้ำจากคลองระบายนำเพื่อให้น้ำแก่ข้าว องค์ประกอบหลักของการ
จัดการน้ำที่เหมาะสมที่สุดมีดังนี้ ระดับน้ำในพื้นที่เท่ากับ ๐.๐๘ เมตรสูงกว่าผิวดิน ระดับน้ำนี้จะ
สามารถให้น้ำแก่ข้าวโดยแรงโน้มถ่วงได้ในช่วงแล้งฝนและปริมาณฝนส่วนเกินสามารถระบาย
โดยแรงโน้มถ่วงได้ในช่วงฝนชุก, เปอร์เซ็นต์ของพื้นที่ผิวน้ำเท่ากับ ๑.๑, ความสามารถระบายน้ำ
ออกจากนาเท่ากับ ๖.๗ มิลลิเมตรต่อวัน และความสามารถในการสูบน้ำของสถานีสูบน้ำเท่ากับ
๗.๕ มิลลิเมตรต่อวัน ระดับน้ำจากผิวดินสัมพันธ์กับการ ไหลซึมของน้ำ ระดับน้ำยิ่งลึกการไหล-
ซึมของน้ำจากนาผ่านรอยแตกของดินก็ยิ่งสูงขึ้นเท่านั้น ดังนั้นข้าวอาจสูญเสียมากในช่วงแล้งฝน

ในทางกลับกันไหลซึมของน้ำอาจมีประโยชน์ต่อพืชเนื่องจากระดับน้ำจะไม่สูงมากในกรณีที่มี
ฝนตกจำนวนมากในเวลาอันสั้น ในขณะเดียวกันระดับน้ำยิ่งตื้นจะเพิ่มโอกาสของการเกิดน้ำท่วม
และน้ำล้นข้ามคันนาซึ่งอาจจะทำให้เกิดความเสียหายอย่างมาก

พื้นที่ปิดล้อมที่จำลองปลูกข้าวนาชลประทานในประเทศไทย

โดยทั่วไปองค์ประกอบหลักของการจัดการน้ำมีดังนี้ ระดับน้ำในพื้นที่เท่ากับ ๐.๘๐ เมตรจาก
ผิวดิน เปอร์เซ็นต์ของพื้นที่ผิวน้ำเท่ากับ ๒.๔ ความสามารถระบายน้ำออกจากนาเท่ากับ ๔๖.๐
มิลลิเมตรต่อวัน และความสามารถในการสูบน้ำของสถานีสูบน้ำเท่ากับ ๒๖.๐ มิลลิเมตรต่อวัน
องค์ประกอบหลักของการจัดการน้ำที่เหมาะสมที่สุดมีดังนี้ ระดับน้ำในพื้นที่เท่ากับ ๐.๘๔ เมตร
จากผิวดิน, เปอร์เซ็นต์ของพื้นที่ผิวน้ำเท่ากับ ๒.๕, ความสามารถระบายน้ำออกจากนา เท่ากับ
๔๔.๙ มิลลิเมตรต่อวัน และความสามารถในการสูบน้ำของสถานีสูบน้ำเท่ากับ ๓๒.๕ มิลลิเมตร
ต่อวัน ระดับน้ำในสภาวะที่เหมาะสมที่สุดและระดับทั่วไปของการจัดการน้ำการให้น้ำชลประ-
ทานจะต้องทำการสูบน้ำเท่านั้น ในพื้นที่ปิดล้อมนาชลประทาน ระดับน้ำในการจำลองนี้จะอยู่
ในระดับที่กำหนดเสมอเนื่องจากมีการส่งน้ำเข้าสู่ระบบคลองโดยสม่ำเสมอ ความสามารถใน
การสูบน้ำของสถานีสูบน้ำที่เหมาะสมที่สุดมีค่ามากกว่าค่าโดยทั่วไป สำหรับค่าองค์ประกอบ
หลักของการจัดการน้ำอื่นมีค่าไม่แตกต่างกันมากนัก การเปลี่ยนความสามารถในการสูบน้ำของ
สถานีสูบน้ำและระดับน้ำในพื้นที่เป็นองค์ประกอบที่มีอิทธิพลมากต่อกับมูลค่าเทียบเท่าทั้งหมด
ของระบบต่อปี การพิจารณาเพิ่มความสามารถในการสูบน้ำอาจต้องกระทำ

พื้นที่ปิดล้อมที่จำลองปลูกผักแบบยกร่องในประเทศไทย

โดยทั่วไปองค์ประกอบหลักของการจัดการน้ำมีดังนี้ ความลึกของระดับน้ำในร่องเท่ากับ ๐.๖๐
เมตรจากผิวดิน, ระยะห่างระหว่างร่องเท่ากับ ๖ ถึง ๘ เมตร ระดับน้ำในพื้นที่เท่ากับ ๐.๕๐ เมตร
จากผิวดิน เปอร์เซ็นต์ของพื้นที่ผิวน้ำเท่ากับ ๑.๑ ความสามารถระบายน้ำออกจากสนามเท่ากับ
๔๖.๐ มิลลิเมตรต่อวันและความสามารถในการสูบน้ำของสถานีสูบน้ำเท่ากับ ๒๖.๐ มิลลิเมตร
ต่อวัน องค์ประกอบหลักของการจัดการน้ำที่เหมาะสมที่สุดมีดังนี้ ความลึกของระดับน้ำในร่อง
เท่ากับ ๐.๗๔ เมตรจากผิวดิน, ระยะห่างระหว่างร่องเท่ากับ ๘.๔๕ เมตร ระดับน้ำในพื้นที่เท่ากับ
๐.๐๖ เมตรสูงกว่าผิวดิน เปอร์เซ็นต์ของพื้นที่ผิวน้ำเท่ากับ ๒.๔ ความสามารถระบายน้ำออกจาก
สนามเท่ากับ ๓๔.๙ มิลลิเมตรต่อวันและความสามารถในการสูบน้ำของสถานีสูบน้ำเท่ากับ
๓๖.๕ มิลลิเมตรต่อวัน ระดับน้ำในสภาวะที่เหมาะสมที่สุดสูงกว่าผิวดิน ที่ระดับน้ำนี้จะสามารถ
ให้น้ำแก่พื้นที่เพาะปลูกโดยแรงโน้มถ่วงได้ ความสามารถในการสูบน้ำของสถานีสูบน้ำที่เหมาะ

สมที่สุดมีค่ามากกว่าค่าโดยทั่วไป สำหรับค่าองค์ประกอบหลักของการจัดการน้ำอื่นมีค่าไม่แตก
ต่างกันมากนัก การเปลี่ยนความลึกของระดับน้ำในร่องและระยะห่างระหว่างร่องในพื้นที่เป็น
องค์ประกอบที่มีอิทธิพลมากต่อกับมูลค่าเทียบเท่าทั้งหมดของระบบต่อปี การพิจารณาลดระดับ
น้ำในร่องเท่ากับจาก ๐.๖๐ เป็น ๐.๗๔ เมตรจากผิวดินในบริเวณที่มีการก่อสร้างแล้ว สำหรับ
ในบริเวณที่มีการก่อสร้างใหม่ระยะห่างระหว่างร่องสามารถเพิ่มจาก ๖ เป็น ๘.๕ เมตร การเพิ่ม
ความสามารถในการสูบน้ำอาจต้องกระทำ ระดับน้ำในช่วงแล้งสามารถตื้นกว่าระดับก้นร่องระ-
บายน้ำเพื่อที่สามารถส่งน้ำเข้าแปลงโดยแรงโน้มถ่วง

พื้นที่ปิดล้อมในพื้นที่ชนบทในประเทศไทย

พื้นที่ของพื้นที่ปิดล้อมนี้เท่ากับ ๑๘,๗๖๐ เฮกแตร์ ซึ่งประกอบด้วยพื้นที่เมือง ๓,๓๖๐ เฮกแตร์
และพื้นที่การเกษตร ๑๑,๓๗๐ เฮกแตร์ ซึ่งประกอบด้วยนาข้าวร้อยละ ๑๒.๘ และบ่อปลาร้อยละ
๘๗.๒ การจัดการน้ำในพื้นที่เรียกว่า 'การจัดการน้ำแบบอนุรักษ์' โดยน้ำจะถูกเก็บไว้ในระบบ
คลองก่อนที่จะสิ้นฤดูฝนเพื่อการชลประทานและอื่นๆ ชาวนาจะต้องสูบน้ำจากคลองสายหลักสู่
คลองซอยและสูบเข้าสู่แปลงนาเพื่อที่จะให้น้ำแก่พืช โดยทั่วไปองค์ประกอบหลักของการจัดการ
น้ำมีดังนี้ ระดับน้ำในพื้นที่เท่ากับ ๐.๘๐ เมตรจากผิวดิน เปอร์เซ็นต์ของพื้นที่ผิวน้ำเท่ากับ ๓๓
ความสามารถระบายน้ำออกจากนาเท่ากับ ๔๖.๐ มิลลิเมตรต่อวันและความสามารถในการสูบน้ำ
ของสถานีสูบน้ำเท่ากับ ๕.๓ มิลลิเมตรต่อวัน องค์ประกอบหลักของการจัดการน้ำที่เหมาะสม
ที่สุดมีดังนี้ ระดับน้ำในพื้นที่เท่ากับ ๐.๒๔ เมตรจากผิวดิน, เปอร์เซ็นต์ของพื้นที่ผิวน้ำเท่ากับ
๒๐, ความสามารถระบายน้ำออกจากนาเท่ากับ ๖๒.๕ มิลลิเมตรต่อวัน และความสามารถใน
การสูบน้ำของสถานีสูบน้ำเท่ากับ ๖.๓ มิลลิเมตรต่อวัน การเปลี่ยนระดับน้ำในพื้นที่และ ความ
สามารถในการสูบน้ำของสถานีสูบน้ำเป็นองค์ประกอบที่มีอิทธิพลมากต่อกับมูลค่าเทียบเท่าทั้ง
หมดของระบบต่อปี การพิจารณาเพิ่มความสามารถในการสูบน้ำอาจต้องกระทำ ระดับน้ำในช่วง
เวลาเก็บกักสามารถตื้นขึ้นอีกเพื่อจะได้ปริมาณเก็บกักมากขึ้นและสามารถลดค่าใช้จ่ายการสูบน้ำ
เข้าแปลงนาและบ่อปลา บ่อปลาสามารถใช้เป็นพื้นที่เก็บกักปริมาณน้ำส่วนเกินโดยการเปลี่ยน
แปลงระยะเวลาการเลี้ยงปลาในช่วงฤดูฝนสามารถช่วยให้การจัดการน้ำเหมาะสมมีค่าใช้จ่ายลด-
ลงซึ่งต้องมีการศึกษาพิ่มขึ้นทั้งเทคนิคและเศรษฐกิจและสังคม

พื้นที่ปิดล้อมเมืองสุขุมวิทในประเทศไทย

พื้นที่กรณีศึกษาตั้งอยู่ทางแถบตะวันออกของกรุงเทพมหานคร โดยอยู่ในความรับผิดชอบของ
กรุงเทพฯพื้นที่ปิดล้อมย่อยดีเอฟของพื้นที่ปิดล้อมสุขุมวิทได้ถูกนำมาศึกษาเนื่องจากความพร้อม

ของข้อมูลและแยกขอบเขตการระบายน้ำค่อนข้างชัดเจน พื้นที่ปิดล้อมย่อยนี้มีพื้นที่ทั้งหมดประ-
มาณ ๓๖๘ เฮคแตร์ ซึ่งประกอบด้วย พื้นที่ผิวแบบปูด้วยวัสดุฉาบผิว ๒๔๘ เฮคแตร์และพื้นที่
แบบไม่ปูด้วยวัสดุฉาบผิว ๑๒๐ เฮคแตร์ องค์ประกอบหลักของการจัดการน้ำโดยประมาณ
ในปัจจุบันมีดังนี้ ขนาดของท่อระบายน้ำเท่ากับ ๐.๖๐ เมตร ระยะห่างระหว่างท่อส่งน้ำ๖๐๐ เมตร
ระดับน้ำในคลองระบายน้ำในช่วงฤดูฝนเท่ากับ ๒.๐๘ เมตรจากผิวดิน, เปอร์เซ็นต์ของพื้นที่ผิว-
น้ำเท่ากับ ๐.๖ และความสามารถในการสูบน้ำ ของสถานีสูบน้ำเท่ากับ ๑๓๗.๑ มิลลิเมตรต่อวัน
องค์ประกอบหลักของการจัดการน้ำที่เหมาะสมที่สุดได้ทำการคำนวณหาที่พื้นที่ของลากอับสวัด-
โพลเดอร์มีดังนี้ ขนาดของท่อระบายน้ำเท่ากับ ๑.๐๐ เมตร ระยะห่างระหว่างคลองระบายน้ำ
เท่ากับ ๑,๔๘๐ เมตร ระดับน้ำในคลองระบายน้ำเท่ากับ ๒.๕๖ เมตรจากผิวดิน, เปอร์เซ็นต์ของ
พื้นที่ผิวน้ำเท่ากับ ๑.๑ และความสามารถในการสูบน้ำของสถานีสูบน้ำเท่ากับ ๑๑๖.๖ มิลลิเมตร
ต่อวัน การเปลี่ยนแปลงค่าขนาดของระดับน้ำในคลองระบายน้ำเป็นองค์ประกอบที่มีอิทธิพล
มากต่อกับมูลค่าเทียบเท่าทั้งหมดของระบบต่อปี การลดระดับน้ำในคลองระบายน้ำในช่วงฤดูฝน
และเพิ่มขนาดของท่อระบายน้ำสามารถลดความเสียหายเนื่องจากน้ำขังบนผิวถนนได้และถ้าสา-
มารถเพิ่มพื้นที่ผิวน้ำจะทำให้มีความจุน้ำเพิ่มขึ้นและลดอัตราการเพิ่มของระดับน้ำในคลอง
ระบายน้ำได้

การประเมินผลที่ได้

การไหลของน้ำเนื่องจากแรงตึงผิวและความชื้นในชั้นดินไม่อิ่มตัว

ภายหลังจากที่พื้นที่ปิดล้อมได้ถูกทำขึ้นรอยแตกของพื้นดินจะเกิดขึ้นเนื่องจากการหดตัวแนวราบ
และดินจะแน่นขึ้นเนื่องจากการหดตัวในแนวดิ่ง ในดินชั้นบนเนื้อดินจะถูกกวนผสมกันเนื่อง
จากการไถพรวนเพื่อการเกษตรกรรมและทำให้รอยแตกหายไป แต่รอยแตกยังคงอยู่ในดินลึก
จากดินชั้นบนอาจจะคงอยู่ ดังนั้นในการจัดทำแบบจำลองรูปตัดตามความลึกของดินจำลองเป็น
สามชั้นดังต่อไปนี้ ดินชั้นที่หนึ่งชั้นไถพรวน, ดินชั้นที่สองชั้นที่มีรอยแตก และชั้นที่สามดินชั้น
ล่างที่มีความทึบน้ำหรือกึ่งทึบน้ำ

　　การไหลของน้ำเนื่องจากแรงตึงผิวในชั้นดินไม่อิ่มตัวหรือใกล้กับชั้นรากพืชได้ถูกจำลองใน
แบบจำลองโดยค่าแรงดูดซับน้ำซึ่งขึ้นกับความชื้นในดินที่บริเวณเขตแบ่ง ได้แก่ ความลึกของน้ำ-
ใต้ดิน คุณสมบัติของดิน เป็นต้น ในสภาพอากาศแบบปานกลางใช่วงฤดูหนาวดินที่อยู่เหนือ
ระดับท่อ/ร่องระบายน้ำมีค่าความชื้นสูงกว่าความชื้นสนาม ดังนั้นจะไม่มีการไหลของน้ำขึ้นด้าน
บนเนื่องจากแรงตึงผิว ในช่วงเวลานี้ความชื้นในดินชั้นที่สองจะมากกว่าชั้นที่หนึ่งเพราะว่าแบบ
จำลองคำนวณความชื้นของดินชั้นที่หนึ่งจากดุลยภาพของความชื้นกับค่าการใช้น้ำของพืชใน

ขณะที่การไหลน้ำขึ้นด้านบนจากชั้นที่สองเนื่องจากแรงตึงผิวมีค่าน้อย ในช่วงฤดูร้อนดินมีค่า
ความชื้นต่ำกว่าความชื้นสนามดังนั้นจะมีการไหลของน้ำขึ้นด้านบนเนื่องจากแรงตึงผิวมีบทบาท
สำคัญ ในช่วงฤดูร้อนความชื้นในดินชั้นที่หนึ่งในดินเหนียวและดินพรุในประเทศเนเธอร์แลนด์
ลดลงเร็วกว่าความชื้นในดินชั้นที่สองอาจเนื่องมาจากค่าการไหลของน้ำขึ้นด้านบนเนื่องจาก
แรงตึงผิวจากดินชั้นที่สองสู่ดินชั้นที่หนึ่งบวกกับค่าการซึมซับน้ำดินชั้นที่หนึ่งลบด้วยความชื้นที่
พืชใช้ มีค่ามากกว่าการไหลของน้ำจากดินชั้นที่หนึ่งลงสู่ด้านล่างลงสู่ดินชั้นที่สองบวกกับค่าการ
ไหลของน้ำขึ้นด้านบนเนื่องจากแรงตึงผิวจากดินชั้นที่สามสู่ดินชั้นที่สอง ดินในเขตร้อนชื้นจะ
แห้งกว่าในสภาพอากาศแบบปานกลางเพราะมีการระเหยสูงตลอดปีในช่วงแล้งดินเหนียวในเขต
ร้อนชื้นจะแตกระแหงเมื่อมีฝนตกทำให้มีการสูญเสียน้ำในนาไหลสู่ระบบคลองโดยเฉพาะในนา
ข้าวน้ำฝนและการไหลของน้ำขึ้นด้านบนมีปริมาณน้อยเนื่องจากระดับน้ำใต้ดินอยู่ลึก ในนาชล-
ประทานดินมีค่าความชื้นเกือบอิ่มตัวจึง ไม่มีการไหลของน้ำขึ้นด้านบน ดังนั้นจึงมีแต่การไหล
ของน้ำลงสู่ด้านล่าง ซึ่งมีบทบาทในการจัด การน้ำในพื้นที่นาชลประทาน ในแปลงปลูกผัก
ยกร่องดินมีค่าความชื้นใกล้เคียงความชื้นสนาม เนื่องจากมีการให้น้ำโดยการพ่นน้ำดินจะแห้งก็
ต่อเมื่อมีการเก็บเกี่ยวแล้วเท่านั้น การไหล ของน้ำขึ้นด้านบนเนื่องจากแรงตึงผิวมีผลประโยชน์
ต่อพืชและชาวนาเพราะทำให้มีการสูบน้ำน้อย ในช่วงมีฝนค่าความชื้นจะสูงกว่าความชื้นสนาม
ดังนั้นจะไม่มีการไหลของน้ำขึ้นด้านบน เนื่องจากแรงตึงผิว

ปริมาณการเก็บกักน้ำในมวลดิน

ปริมาณการเก็บกักน้ำในมวลดินอาจมีบทบาทสำคัญเกี่ยวกับการเปลี่ยนแปลงระดับน้ำใน
คลองในพื้นที่ปิดล้อมเพราะว่าปริมาณการเก็บกักน้ำในมวลดินอาจมีปริมาณมากในระดับที่อยู่
เหนือระดับน้ำใต้ดินโดยเฉพาะดินพรุ ในเขตร้อนชื้นปริมาณการเก็บกักน้ำในมวลดินในแปลง
ปลูกผักยกร่องมีบทบาทสำคัญเกี่ยวกับการเปลี่ยนแปลงระดับน้ำในแปลงถ้ามีปริมาณการเก็บกัก
น้ำในมวลดินน้อยจะมีโอกาสน้ำท่วมแปลงมากขึ้นปริมาณการเก็บกักน้ำในมวลดินอาจมีบทบาท
สำคัญในนาข้าวน้ำฝนเพราะว่าดินจะแห้งและต้องการปริมาณน้ำฝนเพื่อที่จะเติมลงในดินเพื่อที่
จะสามารถปลูกข้าวได้และสามารถเก็บความชื้นไว้ได้

การเปลี่ยนแปลงระดับน้ำใต้ดิน

โดยส่วนมากพื้นที่ปิดล้อมจะสามารถไม่นำค่าความลาดชันของการไหลมาพิจารณา ดังนั้น
แบบจำลองส่วนที่มีน้ำใต้ดินเสมือนเป็นอ่างเก็บน้ำและการไหลของน้ำลงไปสู่ท่อ/ร่องระบายน้ำ
ใต้ดินสามารถหาค่าแปรตามปริมาตรจริงของอ่างเก็บน้ำดังกล่าว ดังนั้นในสภาพอากาศแบบปาน

กลางความสัมพันธ์แบบไม่เป็นเส้นตรงระหว่างการไหลกับอ่างเก็บน้ำดังกล่าวสามารถนำมาใช้
ได้ ในการคำนวณระดับของน้ำใต้ดินในแบบจำลองพบว่าจะอยู่เหนือระดับของท่อ/ร่องระบาย-
น้ำใต้ดินเสมอ แต่ในความเป็นจริงระดับของน้ำใต้ดินสามารถลดลงอยู่ต่ำกว่าระดับของท่อ/ร่อง
ระบายน้ำใต้ดินโดยเฉพาะในดินพรุ ดังนั้นการขาดแคลนน้ำในช่วงฤดูร้อนเนื่องจากการไหลของ
น้ำขึ้นด้านบนเนื่องจากแรงตึงผิวจะมากกว่าค่าที่คำนวณได้จากแบบจำลอง ในเขตร้อนชื้นระดับ-
น้ำจะเกี่ยวข้องทั้งน้ำบนดินและน้ำใต้ดินน้ำใต้ดิน/น้ำบนดินที่วัดได้ต้องทำการเปรียบเทียบกับ
การคำนวณ การเปรียบเทียบระดับระหว่างค่าระดับที่วัดได้กับค่าระดับที่คำนวณพบว่าค่าระดับ
น้ำใต้ดินที่คำนวณได้ใกล้เคียงกับค่าระดับน้ำใต้ดินที่วัดได้น้ำใต้ดิน โดยเฉพาะในแปลงปลูกผัก
ยกร่อง การจำลองน้ำใต้ดินในนาข้าวน้ำฝนให้ผลที่แตกต่างเนื่องจากได้รับอิทธิพลจากระดับน้ำ
ในแม่น้ำใกล้กับที่แปลงที่ทดลอง ในการทดลองน้ำใต้ดินตอบสนองอย่างรวดเร็วต่อฝนที่ตกเนื่อง
จากดินมีค่าความซึมได้สูง นอกจากนี้น้ำใต้ดินในนาข้าวชลประทานและแปลงปลูกผักยกร่องได้
รับผลกระทบจากระดับน้ำในคลองซอยอีกด้วย ในนาข้าวชลประทานไม่เพียงแต่การเปลี่ยน
แปลงระดับน้ำใต้ดินแต่ยังมีการเปลี่ยนระดับน้ำบนดินเนื่องจากมีการสูญเสียน้ำที่ซึมผ่านคันนา
และซึมลึก

ความสัมพันธ์ระหว่างน้ำไหลบนผิวดินกับฝนในเขตเมือง

ท่อระบายน้ำมีหน้าที่หลักสองอย่างในเขตเมืองคือการสะสมน้ำฝนและน้ำเสียจากบ้านและ
โรงงาน ในสภาพแวดล้อมของพื้นที่ปิดล้อมน้ำไหลบนผิวดินมาจากทั้ง พื้นที่ผิวแบบปูด้วย
วัสดุฉาบผิวและพื้นที่แบบไม่ปูด้วยวัสดุฉาบผิว ความสัมพันธ์ระหว่างน้ำไหลบนผิวดินกับฝน
ในสภาพอากาศแบบปานกลางและในเขตร้อนชื้นมีความแตกต่างเนื่องจากอุณหภูมิและลักษณะ
ของฝน ในกรณีศึกษาของพื้นที่ปิดล้อมเมืองในประเทศเนเธอร์แลนด์และในประเทศไทยถ้าเพิ่ม
ขนาดท่อระบายน้ำจะมีน้ำไหลลงสู่คลองในเวลารวดเร็วขึ้นและมีโอกาสที่ความเสียหายเนื่องจาก
ระดับน้ำสูงเพิ่มขึ้น แต่ในประเทศเนเธอร์แลนด์ฝนตกไม่รุนแรงดังนั้นจึงยากที่จะหาความ
สัมพันธ์ระหว่างน้ำไหลบนผิวดินกับฝนและความเสียหายเนื่องจากน้ำบนถนน การท่วมของน้ำ
อย่างรวดเร็วและน้ำบนถนนมีโอกาสเกิดมากกว่าในประเทศไทยเนื่องจากฝนตกรุนแรงในระยะ
เวลาสั้น

ระดับน้ำในระบบระบายน้ำหลัก

สมมุติฐานเป็นอ่างเก็บน้ำสำหรับระบบระบายน้ำหลักในการคำนวณระดับน้ำซึ่งยอมรับได้
สำหรับพื้นที่ปิดล้อมที่ราบ ในสภาพอากาศแบบปานกลางระดับน้ำในระบบระบายน้ำหลัก

สามารถมีผลกระทบต่อการระบายน้ำของระบบระบายน้ำใต้ดินในกรณีระดับน้ำสูงกว่าระดับของ
ระบบระบายน้ำ ระดับน้ำอาจมีผลกระทบต่อความชื้นในดินชั้นบน อย่างไรก็ตามระดับน้ำ
ที่อยู่เหนือระดับของระบบระบายน้ำมีผลกระทบน้อยถ้าเกิดในช่วงเวลาสั้น ในสภาพอากาศแบบ
เขตร้อนชื้นระดับน้ำในระบบระบายน้ำหลักสามารถได้รับผลกระทบต่อความสามารถการระบาย
น้ำออกจากแปลงนาและแปลงผักยกร่อง ในช่วงเวลาที่ระดับน้ำในระบบระบายน้ำหลักสูงใน
ช่วงเวลาที่มีฝนจะต้องทำการสูบน้ำเพราะน้ำไม่สามารถไหลโดยแรงโน้มถ่วงผ่านท่อระบายน้ำ
และสามารถเกิดความเสียหายขึ้นได้ นอกจากนี้โอกาสที่จะเกิดการล้นข้ามคันนาหรือคันดินยัง
สูงขึ้นถ้าระดับน้ำในระบบระบายน้ำหลักสูงขึ้น ในช่วงเวลาแล้งระดับน้ำในระบบระบายน้ำ
หลักสูงจะทำให้สามารถนำน้ำเข้าแปลงได้โดยแรงโน้มถ่วงโดยเปิดให้น้ำไหลผ่านท่อระบายน้ำ
ทำให้การสูบน้ำลดลง ระดับน้ำในระบบระบายน้ำหลักในพื้นปิดล้อมในสภาพอากาศแบบเขต
ร้อนชื้นสามารถเก็บในระดับที่สูงในช่วงปลายฤดูฝนเพื่อที่จะเก็บน้ำไว้ใช้ในการชลประทานใน
ช่วงฤดูแล้ง

พื้นที่ผิวน้ำ

    พื้นที่ผิวน้ำเกี่ยวข้องโดยตรงกับความจุของการเก็บกักในพื้นที่ปิดล้อม พื้นที่ผิวน้ำยิ่งน้อยเท่า
ใดความจุของการเก็บกักยิ่งน้อยเท่านั้น โดยทั่วไปการลดความจุของการเก็บกักจะทำให้ระดับน้ำ
ในพื้นที่ปิดล้อมสูงขึ้นในสภาวะการระบายน้ำเดียวกัน โอกาสที่จะเกิดความเสียหายต่อทรัพย์สิน
ในเขตเมืองเมื่อพื้นที่ผิวน้ำน้อย นอกจากนี้ความต้องการการสูบน้ำยังเพิ่มขึ้นด้วยทั้งในนาข้าว
และแปลงผักยกร่อง ทั้งนี้ยังทำให้โอกาสที่จะเกิดการน้ำล้นข้ามคันนาและคันดินเพิ่มขึ้นเมื่อพื้น
ที่ผิวน้ำลดลง

ความสามารถในการระบายน้ำ

    ส่วนมากการระบายน้ำจากพื้นที่ปิดล้อมจะทำโดยการสูบน้ำ, การระบายผ่านท่อระบายน้ำ
หรือฝายน้ำล้น ความสามารถในการระบายน้ำยิ่งมากขึ้นเท่าใดระดับน้ำที่สูงขึ้นยิ่งน้อยเท่านั้น
ในประเทศเนเธอร์แลนด์ส่วนมากจะทำการระบายน้ำโดยวิธีสูบน้ำในสภาพอากาศแบบเขตร้อน-
ชื้นความสามารถในการระบายน้ำสามารถแบ่งออกได้เป็นสองชนิดคือความสามารถในการระบา
ยน้ำจากแปลงและความสามารถในการระบายน้ำจากระบบคลอง ถ้าลดความสามารถในการ
ระบายน้ำจากแปลงในนาข้าวลงความเสียหายต่อข้าวเนื่องจากระดับน้ำ ในนาสูงจะเพิ่มขึ้น ถ้าลด
ความสามารถในการระบายน้ำจากแปลงผักยกร่องลงโอกาสที่น้ำท่วมแปลงที่ปลูกผักจะสูงขึ้น

ความเสียหายจะสูงมาก นอกจากนี้ในพื้นที่ปิดล้อมแปลงผักยกร่องจะทำให้ระดับน้ำในร่องแปลง
สูงขึ้นทำให้เกิดระดับน้ำใต้ดินสูงซึ่งจะทำให้ความเสียหายเพิ่มขึ้น

ข้อคิดเห็นปิดท้าย

แบบจำลองรวมโอปอลเป็นเครื่องมือที่มีประโยชน์ในการหาค่าองค์ประกอบหลักของการจัดการ
น้ำที่เหมาะสมที่สุดของการจัดการน้ำในพื้นที่ปิดล้อมและสามารถช่วยให้เกิดความเข้าใจพฤฒิ-
ติกรรมของผลกระทบขององค์ประกอบหลักของการจัดการน้ำต่อการลงทุนและความเสียหายอีก
ด้วย ค่าขององค์ประกอบหลักของระบบการจัดการน้ำต้องมีการปรับแก้อีกครั้งตามสภาพความ
เป็นจริงได้แก่ สภาพทางกายภาพ, สภาพการจัดการและบำรุงรักษา, การใช้พื้นที่, การปฏิบัติทาง
การเกษตร, นโยบาย, เทคนิค, ลักษณะของดิน, สิ่งแวดล้อมและทิวทัศน์ ข้อมูลเฉพาะจำเป็นต้อง
ได้จากข้อมูลในพื้นที่

    ค่าของมูลค่ารายปีมีค่าต่ำในพื้นที่ปิดล้อมชนบทและเมือง ดังนั้นการเผื่อค่าความปลอดภัย
ควรจะเพิ่มในองค์ประกอบหลักของการจัดการน้ำซึ่งได้จากแบบจำลองรวมโอปอล

    ค่าของมูลค่ารายปีเพิ่มขึ้นอย่างมากเมื่อองค์ประกอบหลักของการจัดการน้ำมีค่าน้อยกว่าองค์
ประกอบหลักของการจัดการน้ำที่เหมาะสมที่สุด ดังนั้นความเสี่ยงของการออกแบบน้อยกว่าค่า
ดังกล่าวควรจะต้องระมัดระวัง อย่างไรก็ตามการออกแบบควรตระหนักถึง การใช้งานในอนาคต,
อายุการใช้งานของอาคาร, การเจริญเติบโตของเมืองและการเพิ่มของประชากร, การใช้เทคนิคที่
ก้าวหน้า, การเปลี่ยนแปลงของสภาพแวดล้อม, เศรษฐกิจสังคม, นอกจากนี้ยังควรรวมสถิติใน
ทางอุทกวิทยา, การวิเคราะห์ความเสี่ยง เป็นต้น

    ผลที่ได้ในพื้นที่ปิดล้อมชนบท, พื้นที่การเกษตรความลึกของการระบายมีผลกระทบมากที่
สุดต่อมูลค่ารายปีในพื้นที่ปิดล้อมบริเวณที่เป็นดินเหนียวและดินพรุในประเทศเนเธอร์แลนด์ใน
ประเทศไทยระดับน้ำในพื้นที่เปิดในพื้นที่ปิดล้อมมีผลกระทบมากที่สุดต่อมูลค่ารายปีทั้งในพื้นที่
นาข้าวแบบนาน้ำฝนและนาชลประทาน ในพื้นที่นาชลประทาน ไม่เพียงระดับน้ำในพื้นที่เปิดแต่
ความสามารถในการระบายน้ำก็มีผลกระทบมากเช่นกัน ในขณะที่ระดับน้ำในร่องคูในแปลง
มีผลกระทบมากที่สุด ในพื้นที่ปิดล้อมเมืองระดับน้ำในคลองระบายมีผลกระทบมากที่สุดแต่ผล
กระทบไม่มากนักในประเทศเนเธอร์แลนด์

    ในการศึกษาครั้งนี้แสดงให้เห็นว่าการจัดการน้ำที่เหมาะสมที่สุดในพื้นที่ปิดล้อมในสภาพ
อากาศแบบปานกลางและสภาพอากาศแบบร้อนชื้นสามารถหาได้

Curriculum vitae

The author of this dissertation was born at Petchaburi province, Thailand on 17th November 1960. In 1978 he obtained his secondary school degree at Prommanusorn School, Petchaburi province and started his studies in Civil Engineering at Chiang Mai University, Chaing Mai, Thailand. He got his Bachelor degree in Civil Engineering in 1983.

From 1984 till 1998 he worked as government officer in the Office of Water Resource Development as a Civil Engineer, Royal Irrigation Department, Thailand. He was involved in the implementation of large-scale irrigation projects in Thailand. His work concerned construction planning, technical work, contracts and budgeting for irrigation projects. During his work he was member of national and international committees for bidding on several irrigation projects in Thailand. He was also a guest lecturer for training at the Royal Irrigation Department and joint committee on international project implementation between Thailand and Malaysia in the Golok River mouth improvement project. He joined the international course on 'River and Dam Engineering' at Sukuba, Japan in 1990 and joined the international course on 'Land Drainage' at Wageningen, the Netherlands in 1997.

In September 1999 he came to the International Institute for Infrastructural, Hydraulic and Environmental Engineering (IHE) in Delft, the Netherlands. He received his Master of Science degree in Land and Water Development, Hydraulic Engineering in April 2001. The topic of his thesis was 'Some Examples of Optimization of Water Management in the Netherland's Polders'. Directly after obtaining his MSc degree he continued with his PhD research on 'Optimization of Water Management in Polder areas. Some examples for the temperate humid and the humid tropical zone' at UNESCO-IHE. During his research from 2002 to 2004, he carried out several field experiments in Thailand and obtained data from polder areas in the Netherlands and in Thailand.

He attended the Second World Water Forum, March 2000, the Hague, the Netherlands. He participated in two conferences of the International Commission on Irrigation and Drainage (ICID), Utrecht, the Netherlands, 10 - 13 September 2003 and Moscow, Russia, 5 - 12 September 2004. He presented 3 papers related to his research topic in the two ICID conferences.

Printed and bound by CPI Group (UK) Ltd, Croydon, CR0 4YY

22/10/2024

01777530-0005